Communicating Technology

Communicating Technology

Dynamic Processes and Models for Writers

Fred D. White
Santa Clara University

HarperCollins*College*Publishers

Executive Editor: Anne Elizabeth Smith
Project Editor: Brigitte Pelner
Design Manager: Mary Archondes
Text Designer: Robin Hessel Hoffmann
Cover Designer: John Callahan
Art Studio: FineLine, Inc.
Photo Researcher: Karen Koblik
Electronic Production Manager: Valerie A. Sawyer
Desktop Administrator/Electronic Page Makeup: Hilda Koparanian
Manufacturing Manager: Helene G. Landers
Printer and Binder: R.R. Donnelley & Sons Company
Cover Printer: The Lehigh Press, Inc.

For permission to use copyrighted material, grateful acknowledgment is made to the copyright holders on pp. 569–571, which are hereby made part of this copyright page.

Communicating Technology: Dynamic Processes and Models for Writers

Library of Congress Cataloging-in-Publication Data

White, Fred D., 1943-
    Communicating technology: dynamic processes and models for writers / Fred D. White.
        p.   cm.
    Includes index
    ISBN 0-06-500508-2
    1. English language—Technical English—Problems, exercises, etc.
  2. Communication of technical information—Problems, exercises, etc.
  3. English language—Rhetoric—Problems, exercises, etc.
  4. Communication in science—Problems, exercises, etc.
  5. Technology—Authorship—Problems, exercises, etc.   6. Science—
  —Authorship—Problems, exercises, etc.   7. Technical writing—
  —Problems, exercises, etc.   I.   Title.
PE1475.W45   1994
808' .0666–dc20   94-29053
                                                            CIP

95 96 97 98 9 8 7 6 5 4 3 2 1

For Beverly

Contents

Chapter 6
Basic Rhetorical Techniques I: Presenting Information Clearly and Accurately   **109**

Chapter 7
Basic Rhetorical Techniques II: Argument in Technical Writing   **136**

Chapter 8
Editing to Improve Style I: Sentences   **174**

Chapter 13
Proposing a Project   **284**

Chapter 14
Reporting Progress or Activity   **307**

Chapter 15
Reporting Empirical Research   **325**

Chapter 16
Reporting Feasibility Research   **359**

PART III:
SCIENTIFIC AND TECHNICAL JOURNALISM

■   ■   ■

Chapter 20
Writing and Designing Brochures and Fact Sheets   **480**

Chapter 21
Newsletter Writing and Editing   **499**

Preface

This textbook provides a full-spectrum introduction to technical writing. While ideal for the technical communication major, the textbook is also suitable for students of all majors whose careers are likely to involve writing of a technical nature. To be sure, as a result of the widespread incorporation of sophisticated data processing and telecommunications equipment into all areas of business and industry, it is hard to imagine a career in which some form of technical writing would not be done routinely.

A strong assumption underlying this book is that technical writing is academically important as well as vocationally practical, and that it can enrich a liberal-arts education, just as a liberal education can enrich the experience of technical writing. "We should resist teaching skills without intellectual content," states Theresa Enos. Teachers of scientific and technical communication can do much to help their students as professionals in the making to realize "that imagination, expression, and rationality are part of all of us as human beings."[1]

---

[1] Theresa Enos, "Rhetoric and the Discourse of Technology." In *New Worlds of Writing*, ed. Carolyn B. Matalene. New York: Random House, 1989: 97.

Three principles have guided the author's approach to the subject matter:

1. To master the art of technical communication, one should understand the milieu in which such communication takes place. Technical writing in the workplace, for example, is rarely done in solitude. Writers—that is, anyone who must write, not just technical writers per se—continually interact with managers, clientele, technicians, programmers, engineers, and other writers, at many stages of the writing project's development. Also, students need to understand the nature of such interaction: the realities, both positive and negative, of deadline pressure; of locating the right experts for background information; the competition that underlies proposal writing; the way well-written documents such as user manuals and product descriptions are important to product marketability—these too must be factored into the experience of writing on the job.

2. "Technical" is not subject-specific. All subjects have their technical side; hence, all subjects include technical writing. This book will give students a sense of the interdisciplinary character of technical communication.

3. Technology and science are as discourse-driven, hence as *rhetorical,* as the traditional humanistic disciplines of literature, history, and philosophy. Scientific research, engineering, industrial and agricultural pursuits may seem like the unlikeliest of activities to depend on interpretive or argumentative discourse, but a closer look reveals situations such as these:

- A term is defined in great detail for one audience, but not defined at all for another.
- Cautionary remarks in a procedural piece are made especially emphatic and conspicuous on the page to reduce likelihood of an accident.
- In an empirical research report, certain erosion phenomena are meticulously described, while other phenomena are described briefly or ignored completely. A second researcher on the scene gives much more emphasis to the phenomena slighted by the first researcher.

The social context in which these writing situations occur greatly influences the way the subject matter (no matter how "factual") is processed. For this reason, it is important for students to examine technical discourse from a rhetorical perspective. In his seminal study of scientific discourse, Charles Bazerman points out that

> A rhetorical approach to writing well in science would not set forth a set of formal prescriptions to be followed for propriety's sake, nor would it suggest a set of universally advisable procedures. A rhetorical approach would attend to the

range and meaning of current practices and then suggest how to deploy them appropriately and effectively within specific contexts.[2]

Communicating Technology has been written with Bazerman's rationale in mind.

ORGANIZATION OF THE TEXT

Communicating Technology is organized into three sections:

1. *Foundation Skills.* Chapters 1 through 11 are devoted to the fundamentals of technical writing, which include:

- Introducing the field and some of its broad applications (Chapter 1)
- Identifying and addressing different readers (Chapter 2)
- Interacting and collaborating on writing projects in the workplace as well as in the classroom (Chapter 3)
- Gathering information (including interviewing and survey-taking techniques); incorporating information into a report; understanding the dynamic process of researching a technical subject (Chapter 4)
- Composing dynamically; using invention, outlining and storyboarding, and revision strategies in technical-writing contexts (Chapter 5)
- Using the basic rhetorical modes of definition, description, narration, analysis, and argument in scientific and technical writing (Chapters 6 and 7)
- Editing to improve style: strengthening syntax and word choice; improving clarity, emphasis, coherence (Chapters 8 and 9)
- Practicing document-design techniques (Chapter 10)
- Creating and incorporating visual elements (Chapter 11)

2. *Writing and Speaking in Work Environments.* Chapters 12 through 18 focus on the standard technical-writing tasks performed on the job:

- Preparing correspondence (letters and memoranda) for a variety of purposes in business and industry (Chapter 12)

---

[2] Charles Bazerman, *Shaping Written Knowledge*. Madison: Univ. of Wisconsin Press, 1988: 323.

- Formally proposing technical projects (Chapter 13)
- Reporting progress, research, and feasibility (Chapters 14 through 16)
- Writing procedures and guidelines for assembly, operation, maintenance, repair, and computation (Chapter 17)
- Preparing effective oral presentations; using audiovisual media (Chapter 18)

Emphasis here is given to the writing or speaking activity rather than to the product; in other words, to the dynamics of "proposing a project" more than to "a proposal." Different genres of technical writing certainly do exist, and their general attributes are fully presented in this book; but they are not the inflexible, universal molds they are sometimes assumed to be.

3. *Scientific and Technical Journalism.* Chapters 19 through 21 provide guidelines for writing articles and news stories, brochures and newsletters—which technical writers often prepare for general audiences as well as for the larger workplace community. These include readers who need to be informed of developments outside of their areas of expertise. Such documents include:

- Feature articles and editorials for metropolitan newspapers; features on technical subjects for widely circulating periodicals (Chapter 19)
- Public-service documents, such as brochures and fact sheets (Chapter 20)
- Organization-related features for newsletters (Chapter 21)

Instruction in the writing of such documents contributes to scientific and technical literacy, which, in the United States especially, continues to be a critical problem. Technical writers, because of their versatile backgrounds in communication as well as the natural and applied sciences, are best prepared to take advantage of what is rapidly becoming a high-demand resource.

PEDAGOGICAL ELEMENTS

Students in any introductory course appreciate a textbook that helps them master the material efficiently, by including interesting and challenging exercises. Technical-writing textbooks often are written with an antiseptic formality that belies the dynamic and fascinating character of technical communication. Accordingly, the author has tried to kindle enthusiasm for the subject matter by making the chapters as engaging as possible.

Pedagogical features include, for each chapter:

- *A list of topics to be covered,* so that students have a clear sense of direction when studying the material.

- *End-of-chapter checklists,* which can help students prepare a more thorough first draft; they can also work as heuristics for getting the draft under way.
- *Chapter summaries,* which encourage students to review the key ideas presented in each chapter.
- *Topics for classroom discussions and activities.* Most of these topics give students an opportunity to apply their own values and experiences to the subject matter.
- *Notebook suggestions.* Writing skills are developed more readily when writing becomes part of one's daily routine. Self-directed notebook exercises can help instill this important habit.
- *Suggestions for writing projects.* The goal here is to stimulate students' enthusiasm by making the writing tasks real-world practical and challenging, but also interesting.

Three reference sections are included: Appendix A for grammar, mechanics, and usage; Appendix B for documentation formats (American Psychological Association (APA), the Council of Biology Editors (CBE), and the Modern Language Association (MLA); and Appendix C for guides to formatting traditional report elements, such as a letter of transmittal, abstract, table of contents, collected data section, and references page.

INSTRUCTIONAL SUPPLEMENTS

A detailed Instructor's Manual, written by the author, is available to each adopter of *Communicating Technology.* Intended as a resource for teachers, the manual includes an overview of the textbook; sample syllabli for both quarter system and semester schools; a guide to using the reference manual; and for each chapter in the text, the Instructor's Manual includes points to emphasize ways for using the discussion topics, and suggestions for further readings. In addition, a comprehensive set of transparency masters of numerous images is also available.

ACKNOWLEDGMENTS

I am happy to express my gratitude to the many individuals whose expert advice and support helped me to accomplish what I could not have accomplished alone: Professor Bruce S. Henderson, Santa Clara University, for sharing his vast knowledge of technical communication and the technical workplace with me over the years, and for his willingness to read large chunks of manuscript on short notice; Professor Timothy Gracyk, Santa Clara University, for casting a sharp eye on the style and continuity of several chapters; Professors Claudia Mon Pere McIsaac and Ann Brady Aschauer, Santa Clara University, whose knowledge of writing in corporations has been a valuable resource for me; my Chair, Professor Diane E. Dreher, Santa Clara University, for her enthusiastic support of my project during the four years it has taken me to complete it; and of course, my technical-writing students—all fourteen years' worth—especially Lisa Bonnell-Sprenger, Paul Freitas, Daniel Kiely, and Tristen Moors who graciously permitted me to use their work as models.

I owe an equally great debt to the reviewers who have given me rich and abundant suggestions: David Porush, Rensselaer Polytechnic Institute; Carol Cyganowski, DePaul University; Louie Edmundson, Chattanooga State Technical Community College; Beverly Sauer, University of Maine at Orono; Carol Senf, Georgia Institute of Technology; Sheila Honig, University of Missouri-Kansas City; Jennings Mace, Morehead State University (Kentucky); Thomas Warren, Oklahoma State University-Stillwater; Judith Kaufman, Eastern Washington University; Billie Wahlstrom, University of Minnesota.

And how can I fully thank the extraordinary team at HarperCollins, the best associates and friends a writer could have? To Michelle Beese, who was the first person to goad me into turning an idea into a proposal; to Constance Rajala, and Jane Kinney, who believed in my book when it was scarcely more than a shapeless mass; to my project editor Brigitte Pelner and my virtuoso copyeditor, Lee Paradise, I express boundless gratitude.

And finally, to my wife Beverly, and to my son and daughter Michael and Laura, I would not have been able to sustain the energy I needed to get through this task without their steady encouragement.

Fred D. White

I

The Elements of Scientific and Technical Writing

1

Communicating Science and Technology: The Big Picture

The Nature and Scope of Technical Writing

Technical Writing and the Advancement of Knowledge

Technical Writing as a Profession

Technical Literacy as Cultural Literacy

Freelance Technical Writing

L
ike any complex subject, technical writing is often only partially understood, and tends to be easily stereotyped. This chapter will introduce you
to the multifaceted nature of technical writing and its importance in a liberal arts education; to career possibilities for technical writers; and the nature
of writing in the workplace that virtually all professional people (not just technical writers) must do to some degree.

THE NATURE AND SCOPE
OF TECHNICAL WRITING

In one sense, "technical writing" is an ambiguous term. David Dobrin, a technical communications theorist, observes that while "medical writing" or "legal
writing" is writing about medicine, or law, "there is . . . no discipline of 'tech-
nics.'"[1] So what *does* "technical writing" refer to?

In its most common usage, *technical writing* refers to the communicating
of specialized information in any field (particularly in industry), read by technicians, technical managers, owner-operators of machines, and scientific researchers to perform a certain task. In short, technical writing is writing that is
put to use: regarded as a means to an end rather than as an end in itself or pur-
sued for its own enjoyment. Whereas one would read a poem, a novel, or a personal essay mainly for aesthetic reasons (relishing the writing itself as much as
the imagery, story, or ideas it conveys), one would read instructions for installing a new showerhead, or a feasibility study on the potential health-risk factors of a newly developed pesticide, for one reason only: to install the mechanism correctly; to decide whether it is safe to use the pesticide on produce for
human consumption. The writing itself is strictly a means of accomplishing
those end results.

Although no century has seen such a proliferation of technical communication as ours, technical writing is virtually as old as civilization itself. One of
the oldest examples of technical writing ever discovered dates from the seven-
teeth century B.C. (and is a copy of a document two hundred years older than
that): a papyrus fragment, in Egyptian hieratic script, examining triangles, perhaps to aid in the construction of pyramids or other structures (see Figure 1.1).
In ancient and medieval times, priests, farmers, seafarers, recorded their observations of the stars, the weather, the tides, the behavior of rivers, studied mathematical relationships, for practical reasons: to establish one's bearings at sea,

---

[1]David Dobrin, *Writing and Technique.* Urbana: NCTE, 1989: 29.

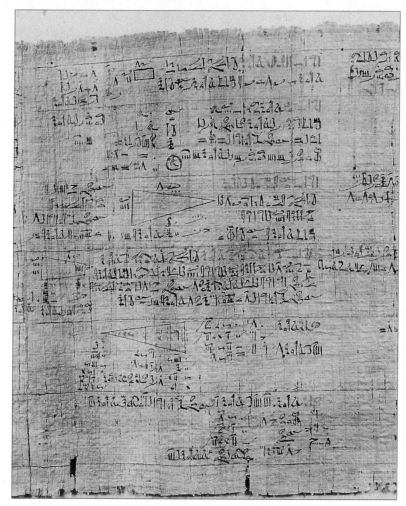

FIGURE 1.1   Technical document from Ancient Egypt.

to know when to plant or harvest crops, to know when and whom to marry (no distinction was yet made between astrology and astronomy!), to erect buildings. Between the fifteenth and eighteenth centuries, during the rise of empirical science and mechanical innovation, artist-scientists like Leonardo da Vinci explored animal and human anatomy (see Figure 11.2); astronomers like Copernicus, Galileo, and Kepler filled their notebooks with observations of the heavens; inventors designed clocks (Figure 1.2) and developed vastly improved instrumentation for scientific work by designing telescopes and microscopes.

One of the goals of this book is to help you develop the skills you will need for writing means-to-an-end technical documents. The first step is to

FIGURE 1.2 Technical drawing of a pendulum clock, seventeenth century.

become familiar with the range of such documents. There are many kinds of technical documents, but we may organize them into four categories: (1) Procedural; (2) Descriptive/Explanatory; (3) Persuasive/Evaluative; and (4) Investigative.

1. Procedural Documents
   - Instructions for assembly, such as for furniture, athletic equipment, toys.
   - Instructions for operation, such as for software, appliances, power tools.
   - How-to articles
2. Descriptive and Explanatory Documents
   - Descriptions of products or services, such as those found in catalogues and on packages.

- Explanations of processes, concepts, discoveries: "How food is digested," "What is the nature of light?" "How Watson, Crick, and others discovered the structure of DNA."
- Reports of progress, in which the work completed on a scheduled project is described, and costs tabulated and explained.

3. Persuasive or Evaluative Documents
- Proposals for research or for engineering or construction projects, to be funded by the agency soliciting proposals from potential clients in competition with each other.
- Product or service evaluations.
- Reports whose main purpose is to recommend an action or policy out of any number of possibilities.

4. Investigative Documents
- Reports whose main purpose is to reveal new knowledge about a phenomenon without necessarily having a utilitarian purpose in mind.
- Articles that examine a mechanism or phenomenon to determine its function or behavior.

This book also introduces you to another facet of technical writing, one that is growing rapidly in importance: *technical journalism,* the writing of articles and brochures on scientific and technical subjects, aimed at a nonspecialized readership. Technical articles, brochures, and newsletters are certainly utilitarian in that they often convey information that can be put to use; but they do more: they attempt to present the information simply and interestingly enough so that nonspecialists can understand it.

Now *everyone,* it should be emphasized, is a nonspecialist outside of his or her area of expertise. Moreover, the distinction between writing that is work-related or discipline-specific and writing that isn't may not always be clear. It is often the case that a news story about a discovery in one field—say, from physics: the decay-rate of a certain radioactive isotope such as carbon 14—could shed unexpected light on a research investigation in another field—say, from archaeology: the dating of a prehistoric encampment based on the extent of decay of the carbon 14 found at the site. That is why it is important to communicate information across the disciplines, in language accessible to all.

Another reason for the importance of technical journalism is that general readers are eager to heighten their "science literacy" aptitude, not only to cope with technical concepts that have woven their way into the workplace (and into the fabric of modern life in general) but to heighten their understanding of the scientific discoveries and technological innovations occurring virtually on a daily basis. Being scientifically and technically literate is gaining importance as a fundamental way to "keep abreast of the issues" in a world where technological change has reached revolutionary proportions.

Because of this extreme rate of technological advance, the demand for scientific and technical writing has also increased dramatically, and it involves the technical aspects of virtually every subject. This is even as true for the arts as it is for the sciences. Amateur photographers, for example, are eager to learn skills such as controlling depth-of-field, knowing which type of film or photographic paper to use, mixing the proper chemicals for developing, learning to use lens filters for special effects, and so on. People volunteering to assist a theater group with constructing a stage set need to understand principles like sloping risers or track lighting. A first-time sculptor needs to be informed of different types of clays and glazes; of casting techniques such as cire perdue (lost wax) and sand-casting; or of sculpting tools such as chisels, bow drills, and rasps. Anyone who enjoys keeping a vegetable garden may find it necessary to keep abreast of, say, environmentally safe pesticides such as chemosterilants.

Now you may be thinking, "Sloping risers?" "Cire perdue casting?" "Chemosterilants?" Isn't that getting *too* technical for nonspecialists? The question is meaningless outside the context of a particular *rhetorical situation;* that is to say, the answer to the question depends on three main concerns:

- The purpose of the document to be written.
- The needs of the target audience.
- The author's experience and knowledge of the material.

The following situation will make this clearer: Assume that you are a gardener and that you have been asked to write about techniques of vegetable gardening for two different audiences: (1) eighth graders who have been asked to grow vegetables for a science project; (2) experienced vegetable gardeners. You realize at once that the two articles will be very different from each other because the backgrounds of your audiences are so different. The amount of technical detail you present, and the way you present it, will be the crucial distinguishing characteristics. What would be familiar to the experienced gardeners would be new and unfamiliar to the eighth graders, who would have to be provided with clear definitions, explanations, analogies. Compare the following passages:

1. The three primary plant nutrients are nitrogen, phosphorous, and potassium. Plants also need three secondary nutrients—calcium, magnesium, and sulfur— and very small quantities of trace nutrients, including iron, manganese, and zinc.

—*All About Growing Fruits and Berries.* San Francisco: Ortho Books, 1982: 26.

2. Plants need nutrition to grow healthy, just as people do. Just as a person needs calcium for strong, healthy bones, so does a plant need calcium for strong, healthy stalks and roots. . . .

Just as technical writing can be adapted for different age groups, it can also be adapted for different purposes. As already mentioned, "technical writing" conventionally refers to writing that is used, virtually as a kind of tool, for getting work done. But technical writing can also include writing that helps people understand and appreciate complex ideas and functions. The typically "technical" elements in such a document may be muted; the writing may even be expressive and metaphoric, emphasizing the awesomeness of the phenomenon over its objective properties. In the following passages, both of which discuss the birth of stars, notice how one author emphasizes the aesthetic over the practical, while the other emphasizes the practical over the aesthetic.

> 1. Stars like the Sun are born in batches, in great compressed cloud complexes such as the Orion Nebula. Seen from the outside, such clouds seem dark and gloomy. But inside, they are brilliantly illuminated by the hot newborn stars. Later, the stars wander out of their nursery to seek their fortunes in the Milky Way, stellar adolescents still surrounded by tufts of glowing nebulosity, residues still gravitationally attached of their amniotic gas.
>
> —Carl Sagan, *Cosmos*. New York: Random House, 1980: 226.

> 2. It is virtually certain that the stars in a cluster have a common origin, and are of about the same age. . . . The best means of [investigating stellar evolution] is to measure the colour indexes and luminosities of individual stars. When colour index is plotted on the horizontal axis against luminosity on the vertical axis, it is clear that the distribution is not random.
>
> —E. L. Schatzman, *The Structure of the Universe*. Trans. Patrick Moore. New York: McGraw-Hill, 1968: 97.

Clearly, the second passage emphasizes the practical; but what specific characteristics make it so? Whereas Sagan attempts to relate the birth of a star to the birth of a human being (by using the metaphor of "nursery") in order to generate a sense of delight and philosophical reflection on the interrelatedness of stars and humans, Schatzman is more interested in communicating the methodology used to acquire insight into the phenomenon of stellar evolution. Schatzman's approach is intended to be accessible to lay readers (at least to those with a background in high school physics), but it probably would be of greater interest—indeed, of greater *usefulness*—to students of astronomy.

It should be pointed out that the purpose of these comparisons is not to establish a clear-cut distinction between "technical" and "nontechnical" writing; that is neither possible nor useful. The purpose, rather, is to demonstrate the wide range of technical communication, servicing a wide range of audience needs.

Let's look at another example, a description of the chemical properties of hydrogen, intended for beginning students of chemistry:

> Hydrogen is not very reactive at room temperature, but at higher temperatures it burns vigorously, and often explosively, in air or oxygen to form water.

$$2 H_2 + O_2 = 2H_2O$$

Hydrogen is a moderately strong reducing agent. . . . For example, it reduces the oxides of metals less active than manganese to the free metals:

$$H_2 + CuO = H_2O + Cu$$

Hydrogen will combine directly with the very active metals to form compounds called ionic hydrides, in which the valence number of hydrogen is −1.

$$Ca + H_2 = CaH_2$$

—Fred C. Hess, *Chemistry Made Simple*. Rev. ed. by Arthur L. Thomas. Garden City: Doubleday, 1984: 105.

Once again, utility is the principal consideration. The goal is not only to enable the reader to understand the properties of hydrogen but also to ensure that this knowledge can be properly put to use in the laboratory during experiments with hydrogen. The sentences are concise, the terminology exact. Note that, as is so frequent in technical communication, the text is supplemented with a visual aid—in this case with chemical formulae. Visual aids, be they equations, charts, graphs, tables, photographs, or diagrams (see Chapter 11), are essential for reinforcing understanding and applicability.

Audience-Specific Writing

One more point to make about the characteristics of technical writing is that it is composed with a specific audience in mind. Technical writers need to have some idea of the designated readers' needs for the document and their familiarity with the subject matter. In the preceding chemistry passage, the author knew he was addressing beginning students—most likely those concurrently enrolled in a high school course in physical chemistry. But what about the intended readers of the following passage, from another chemistry text?

When ignited, hydrogen burns in air with a blue or colorless, nonluminous flame, yielding H_2O. When mixed with air, the flammability limit is 4–74% hydrogen. When mixed with oxygen, the flammability limit is 4–94% hydrogen. . . . In sunlight or magnesium light, hydrogen combines with chlorine with violent release of energy, forming hydrogen chloride HCl. When hydrogen is heated with sodium, calcium, and several other metals, the corresponding hydride is formed. In the presence of a catalyst, hydrogen reacts with nitrogen to form ammonia NH_3.

—*Van Nostrand Reinhold Encyclopedia of Chemistry,* 4th ed. New York: Van Nostrand Reinhold, 1984: 471.

The intended readers for this passage are those who already have basic familiarity with chemistry: key terms like *hydride* are not defined in context, and the style (vocabulary and syntax) is more intricate than in the earlier passage. In Chapter 2 we examine these and other techniques of audience targeting in detail.

Table 1.1 Technical Writing Possibilities Across the Disciplines

Discipline	Possible Topic
Anthropology	Comparative cranial anatomy of Neanderthals and *Homo sapiens*
Archaeology	Using plant remains to determine chronology
Architecture	Structure of Gothic vaulted arches
Art (Photography)	Choosing the right film for your needs
Art (Painting)	Exploiting the chemical properties of certain pigments
Art (Music)	African harmonic patterns in progressive jazz
Communication	How communication satellites receive and transmit radio and TV signals
Engineering (Civil)	Reinforcing freeway overpasses in earthquake-prone regions
Engineering (Electrical)	Designing digital processing chips for voice recognition
Geology	Dynamics of river-delta formation
Health	Effects of alcohol on fetal development
Linguistics	Compound-word formation in Old English
Marketing	Feasibility of increasing manufacture and promotion of biodegradable products as a result of increased ecological awareness
Psychology	Investigation of the relationship between stress and depression

TECHNICAL WRITING AND THE ADVANCEMENT OF KNOWLEDGE

Technical writing not only directs the assembly and operation of equipment, not only articulates plans for scientific and technological endeavors; it also advances our knowledge and understanding of the subject matter in question. Writing (and reading) about machines, for example, enables us to understand their functions; but just as importantly it enables us to understand their role in modern culture, to understand their strengths and limitations. A technical writer working on an operator's manual for a computer system, or writing about the usefulness of such a system for his or her company's newsletter, is going to learn an enormous amount about that system in the course of researching and writing the document. Technical writing, that is to say, is always more than *writing:* it is researching, learning, and experiencing an entire technological milieu.

As stated earlier, every subject has its technical facets. Table 1.1 gives you a quick sense of the range of technical writing possibilities across the disciplines.

What makes each topic "technical" is its utilitarian, specialized focus. Writing that translates the *theory* of, say, ideal freeway-overpass reinforcement into a viable *technique* for constructing such reinforcement is truly a dramatic advancement in learning from the abstract-theoretical to the concrete-actual. The very effort to produce such a translation enables the engineers or scientists to fine-tune a theory. What may seem to work on paper often proves to be unworkable when rendered as a set of procedures. As in computer programming, the bugs need to be identified and eliminated; but this cannot be achieved until the actual work (directed by the writing) gets underway.

TECHNICAL WRITING AS A PROFESSION

Nearly all professional people must write to some extent as part of their jobs. Many companies also hire people whose exclusive duties are to write documents to accompany the products they manufacture (such as instruction sheets or user manuals), or to write proposals, write and edit newsletters, publicity releases, and so on. (For a sampling of job notices for technical writers, see Figure 1.3.) Such technical writers must not only be skillful with language; they must also be knowledgeable about the products they are writing documents for, as well as be enthusiastic about science and technology in general. Alas, such a combination of interests is not easy to come by. This predicament has more to do with antiquated assumptions about education than it does with genetics. In 1925, the great philosopher of education Alfred North Whitehead asserted that one's liberal arts education cannot be complete without technical training; nor can one's technical training be complete without a liberal arts education. Many liberal arts colleges are finally taking Whitehead seriously by introducing courses, as well as complete programs, in writing for the professions (what Whitehead, in his use of the word "technical," was essentially referring to). Many corporations have established technical writing apprenticeships and eagerly hire new college graduates with the requisite coursework and drive to succeed in this demanding profession.

What is it like to work as a technical writer? There are many misconceptions, such as that the work is tedious, or that it isn't really "writing"—but merely recording of facts, equations, engineering specs. The truth of the matter is that the work is exceedingly interesting and challenging. Here are some general characteristics of the technical-writing profession:

1. *Corporate technical writing is, directly or indirectly, an aspect of product or service marketing.* When the directions for using and maintaining a product such as a cellular telephone are presented with clarity and precision; when

Senior Technical Writer

You will work closely with engineering and marketing teams to research and write technical product documentation in support of disk drive product line. Documentation requirements include product specifications, product manuals, configuration and installation guides.

Candidates must have a bachelor's degree in a technical discipline or communications with strong technical aptitude and 5-8 years of experience in technical writing, technical support or an equivalent combination of education, training or work experience. Must have the ability to understand technical subject matter and communicate it clearly both verbally and in writing. Ability to understand and write from schematics is desired. Experience in the disk drive industry preferred. Excellent editing, organizational, and communications skills are essential. Experience with Mac desktop publishing systems is desirable; proficiency with FrameMaker is a plus.

An Element of POWER

Science Writer

In this challenging and highly visible marketing communications positions, you will write, edit, and produce a bimonthly customer newsletter and a quarterly employee newsletter. You will also develop, write, and edit marketing/sales collateral materials including datasheets, brochures, application pamphlets, and technical articles; maintain and update existing literature; assist with writing and editing of direct mail, ads, press releases, articles, and presentations; and serve as an editorial resource for Marketing. Naturally, this position requires excellent writing, editing, verbal communication, and print production skills, with a special emphasis on science writing. An understanding of chemistry and a working knowledge of computers (especially word processing and graphics applications) are also required. The ability to work in a high-energy environment, meet tight deadlines, maintain flexibility, and juggle multiple tasks is essential. A BA in English, 2 or more years' experience as a science or marketing writer, knowledge of experience with marketing communications and the software industry, and familiarity with desktop publishing would all be advantages.

Technical Writer

Put your talent to work researching, developing and writing high-quality user manuals for our networking products. Duties include formulating writing project schedules, creating new and rewriting existing product manuals and interfacing with graphic artists. You should have a technical writing certificate and a degree in English or journalism is preferred. In addition, you must have at least 3 years' technical writing experience in related industry, knowledge of LAN and interpersonal skills. A thorough understanding of Unix is desired.

MARKETING COMMUNICATIONS WRITER

You're a word wizard. You respect the English language. You eschew cliches. You know the difference between flout and flaunt. Your writing informs and persuades -- smoothly, efficiently, elegantly. You're knowledgeable about software and computer technology.

You have a background in marketing communications and can write white papers, datasheets, customer profiles, high-end brochures -- and more. You've had newsletter or magazine experience, and thrive under pressure, routinely meeting ridiculous deadlines.

Join our marketing communications team and help us produce the best marketing collateral in the software industry.

WRITER

Responsibilities include writing manuals for PC & video games and monitoring the accurate development of manuals written by contract employees. Must have 2 years manual writing experience with games on Super NES, Sega Genesis and IBM PC. Ability to work under deadline pressure a must.

MEDICAL/ TECHNICAL WRITER

Become a member of our dynamic Technical Publications Department! Research, plan, write and edit FDA-regulated product labeling and literature, including product user manuals, package inserts, and technical papers. Requires a BS degree (preferably in a medical discipline), or equivalent, plus a minimum of 8 years experience as a medical/technical writer. Working knowledge of Macintosh computer and Microsoft Word required.

FIGURE 1.3   Sample technical-writing job notices.

software documentation is easily referrable because of coherent formatting and detailed indexing; when, in a feasibility study, the advantages and disadvantages of a robotic system for an automobile assembly plant are so thoroughly analyzed that the CEO can make a prompt and confident decision to purchase the system, you are looking at writing performed on behalf of a product or a service.

2. *Corporate technical writing is inherently interactive.* A technical writer almost always must interact with other kinds of professionals within the corporation: engineers, managers, draftspeople, budget directors, marketing specialists, computer programmers, and other technical writers. With some types of documents, such as user manuals, it may also be necessary to interact with individuals outside the corporation—product reviewers, for example, who will test the usability of the document in question. Continual review and testing procedures are as necessary before the document can be published as before the hardware can be marketed.

3. *Corporate technical writing is audience-specific writing.* Almost all kinds of expository-writing contexts require authors to have a clear sense of whom they're writing for; but writing in a corporate environment is done to satisfy the needs of *designated* readers—readers you might actually be required to meet. Audience targeting is a methodical and, when done properly, an effective first step in shaping a technical document.

Minimal Skills

What are the skills generally expected of corporate technical writers? Of course, this varies with the kind of writing tasks a firm will assign a technical writer. Generally, however, technical writers would be expected to possess most of the following:

- A bachelor's degree in communication, English (often with emphasis on writing in the professions), or in technical communication itself.
- Course work in computer science.
- Familiarity with a wide range of computer systems and capabilities.
- Familiarity with the mechanical and electronic systems about which one is writing.
- Knowledge of at least one programming language.
- Familiarity with desktop publishing and graphic design.
- Strong oral-communication skills.
- Good interviewing skills (in person or via telephone).
- Excellent command of written English; familiarity with report writing, letter and memorandum writing.
- Knowledge of and skills in information gathering.

In addition, technical writers are expected to have strong social skills and to be able to initiate key contacts on their own. There is nothing sedate about the profession of technical writing!

The Technical Writer in Action

Let us look in on a segment of a technical writer's workday. Ann Martinez, who received her B.A. in English (emphasis on technical communication) with a minor in computer science, is employed by a large software manufacturer; her principal responsibilities include: (*a*) designing and writing user manuals for new software, and (*b*) writing descriptions of the new software for the company newsletter and for the marketing department. On the day you meet her, Ms. Martinez has been working on a user manual for a new word-processing program; she has just completed a first draft of the chapter that explains how to insert graphics into text. Although reasonably confident that her step-by-step directions are clear, she is concerned that she might have omitted something important and that some of the instructions she wrote could conceivably be misinterpreted.

Ms. Martinez decides to make an appointment with the programmer (who created this capability) to discuss these concerns. The programmer has proven to be an important resource for her during the planning and drafting of her document. Now, as a result of her latest interaction with the programmer, Ann realizes what she needs to add to her draft; she also realizes that a complete restructuring of the entire chapter will be necessary.

Other consultations may well be necessary. She may need to discuss the new changes with her manager or with fellow technical writers with whom she may be collaborating, or even to contact experts outside the company.

Sound frustrating? It isn't really; that is how good writing gets done. Without such collaborative interaction, the risk of leaving out crucial information would be greatly increased—along with *real* frustration!

TECHNICAL LITERACY AS
CULTURAL LITERACY

We live in an age in which a comprehensive understanding of and ability to discuss and argue about issues in science and technology is rapidly becoming essential for everyone; technical writers in particular need to keep abreast of these issues. Concepts like ozone depletion, acid rain, CFCs, CAT-scan, in-vitro fertilization, human immunodeficiency virus (HIV), biomass, genetic screening,

FRANK & ERNEST® by Bob Thaves

AMERICAN SOCIETY
OF PHYSICISTS

ELEMENTARY
PARTICLE
SALE

ELECTRONS.. -1
PROTONS ... +1
QUARKS .... ±1/3
NEUTRONS.. FREE OF
CHARGE !

Science and technology in the comics.

FRANK & ERNEST reprinted by permission of NEA, Inc.

and interactive programming have by now been assimilated into your daily lives. Even the comic strips reflect this assimilation! Open this morning's newspaper and most likely you will encounter news relating to science and technology; and it probably won't be limited just to a "Science News" section, either. To give a nearly random example, during a single week in February 1992, the following articles appeared in the *San Jose Mercury News*:

- Senate bill a step forward toward a national energy policy. (Feb. 23)
- FDA cracks down on untested medical devices. (Feb. 24)
- Chemical alternative to CFCs also harms ozone. (Feb. 24)
- Former airline mechanic invents 2-wheel drive bicycle. (Feb. 24)
- Computer provides 3-D images of Earth's interior. (Feb. 25)
- New surgical technique on leg bones makes dwarfs taller. (Feb. 25)
- Auto-programmable radios (working like VCRs) available. (Feb. 25)
- Biologists uncover evidence that male-female behavioral differences are based on genetic and brain differences. (Feb. 26)
- Hewlett-Packard to build an interactive television appliance (ITA) that would connect TV sets to national videogame, information, shopping, and education networks. (Feb. 27)
- FDA discovers that 20% of Pacific seafood is contaminated. (Feb. 27)
- Research and development in the U.S. is being seriously cut back. (Feb. 27)
- Blindfolded moms able to identify babies by touch. (Feb. 28)
- Gender-testing for athletes. (Feb. 29)
- Scientists urge screenings to test for "good" cholesterol. (Feb. 29)

Contemporary culture, indeed, is dominated by rapid-fire developments in communication and information systems, in medicine, in space sciences, in the rapidly burgeoning environmental sciences and technologies. Yet alarming numbers of people—including college students—continue to be seriously underinformed in these areas.

In a recent speech to an annual meeting of the American Association for the Advancement of Science, Donald Kennedy, the former president of Stanford University and a biologist, called on all scientists to try harder to communicate with the public, to help them understand and appreciate their work. One could extend this calling to include everyone who can translate technical data into easily understandable instructions or explanations. In other words, technical writers possess not only the competence but the ethical responsibility for reaching nonspecialized audiences. Without communication, there can be no public support of or interest in science and technology, and that can only lead to distrust. The public will not support what it distrusts and cannot understand; as a result, crucial research that could have long-term benefits for human welfare would not get funded.

What is more, according to researchers in a Northern Illinois University study, "Only one in eighteen American adults is scientifically literate"—that is, only 6 percent of those surveyed "had both a general understanding of how science is conducted and at least a minimal knowledge of scientific concepts."[2] Efforts are underway in public schools to reverse the misconceptions associated with science—that it is dull, impersonal, devoid of humanistic concerns, consisting of memorization of facts; or that it is really only for boys. To give one example of how the latter misconception is being combatted, in 1992 the American Association of University Women published a report which made it clear that girls are often subtly discouraged from courses (not to mention careers) in math and science.

Clearly, scientific and technical illiteracy must be combatted as forcefully as illiteracy in its root sense (the inability to read or write). As Mark Twain once said, "The person who does not read has no advantage over the person who cannot read." Writers who can communicate ideas in science and technology are performing a valuable public service. Communicating such knowledge accomplishes two things: (1) it brings about a "discourse community" wherein ideas in science and technology are digested, criticized, refined by persons with a wide range of backgrounds; and (2) it ultimately gets work going; that is the utility factor which lies at the heart of all technical discourse.

FREELANCE TECHNICAL WRITING

The past ten years have seen an astonishing proliferation of magazines and newsletters devoted to virtually every aspect of science and technology. One can find, at most newsstands, magazines devoted to oceanography, the environ-

---

[2]*Chronicle of Higher Education,* Jan. 25, 1989: A7; see also For Discussion topic 4 at the end of this chapter.

ment, archaeology, astronomy, and computer science—in addition to the general-science magazines like *Scientific American, Technology Review, Discover,* and *Omni*. Also, many newspapers such as *The New York Times,* the *San Jose Mercury News,* and the *Wall Street Journal* have weekly or daily sections devoted to science, technology, medicine and health—as do the major weekly newsmagazines such as *Newsweek, Time,* and *U.S. News and World Report.*

This means that people have an insatiable appetite for clearly and engagingly written articles about science and technology: articles that, for example, explain why certain chemicals are dangerous to sea life, or how to install a water-filtration system in your kitchen (see Chapter 19 for discussion of how to write such articles). Anyone who learns to write well in these areas will find a widespread, receptive readership. One can earn a substantial secondary income as a freelancer writing articles to mass-circulation periodicals.

Finally, several "science-literacy" books have been published in the last few years; these are designed to introduce general readers to fundamental concepts in science and technology. One such book, *Science Matters: Achieving Scientific Literacy,* by Robert M. Hazen (a geology professor) and James Trefil (a physics professor), clearly introduces 150 concepts such as:

absolute zero	gamma radiation
alpha decay	half-life
amino acid	ion
conduction	laser
DNA	microwaves
electromagnet	RNA
entropy	transistor

Anyone who wishes to be a writer must read widely and find time to reflect—in writing—on what is being read. Keep a notebook and use it to summarize articles and newly learned concepts, to practice writing descriptions of objects such as tools and mechanisms, to record questions about matters that are confusing to you, and so forth. Check the end of each chapter in this book for additional notebook-writing suggestions.

■ CHAPTER SUMMARY

Technical writing, which can be traced to ancient times, is basically utilitarian writing—writing that is used to perform a task, make a policy-changing decision, explain the results of a laboratory experiment, or convey the significance of a new discovery.

Technical documents in the workplace may be procedural (directing readers to perform a certain task); investigative (providing readers with in-depth analysis of observed activity such as laboratory experiments); persuasive (urging readers to adopt a procedure or policy); or descriptive/explanatory (explaining principles, describing and defining concepts and processes).

Technical writing can also be addressed to general audiences, by way of articles and brochures, for conveying knowledge about important issues in science and technology—a necessary dimension of a modern liberal education, and important preparation for understanding some of the major issues of contemporary culture.

As a profession, technical writing is highly interactive, involving consultation with other writers, technicians, executives. Many corporations like to hire new liberal arts graduates with coursework in technical communication and to train them, through their own formal apprenticeship programs, to become skilled in a specialized kind of technical writing, such as user documentation for a particular kind of computer system. Opportunities also exist for freelance technical writers to publish features on science and technology in mass-circulating periodicals.

■ FOR DISCUSSION

1. What are some of the specialized aspects of your major field of study? In what way are they "technical?" Suggest a topic based on one of these specialized concerns that would be appropriate for a technical document.

2. Hobbies often have a technical side to them. Think about one of your hobbies: what aspects of it might be confusing to an outsider? Explain one of these aspects simply enough for a novice to understand.

3. Find a magazine article on a technical or scientific subject and prepare to do the following with it in class:

    a. Summarize the article in just one paragraph.
    b. If you were originally unfamiliar with the topic, explain what the author did to make the information understandable to you. Or, if you had a difficult time with the article, speculate on what the author might have done to make it more understandable.

4. According to a national "technological literacy" survey conducted in 1986 by the Public Opinion Laboratory at Northern Illinois University, a major percentage of the U.S. population is technologically illiterate. The survey, which defined technological literacy as "an understanding of the applica-

tion of science and engineering to the solution of concrete problems," revealed that

70% do not understand radiation
40% think space rocket launchings change the weather
80% do not understand how a telephone works
40% believe some UFOs are actually visitors from other worlds

Prepare a set of five questions that you consider important indications of technological literacy, and conduct your own survey (see Chapter 4, pages 79–80, 83–85). Report your findings to the class.

■ FOR YOUR NOTEBOOK

1. List concepts from one scientific or technical subject which seem to you important enough for everyone to be familiar with. Justify your choices.

2. List possible ways that some people (perhaps yourself) get "turned off" or "turned on" by topics in science or math.

3. Write an imaginary dialogue between you, an expert in some scientific subject (e.g., "holographic imaging" or "the hazards of certain chemical additives in food"), and a friend who knows nothing about this subject. Demonstrate, in the dialogue, how you bring your friend toward an understanding of the subject.

■ WRITING PROJECTS

1. Check the classified ads of three or four issues of your local newspaper for "Writers" or "Technical Writers." Write a 2–3-page paper in which you discuss the kinds of technical writing skills, education, and work experience required. What are the most sought-after qualifications? What seems to be the best kind of educational preparation for these writing jobs?

2. Locate two technical articles on the same topic, and summarize the similarities and differences between them. Is one more technical than the other? In what way? Do the authors emphasize different things? What is the rationale in each case? Which article seems better organized? Which seems easier to follow, and why? Discuss your findings in a 2–3-page essay.

3. In the Introduction to their book *Science Matters,* Robert M. Hazen and James Trefil state, "Every citizen will be faced with public issues whose discussion requires some scientific background." Drawing from your own experience, write an essay in which you identify and discuss such issues.

2

Identifying and Addressing Appropriate Audiences

Readers of Technical Writing

Types of Readers

Determining Readers' Needs

Writing for Multiple Audiences

Writing for International Readers

Reader Awareness at the Drafting Stage

Audience Targeting: A Checklist

To understand how readers read can shed much light on what it takes to make a writing task succeed. How readers process information and acquire meaning from written language; and how they are most attentive to matters that involve them or that best represent their values—these cannot be emphasized enough in technical writing. Here the writer needs to convey information that is not merely to be read for edification but is to be actively put to use. The information therefore must be presented with utmost clarity, thoroughness, and precision. Of course, it is easy to say, "Be clear!" To achieve clarity, however, requires a much greater effort. "Clarity" is a relative term: its variables are the complexity of the subject matter, the circumstances that give rise to the writing task, the needs and the preparedness of the readers, and the skill of the writer. But awareness of these variables is part of a writer's responsibility toward his or her readers. Understanding these variables can do much to ensure a document's success.

Perhaps the best way to begin thinking about readers' habits and needs is with your own experiences as a reader. Surely you have felt the frustrations of trying to make sense of vague, convoluted, or poorly structured writing. Most of the time the fault is not yours but the writer's; yet all too often, you are likely to blame yourself for the confusion—as if the writer could do no wrong!

Consider the following excerpt from a description of computer disks, intended for an inexperienced reader:

> Computer disks consist of an iron-oxide coating on a substrate. In the case of floppy disks, the substrate is a flexible plastic. On hard disks, the substrate is made of aluminum and is called a platter.

To those familiar with disk technology, this may be a perfectly clear description. But what if the readers are novices? Before reading on, try to imagine yourself as the beginner you once were. What would you need to have clarified in order to make sense of the passage?

Certainly, the term *substrate* needs to be defined. Also, the distinction between *floppy disk* and *hard disk* may not be sufficiently clear to beginners. Here, then, is the passage as it was originally written:

> All computer disks consist of an iron-oxide coating on a *substrate.* In the case of floppy disks, the substrate is a flexible plastic similar to recording tape. Inside the rectangular, hard-plastic case, it's doughnut-shaped—round with a hole in the center. (*Floppies* get their name from the flexibility of the substrate.) On hard disks, the substrate is . . . usually made of aluminum and is called a *platter.*
>
> —Sharon Zardetto et al., *The Macintosh Bible,* 3rd. ed. Berkeley: Goldstein & Blair, 1991: 137–138.

It is also important to be aware of your reader's need for a logical and coherent structure to whatever document you are writing. Always ask yourself four questions related to organization when you are planning or drafting a document:

1. Have I revealed the purpose of the document in the introduction?
2. Have I divided my topic into logical segments and presented each segment in a logical sequence?
3. Does the discussion progress smoothly, coherently, from one segment to the next?
4. Do my paragraphs have internal coherence? That is, does each sentence relate clearly to the one preceding it? Does every word within each sentence serve to build the overall meaning?

According to science-writing specialists George D. Gopen and Judith A. Swan, "Readers make many of their most important interpretive decisions about the substance of prose based on clues they receive from its structure."[1]

It is helpful to remember that to write means to share ideas with others, to make them public—the root meaning of *publish*. This sharing is a give-and-take relationship: writers give readers information and ideas that the readers find useful and/or entertaining and which they can understand. Readers in turn give the writers their undivided attention, trust, and support.

Writers who do not consider the needs of their readers risk falling prey to what might be called the Humpty Dumpty syndrome. You may recall from Lewis Carroll's *Through the Looking Glass* that H. D. is a rather arrogant sort (foolhardy for an egg), who boasts that he can make words mean anything he pleases. But unless he accompanies those words with clear definitions (as he did when explaining the Jabberwocky poem to Alice), audiences will turn away in bafflement.

Technical writing is especially reader-oriented. Unlike, say, fiction, the readers of which have no direct say in what the writer produces (other than to expect an absorbing plot-structure and memorable characters), technical writing is often writing-to-order.

READERS OF TECHNICAL WRITING

Nearly everyone who can read does have need for technical writing of some kind at one time or another, procedural documents (such as assembly instructions) being perhaps the most common. When nonspecialized readers have trouble understanding such documents, they may blame themselves for not being

---

[1] George D. Gopen and Judith A. Swan, "The Science of Scientific Writing." *American Scientist* 78 (1990): 6.

able to "absorb" technical information; but quite often the real cause is the writer's forgetting to provide background information that a nonspecialized audience needs in order to understand the material. This is not mere carelessness on the writer's part: as one's knowledge of a subject advances, the distinction between "basic" and "specialized" knowledge begins to blur.

Take, for example, a water-quality specialist whose job is to report, to the general public, the quantities of contaminants detected in the local water supply. This writer may assume that the data are sufficiently clear "to speak for themselves," as in the following table.

Constituent	Units	Maximum Contaminant Level	Great Oaks Water Results
Inorganic Chemicals			
Aluminum	mg/l	1.0	ND
Barium	mg/l	1.0	ND
Chloride	mg/l	500.0	27.0–84.0
Fluoride	mg/l	1.4–2.4	.3–.82
Lead	mg/l	.050	ND
Radioactivity			
Gross Alpha R.	pCi/l	15.0	1.4–1.5
Gross Beta R.	pCi/l	50.0	.5–7.2
Additional Constituents Analyzed			
Hardness	mg/l	NS	272–388
Calcium	mg/l	NS	43–110
Alkalinity	mg/l	NS	200–302

Adapted from Great Oaks Water Co. (San Jose, CA), Annual Water Quarterly Report: 1991 Test Results; Feb. 24, 1992; p. 5.

For this data to be made meaningful to outsiders, some explaining would be necessary—not just to translate the abbreviations used (mg/l = milligrams per liter; pCi/l = picocuries per liter; ND = not determined; NS = no standard), but to explain the degree of water safety in terms of the ratio of maximum contaminant levels to the contaminant levels actually detected in the current test. Also, readers would want to know more about the dangers of particular contaminants detected in the water supply, however slight their presence may be.

Also, at the start, merely the appearance of this tabulated chemical data can be off-putting to nonspecialists. Technical writers who need to address a general audience would want to cover the material in more-or-less conventional prose. For example:

> Among the inorganic chemicals (aluminum, barium, fluoride, chloride, and lead) detected in the Great Oaks test, chloride was detected in 27–84 milligrams-per-liter concentrations, and fluoride in 0.3–0.8 milligrams-per-liter concentrations. The maximum permissible contaminant levels determined for chloride and fluoride are 500 milligrams per liter and 1.4–2.4 milligrams per liter respectively. This suggests that while there may be no immediate danger, there is enough presence of dangerous contaminants in the drinking water to recommend careful and continuous monitoring.

By being aware that you are presenting technical information to a lay readership rather than a specialized one, you will be more likely to include some or all of the following elements that enhance reader comprehension:

- Definition of terms.
- Use of examples.
- Explanation of unfamiliar concepts.
- Discussion of implications.

TYPES OF READERS

Readers of technical documents are most commonly divided into "specialists" and "nonspecialists" or "laypersons"; and while this is a convenient dichotomy to make, it may be too simplified for some contexts. Some nonspecialists are more closely involved with a project than others. For example, a project manager may have specialized knowledge of the uses of particular computer systems for particular work environments, but he or she may not have a computer engineer's understanding of those systems' architectures. It would be more useful, then, to associate types of readers with job function.

Study the table on page 26, which categorizes the different types of readers according to job function and lists their characteristics. You will then be prepared to consider some useful strategies for meeting the needs of each audience type.

Keep in mind that these are rough distinctions at best; in actuality, the characteristics of one type of reader often overlap with those of another. For example, some corporate executives have enough technical expertise in the products they manufacture to understand a good deal of the technical information in the feasibility reports they request. It is not so much a difference in readers per se as it is a difference in the roles a given reader may assume—usually out of necessity rather than by choice—when consulting a document.

Type of Reader	Characteristics
Technical specialists	Engineers, designers, programmers, testers, who consult the document to assemble or operate the system being described in the document. Each specialist represents a different type or level of expertise, which means that a mixed audience exists within this category alone.
Production and Promotion Managers	Those who decide what the production and marketing strategies of a system should be. Managers read documents that help them understand a system well enough to understand how it can be utilized for a specific market—a new type of photocopier for educational institutions, for example.
Administrators	Supervisors, vice presidents, board members, who oversee operations, launch projects, determine budgets, and make policy decisions based on the written documents they receive.
Clients; Stockholders	Those outside the workplace, involved in corporate activities and decision making, who need to understand the potential of the technical systems being produced.
Laypersons	The general public: readers with a practical or intrinsic interest in technology and science, who want to keep abreast of latest developments by reading magazine articles for the nonspecialist.

DETERMINING READERS' NEEDS

Readers of technical writing possess certain expectations about—and limitations to understanding—particular documents. Although writers cannot be completely familiar with their readers' needs and expectations—nor would such information necessarily be advantageous—it is wise to make some basic assumptions to reduce the risk of omitting material that is important to one type of readership; or of adding material that is unimportant to another type of readership. Among the important concerns are whether or not to define key concepts; to use technical terms (which specialized readers will expect); or to use more universal terms (which nonspecialists require). The table on page 27 lists the typical needs and expectations of each reader type.

WRITING FOR MULTIPLE AUDIENCES

Another important point to keep in mind is that technical documents are usually going to be read by more than one type of reader, and that each reader will

Type of Reader	Principal Needs
Technical Specialists	Need precise data: exact quantities, weights, ratios, specific directions for performing tasks or reproducing experiments. Specialists usually possess the same level of knowledge about the subject as the writer; thus the writer can assume familiarity with technical language and procedures.
Production and Promotion Managers	Need to understand what potential a product or service can have on the market; what new markets can be secured; how production or service should proceed.
Administrators	Need "bottom line" information, such as costs, long-range benefits, liabilities. As policy makers, they need to understand the technical aspects of the subject matter well enough to determine budgetary and legal implications.
Clients; Stockholders	Need information relevant to assembly or operation, to the practical uses and benefits of the product or service, or to its investment potential.
Laypersons	Need to see the product or service placed in larger business and social contexts.

read only those sections of the report most relevant to his or her needs. This is why it is important to subdivide (with succinct, informative, and conspicuous subheads) such documents. Knowing specifically who will be reading your report will determine how much technical or nontechnical elaboration to give to a particular subsection. The following example shows how multiple-audience accommodation can be employed in a technical report.

In 1992 the World Wildlife Fund published a 142-page report, *Getting At the Source: Strategies for Reducing Municipal Solid Waste,* an investigation of ways in which the quantity and toxicity of municipal solid waste might be significantly reduced, in light of rapidly disappearing landfill sites. Its audience, like that of the steering committee which produced the report, consists of specialists, managers, and administrators from industry, government, and public-interest groups. Each of the obstacles to solving the problem—technical, informational, economic, political, and so on—requires strategies that reflect understanding of the specific constraints involved in each situation. Some of the readers will be better able to deal with a given obstacle than others.

The writers who prepared this report wisely subdivided their discussion of obstacles so that readers, regardless of their respective areas of expertise, could peruse the report efficiently, and locate quickly those specialized areas they would be best qualified to deal with. For example, specialists in public policy

Obstacles to Source Reduction Options

This tool is a discussion of various obstacles that can impede adoption of a source reduction option and corresponding questions the analyst should address. In summary, the suggested criteria are:

1. Technical obstacles
2. Information obstacles
3. Economic obstacles
4. Public policy obstacles
5. Consumer preference obstacles
6. Institutional obstacles

These obstacles are described in more detail below.

The need to first understand obstacles before selecting implementation strategies is based on the rationale that the option probably would have been implemented already, or implemented more successfully, were it not for the existence of such impediments as lack of information or consumer preferences. Strategies should be selected, therefore, according to their ability to overcome these types of obstacles.

1. TECHNICAL OBSTACLES

Adoption of source reduction alternatives may be impeded by a variety of technical obstacles. More research and development may need to be done before alternatives are developed or commercially available. How much will this cost, and how long is it projected to take? If research needs are small, further study may be sufficient to solve this obstacle. If the best alternative will not be available for a number of years, it may make more sense to encourage the adoption of other alternatives in the interim.

The raw materials, production capacity necessary to switch to alternatives, etc., may not be available. How long will it take to overcome these obstacles, and at what cost?

There may be physical constraints that limit the adoption of alternatives. For example, a cafeteria may be unable to purchase reusable dinnerware because it lacks the space or fittings for a dishwasher; a particular store may lack the storage space for refillable bottles returned by customers. What will it cost to correct these obstacles, or what other accommodations will have to be made?

(continued)

2. INFORMATION OBSTACLES

In some cases, manufacturers and consumers may not have adopted source reduction options because they lack the facts to make informed decisions. Their information may be incorrect, out of date, or simply unknown. The obstacles to obtaining information can be the proprietary nature of the information, the expense of obtaining the information, the lack of a means to transmit the information. It is often the case that the necessary research has not been done, so that the information does not exist. For instance, there may be insufficient information on alternatives for toxics, or a life-cycle assessment needed to compare two products may not have been performed. Some examples of information that manufacturers and/or consumers may lack include the volume of waste generated and the cost of disposal; the problems that production, use, or disposal of this product creates; and the availability, performance, costs or savings, and size of the market for alternatives.

3. ECONOMIC OBSTACLES

There may be a variety of economic obstacles to a source reduction option. One common obstacle is that consumers reject the alternatives as too expensive. This can be due to the maintenance, repair, or labor costs, as well as the purchase price.

Different economic obstacles can also impede the manufacture of alternative products. These can include: producing an alternative product would place a company at a competitive disadvantage with other firms in the industry; the alternative is produced by a different industry entirely; the profit margin or return on investment is smaller for the alternative; market studies or experience show that consumers will not buy the alternative; demand for alternatives is too small for manufacturers to produce them profitably (it is below the minimum efficient scale); the firm lacks the funds for the initial capital investment required to begin producing the alternative; and producing an alternative product may increase a manufacturer's product liability costs (for instance, by making the product less resistant to product tampering).

Another economic obstacle can be that the waste generator is not responsible for the full cost of waste disposal, and thus does not have an economic incentive to include these costs in decision making. For example, senders of unsolicited mail do not pay for its disposal. School cafeterias may not pay for water, electricity, or disposal costs out of their

(continued)

own budget. (This may also be an information obstacle, since no one may even know what these costs are, or an institutional obstacle, as noted below.)

4. PUBLIC POLICY OBSTACLES

Government standards (for safety, environmental protection, etc.) may limit adoption of alternatives, or force the substitution of less environmentally benign alternatives. In some cases this takes the place of a trade-off. For instance, paint manufacturers may comply with regulations to control smog by replacing such solvents as mineral spirits, toluene, and xylene with 1,1,1 trichloroethane, which contributes to stratospheric ozone depletion and may be classified as an air toxic.

There are many cases where such standards are implemented for good reason. For instance, refrigerator doors using catches and flexible gaskets seal effectively, and are thus energy efficient. However, catches were replaced by magnetic seals to prevent children from locking themselves in and suffocating. Energy efficiency was reduced until more effective magnetic seals could be developed.

In addition to standards, other government policies (taxes, subsidies, import restrictions, etc.) may provide contradictory incentives or otherwise present obstacles to the adoption of alternatives. For instance, there is a limitation on imports of cotton diapers from China (designed to protect an American manufacturer of cloth diapers) that is making it difficult for diaper services to expand. Those who believe that cloth diapers are better for the environment than disposables would see this import restriction as creating an obstacle.

5. CONSUMER PREFERENCE OBSTACLES

There are many factors other than cost (which is covered in "Economic Obstacles") that affect consumers' purchasing decisions. In the absence of a strategy to encourage their purchase, consumers might not buy alternative products that do not meet these needs. For instance, the alternative may require a change in lifestyle, or there may be other behavioral changes necessary. It may take more time or effort to use, or be less convenient in some other way. There may be other intangible factors that consumers value, such as the look or feel of a particular

(continued)

product. Or they may refuse to buy products that are hard to find in the marketplace, such as those that must be purchased at specialty shops or through the mail. Market research may indicate what consumers think and whether an option will meet resistance or prove acceptable.

The increasing trend toward shorter product life spans ("planned obsolescence") reflects consumer preferences for new styles or for products that cost less than more durable alternatives and that they can replace rather than maintain or repair. Others might believe such trends are beneficial largely to manufacturers. In either case, consumer obstacles may be a factor in switching to product alternatives or modifying lifestyles.

6. INSTITUTIONAL OBSTACLES

Institutional factors can prevent manufacturers and consumers from implementing source reduction. Such obstacles include organizational structure, the lack of full-cost accounting, and the need for the necessary infrastructure.

The lack of an institutional structure to encourage source reduction may prevent it from happening. An organization, in both its capacity as a producer and a consumer, may not have a clear policy or guidelines on source reduction. There may be no one responsible for implementing such a policy, or sufficient resources may not have been devoted to it. A formal organizational structure to reduce the amount or toxicity of individual products, either in individual companies or in the industry as a whole, may serve to encourage source reduction.

Another institutional obstacle may be the absence of a mechanism to make individuals, organizations, or parts of organizations accountable for their actions by linking the costs or benefits of source reduction to purchasing and use decisions. This applies to both manufacturers and consumers. For instance, as has already been noted, the electricity, water, and waste disposal costs for school cafeterias are often paid for out of the school's general fund, instead of by the cafeteria itself. In fact, no one may even know what these costs are. Thus, the cafeteria may not have an incentive to conserve resources. Similarly, in communities where trash collection is paid for through property taxes or flat fees, as opposed to volume-based rates, consumers do not face the correct economic signals.

A final obstacle may be the lack of the necessary infrastructure to support source reduction. Some examples are: a product that could be

(continued)

redesigned to achieve source reduction, but the design changes would mean that most of the service personnel in repair shops around the country would need to learn how to repair the revised product; an organization that uses disposable products because no employee has the responsibility of collecting and cleaning reusable items; alternatively fueled cars that might not be driven outside of a major metropolitan area because there would be no facilities to refuel.

would pay special attention to Section 4, "Public Policy Obstacles," because they would be the likeliest readers to, say, suggest possible solutions to the dilemma faced by manufacturers who had to compromise one environmental objective in order to comply with another.

WRITING FOR INTERNATIONAL READERS

The contemporary workplace is globally interactive, multinational and multicultural. The diversity of peoples and customs makes global interaction a rich and exciting experience. But it is important to be aware of the following principles:

1. The manner of conveying information can differ greatly between one culture and another. What might be complimentary to one culture may be offensive to another. What might be crystal clear to one culture might be incomprehensible to another. Researcher Emily A. Thrush tells of her experience teaching English in Saudi Arabia and using a U.S.-published textbook that kept alluding to vending machines. "None of my students had ever seen a vending machine and they all found the concept of food coming from vending machines so bizarre that I had to throw away the books and find other materials to use as teaching aids."[2]

2. Information is sometimes processed differently in different cultures. Hierarchical arrangement, for example, would be neither appreciated nor understood in some countries.

---

[2]Emily A. Thrush, "Bridging the Gaps: Technical Communication in an International and Multicultural Society." *Technical Communication Quarterly* 2 (Summer 1993): 275.

3. In some cultures, proper forms of address in business correspondence go considerably beyond a mere formulaic greeting ("Dear Ms. Smith," etc.): the recipient must be accorded explicit respect throughout the letter. What might pass as conciseness in an American context might pass as rudeness" in, say, a Japanese or a French context.

If you find yourself writing for a reader whose cultural traditions are unfamiliar to you, what do you do? The answer is easy: Ask for guidance from those who might know. Or search out model documents from the culture in question. Above all, be sensitive to the needs and expectations of different traditions. Interaction no longer implies assimilation.

READER AWARENESS AT THE DRAFTING STAGE

Writing, virtually by definition, is a symbiotic relationship of author-based goals with reader-based expectations. This especially holds true for technical writing. One could say that, ideally, the author's goals are indistinguishable from the reader's expectations (i.e., What I wish to accomplish in this document is exactly what you hope that I accomplish). The reader, after all, must put the document to use; and for a document to be usable, the author must know what the reader is looking for and must have a clear sense of what he or she is capable of understanding.

But how does the author come to know these things? Sometimes the answer is startlingly easy: in the workplace, the reader literally will *tell* the author what he or she expects to see in the finished document. Aside from that, it is difficult to generalize, for every writing task constitutes a unique set of variables: the nature and purpose of the document in question; the circumstances that occasioned the document; the explicitly articulated needs of the audience for whom the document is principally intended. By recognizing the complexity of the rhetorical situation, writers will be less apt to make stereotyped judgments about their target readership.

To give an idea of how a writer uses audience awareness to help shape the draft of a technical document, the examples on pages 34 and 35 represent three situations—one for each type of reader.

In empirical-research documents such as this, the writer is a technical expert writing to other technical experts. In this case, she makes several assumptions at the outset:

1. These experts are already aware that ice impedes water flow in rivers.

Type of document:	Empirical research report.
Purpose of proposal:	To establish a means of determining the flow resistance of river-ice covers.
Type of specialists:	Hydraulic engineers.
Needs of these readers:	(1) a reliable model for determining flow resistance of river ice; (2) detailed procedures for establishing the model; (3) description of the river system used as a model, and why it can be regarded as representative; (4) statement of results.

2. They may need to be reminded that the underside of newly formed river ice is rough, but it gradually smooths out as the ice thickens.

3. They understand the mathematics required to compute such principles as resistance coefficients.

4. They understand the techniques of simulation modeling.

Here are two excerpts from the report:

> Modeling flow conditions in a large, ice-covered river requires the simulation of time-dependent variation of the undersurface resistance of the ice cover. It is well recognized that the roughness of the underside of the ice cover varies over a wide range throughout the winter [bibliographic references given].

> * * *

> It has been observed [references given] that the resistance to flow under an ice cover generally decreases from a high initial value to a low value before the beginning of the melting period, and then increases until the breakup and disintegration of the ice cover at the end of the season. Based on these observations, and the data in the St. Lawrence River, an empirical model is developed for the simulation of the time-dependent variation of n_i. In this model, the resistance coefficient for the ice cover is considered to consist of three components, i.e., $n_i = n_d + n_t + \bar{n}. \ldots$ The first component, n_d, is a function of time that increases monotonically during the freeze-up period and decreases monotonically during the rest of the stable ice-covered period. The component n_t represents the increase in ice resistance due to the formation of ice ripples during the melting period.

> —Hung Tao Shen and Poojitha D. Yapa, "Flow Resistance of River Ice Cover," *Journal of Hydraulic Engineering* 112 (Feb. 1986): 142; 145.

It is clear from the preceding excerpts that the authors are addressing an audience of fellow specialists. The writing is impersonal, precise. No effort is made to explain concepts to nonspecialists, because this report is not intended for—nor published in a journal for—nonspecialists. However, explanations of points, such as newly devised equations, that the *specialists* might not be familiar with, are carefully presented.

Situation B: Document for Administrators

Type of document:	Progress report (testing of plant foods).
Purpose of report:	To account for amount of work completed within a specific time-frame, expenses accrued within that time, an estimation of work and expenses remaining, and the differences, if any, between these estimates and the original estimates before the work began.
Type of administrator:	Project supervisor for a commercial nursery.
Needs of this reader:	(1) Detailed narrative of work completed (number of plant species used, which plant foods used for each species, any unexpected problems and how they were resolved); (2) list of materials or equipment purchased and used; (3) record of all expenditures, matched against original budget allotments.

Writers of progress reports need to be aware that their readers are as accountable as they are, so they must report the progress of their work in such detail as to leave no question about how time and funds were spent. This is not merely a "watchdog" situation, either: the information derived from such reports is used to estimate funding and time allotments for future projects, as well as to help determine retail pricing—in this case, how much more to charge for the plants with the plant food(s) selected via the testing. More information about writing progress reports can be found in Chapter 14.

Situation C: Document for Laypersons

Type of document:	Brochure on influenza.
Purpose of brochure:	To provide clear and accurate information about influenza, how it can be prevented and treated.
Type of laypersons:	Those with a general interest in health and medicine. No specialized knowledge is assumed.
Needs of these readers:	Specific information about (1) causes of the flu; (2) symptoms; (3) seriousness of the disease; (4) who is most susceptible; (5) how it can be prevented; (6) reliability of flu vaccines; (7) effective medications; all of the above explained in clear, nonspecialized language.

Readers of brochures need the most basic and useful information: definition of the concept, how to test for it, how to obtain or avoid it, how to deal with it. The writing must be as concise and accessible as possible. More about writing brochures can be found in Chapter 20.

As you can see in each of these cases, the readers' needs and the document's type and purpose are closely related; one complements the other. However, the attempt to pin down readers' needs on the basis of reader profiles alone may cause you to overlook some of the important elements in the document in progress.

AUDIENCE TARGETING: A CHECKLIST

The following list of questions should be referred to during the initial drafting stages of the document in question:

1. *Workplace-related documents*
   - [ ] Have I determined the technical or administrative expertise of the person or committee requesting the document?
   - [ ] Am I clear about the situation that occasioned this document?
   - [ ] Have I included sufficient background information for this reader?
   - [ ] Have I included appropriate technical data—formulae, equations, statistics, charts, graphs—that my reader would expect me to include?
   - [ ] Will this reader expect a cost or budget analysis?
   - [ ] Have I answered the most important question this reader will expect the document to answer?
   - [ ] Have I structured this document the best way for my target audience?
     - [ ] Informative headings, subheadings?
     - [ ] All facets of the topic covered?
     - [ ] Appropriate, concrete examples?
     - [ ] Clear introduction?
     - [ ] Clear, concise summary?
   - [ ] Is my reader a member of a culture whose discourse traditions are unfamiliar to me? If so, have I checked with experts to see if my organization, wording, or allusions are acceptable, clear, and respectful?

2. *Journalistic Documents*
   - [ ] Have I assumed no prior knowledge about the topic? Or if I have, is my assumption justified for my particular readership?

☐ Have I defined or explained concepts that are likely to be unfamiliar to my readers?

☐ Have I determined what my readers most expect to learn about the topic from reading this document?

☐ Should I include visual aids, such as graphs or drawings? (These are discussed in Chapter 11.)

☐ Will my readers be engaged by the way I approach the topic?

☐ Is the organization of the piece suitable for my intended readers?

   ☐ Does the introduction engage the reader's interest and set the stage for what follows?

   ☐ Have I divided the discussion into logical sections, each introduced by a subhead?

   ☐ Does the ending summarize the key ideas and reach a conclusion, if appropriate?

■ CHAPTER SUMMARY

All writing is to some extent an interaction between a writer's intentions and readers' needs. In work-related contexts, technical writers usually receive their assignments from the readers themselves: technicians, project managers, executives, or some outside agency in a contractual relationship with the writer's company. Such writers should be aware of their reader's principal needs: the specialist's need for precise data; the manager's need for a product's or service's market potential or production procedure; the administrator's need for cost analysis; the client's need for understanding assembly, operation, or benefits. Outside of work-related contexts, much technical writing is produced in the way of newsletters, brochures, articles, and essays for nonspecialists. Here the writer must be able to make technical concepts and language comprehensible and interesting; such readers are interested in the social implications of the scientific and technical discoveries or progress being discussed.

■ FOR DISCUSSION

1. You have been asked to prepare a report assessing the impact that a cinema complex would have in a suburban neighborhood. Imagine that the finished report would be distributed to the following readers:

a. Residents of the neighborhood in question.
b. Investors in the cinema complex.
c. Small-business owners in the neighborhood.
d. Administrators of the neighborhood's senior citizen residence.

What would be the principal concerns of each audience? What would be the best way for you to accommodate these different readers?

2. Discuss each of the following passages in terms of their ability to communicate to their respective audiences. What, if anything, could each author do to improve clarity?

a. [*military aircraft enthusiasts*] NVGs [night vision goggles] can be divided into two broad types. For the sake of this discussion I'll refer to them as ANVIS and Cat's Eyes. With the ANVIS type, you look directly through the light intensification tube. With the Cat's Eyes, you look at an image that has been piped to your eyes from a remote intensification tube. . . . The distinction is important because the ANVIS type does *not* work well in a fighter cockpit. ANVIS NVGs make it extremely difficult to see cockpit displays. If you are trying to view a FLIR image on the HUD (an image from a LANTIRN pod for example), you must now view the real world through the two levels of less than optimal resolution. Not good.

—Joe Bill Dryden, "Out of the Dark," *Code One* (January 1992): 18.

b. [*students in an introductory calculus course*] A function of a variable x is a rule f that assigns to each value of x a unique number $f(x)$, called the value of the function at x. [We read "$f(x)$" as "f of x."]

—Larry L. Goldstein et al., *Brief Calculus and Its Applications,* 4th ed. Englewood Cliffs: Prentice-Hall, 1987: 5.

c. [*lay readers interested in space flight*] [Q] Would it be possible to have a baby in space? [A] Nobody knows for sure, because it hasn't been done. However, nothing in the discoveries to date has indicated that it would not be possible.

Some puzzling results were obtained on a 1989 shuttle flight which carried thirty-two fertile chicken eggs. Sixteen had been laid two days before launch and sixteen had been laid nine days before. After the flight, the eggs were recovered and the incubation period was continued to completion on Earth. All of the nine-day eggs hatched out but none of the two-day eggs hatched.

There has been much speculation about the cause of this difference. However, the generally accepted explanation is that gravity is required for the early days of incubation. The yolk of the egg must "settle" to the inside wall of the shell where it attaches. Apparently, this contact with the inside surface of the shell is necessary for the proper development of the chick embryo.

—William R. Pogue, *How Do You Go to the Bathroom in Space? All the Answers to All the Questions You Have about Living in Space.* New York: TOR Books, 1991: 108.

3. Locate a set of instructions (e.g., from a computer program's user manual), and analyze the writer's techniques for accommodating his or her intended reader(s). How might the author have made the document more accessible?

4. Get into small groups and exchange drafts of an assignment you had prepared earlier. Critique each other's paragraphs solely on the basis of reader accommodation or the lack of it.

5. Discuss the problems associated with writing a technical report for a mixed audience. Assume that the purpose of the report in question is to demonstrate the urgent need for housing development in a region occupied by an endangered species of bird. Suggest strategies for addressing the concerns of the advocates (environmentalists), the opponents (e.g., a builder wishing to erect a shopping center on the site in question), as well as the taxpaying local residents.

■ FOR YOUR NOTEBOOK

1. Summarize a scholarly article in your major subject for each of the following readers:
   a. another student majoring in the subject
   b. your 12-year-old cousin
   c. your technical-writing instructor

2. Keep a running record of your own efforts to accommodate your target audience in the next assignment you prepare. What comes easiest? What do you find hardest to determine about reaching your audience?

3. Describe your own habits as a reader; as you read a document of fair complexity, try to record everything you do or think about to create meaning out of the words. Do you reread a single sentence—perhaps several times? Do you try reading it aloud? Why? Do you underline or take notes? What do you do when you encounter a confusing expression or explanation?

■ WRITING PROJECTS

1. Write out a detailed critique of an author's reader-targeting strategies in the document you've located for the preceding For Discussion topic 3, or in another technical document. Use the audience-targeting checklist in this chapter for your criteria.

2. Choose one of the following concepts, or a concept from your major field, and write a 1-page explanation to two different types of lay readers: (1) seventh graders; (2) college seniors (humanities majors).
   a. Integrated circuitry.
   b. The Big Bang.
   c. Radioactivity.
   d. Alcohol addiction.
   e. Symbiotic relationships.
   f. Animal hibernation.
   g. The scientific method.
   h. Sonar.

3. Write a letter, imaginary or actual, to a current or former employer, requesting a raise in salary, providing clear justification for your request. Next, write a description of the strategies you used to accommodate this particular reader; include reasons for your choice of strategies.

3

Writing Interactively

An old stereotype about writing is that it is best done in isolation. This reflects, perhaps, the working habits of many literary *artists* (poets, novelists, and the like). But technical writing is not one of the fine arts; and while it is true that solitude is indeed important at certain times for all writers, it is also true that technical writers benefit in important ways from interaction with one another; with representative readers (i.e., with those who can respond objectively and critically to a draft-in-progress); and with managerial and technical personnel, whose feedback can be valuable for job-specific writing projects.

Of course there are no rules to dictate when writers should work alone or when they should work in groups or when they should do both. But the fact remains that many technical writing tasks are done by teams. The more complex the document, the larger the number of writers assigned to the task. This chapter suggests ways in which you can be an efficient member of such a writing team.

THE IMPORTANCE OF WRITING INTERACTIVELY

The benefits of interacting with fellow writers, technicians, managers, and other professionals are many. Because written documents such as proposals, reports, and surveys form the basis for important (and expensive!) corporate decisions, immediate feedback is vital. Like moviemaking, wherein directors and actors view "rushes" (the filming for that day), technical writing benefits by being tested every step of the way. It is not uncommon for substantial losses to be incurred as the result of an insufficiently developed or clumsily written document. This is especially true with regard to proposal writing.

For example, in their study of proposal writing in a Silicon Valley engineering firm, AJI (Atherton Jordan, Inc.—not its real name), researchers Claudia McIsaac and Mary Ann Aschauer provide a vivid picture of what it is like to write interactively in a highly competitive corporate environment:

> Once AJI receives an RFP [Request for Proposal], they must move so quickly to write a proposal that frequent customer contact, even well before the RFP comes out, is critical in winning a proposal. As a result, AJI's business acquisitions representatives will spend one to three years building rapport with a customer and building a customer's needs. Then, when the RFP arrives, AJI not only is expecting it, but may have helped shape it. . . .
>
> In spite of this early preparation, the proposal-writing environment is a virtual pressure cooker. Large proposals exceed 1000 pages, and AJI has only 30–45 days [to complete one]. Putting a large proposal together involves a com-

plex 52-step process. Everything—from conceiving the design to developing outlines, themes, graphics, and various drafts—runs on an extremely tight schedule.

Average proposals involve at least 40 people in the writing process; large proposals involve up to 100 people. . . . Clear and frequent communication is crucial among the engineers as they work together. For example, the engineers working on the technical volume, the most critical part of the proposal, must communicate constantly with one another, for each part of the design affects and is affected by other parts.

—Claudia Mon Pere McIsaac and Mary Ann Aschauer,
"Proposal Writing at Atherton Jordan, Inc.,"
Management Communication Quarterly 3 (May 1990): 534–535.

McIsaac and Aschauer call attention to the difficulty these writer-engineers face with such rhetorical problems as writing for a specific audience, being sufficiently motivated by the project managers, learning to write effective generalizations based on technical specifics, and being persuasive. To deal effectively with these rhetorical problems, AJI created a Proposal Operations Center, staffed by 12 persons "whose educational backgrounds range from engineering and business to journalism and literature" (540).

Your technical-writing course will probably give you a taste of what it is like to work interactively with others on an assignment—not only with classmates but also perhaps with academic specialists (professors and technical support staff) within the college community or even outside of it. The aim here is to introduce you to the way writers—whether they are freelance scientific/technical journalists or company-employed writers—work together on the job.

Because scientific and technical fields consist of so many specialized subdisciplines, it is unlikely that any one person can be sufficiently informed of those specialities, outside his or her own, which often need to be tapped as information sources for a document in progress. When a writer interacts with a specialist, it is important to realize that the specialist usually does a lot more than simply provide the "content" that the writer can then magically transform into pellucid prose. Such razor-sharp separation of form and content is essentially a false dichotomy. The following case in point illustrates why.

A CLOSER LOOK AT WRITING INTERACTIVELY IN THE WORKPLACE

Craig, a junior technical writer responsible for preparing catalogue descriptions, brochures, and other promotional materials for his company's line of computer games, has been asked by his supervisor, Ruth, to write an evaluation

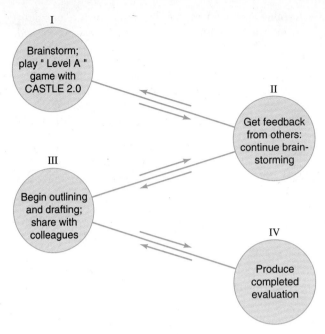

I
Brainstorm; play " Level A " game with CASTLE 2.0

II
Get feedback from others: continue brain-storming

III
Begin outlining and drafting; share with colleagues

IV
Produce completed evaluation

FIGURE 3.1 Craig's four-step procedure for producing his evaluation.

of proposed upgrades for their popular chess program. The new version, CAS-TLE 3.0, Ruth explained to Craig, would basically include new levels of ease for beginning chess players as well as new levels of difficulty for experienced players. But most importantly, it would include a chess tutor, accessible at any level, and for every stage in the game.

Before reading on, try to anticipate Craig's strategy. What steps would you take to complete this project?

To make his task manageable, Craig decided to approach it in four steps—two of them planning-centered, the other two writing-centered (see Figure 3.1 for the flow chart Craig prepared for himself).

Step One

There is such a thing as interacting with others too soon. Craig felt that before he sought anyone else's input, he needed to determine his own. So he sequestered himself in his office, booted up the current version, CASTLE 2.0, and played a game at Level A, to help him recall what it was like when he first learned chess strategy. He knew that it wouldn't be enough for the proposed tutoring capability simply to list the best move for the player to choose in a given situation. No *learning* experience there. Craig jotted down more complex operations:

After computer moves, player might ask for a list of several options to choose from.

Player can ask program to describe the likely effect that a given option can have (say, over next three or four moves).

Player can devise own moves, then ask the program to analyze each one's effectiveness before committing to one of them.

Computer might alert player to reconsider highly unusual moves or apparent blunders.

Step Two

Now Craig felt that he was ready to solicit the responses of two or three others whom he would ask to do the same thing: play chess with CASTLE 2.0; jot down anything you would imagine a novice appreciating or requiring help with. Later in the week, Craig scheduled a brainstorming session. One discovery he made was that each of his respondents came up with some identical suggestions (e.g., a "pull-down" list of desirable moves accompanied by a rationale). To these unanimously agreed-upon, independently arrived-at recommendations, Craig would give top priority. Other recommendations were quickly agreed upon by the group: voice-activated moves; three-dimensional chess capability; a "Great Games" collection, containing some of the famous American Grandmaster and World Chess Competition games, from Capablanca to Bobby Fischer. Discussion of the remaining recommendations—including those that were not mutually agreed upon or that had been conceived of during the brainstorming session—was scheduled for a separate meeting.

Additional meetings are always a good idea, especially with Craig's brainstorming group. During the excitement of that first meeting, the group members understandably made decisions without thinking them through carefully enough. The second meeting was less "brainstormy," more businesslike: each of the recommendations was examined carefully, objectively. "A three-dimensional capability? Were we serious? It would eat up far too much memory, be much too expensive to produce; better shelve that for a separate project five years down the road." By the time that second meeting was over, the long list of add-ons had dwindled down to two or three, each one accompanied by reasons for inclusion.

Steps Three and Four

Finally it was time to start drafting the agreed-upon recommendations. Craig wondered what would be the best way of working interactively on the document. Should he write a complete first draft, then show it to the others? Or should he solicit feedback after finishing one or two sections? Or maybe he should first work up an outline and get reactions to that? Craig decided to follow his own inclination as a writer, which was to produce a complete draft and then let the others take it apart.

While he was working on the draft, however, Craig was reminded of the fact that his technical-specialist colleagues could do a lot more for him than merely download data; they could glance at his outline or draft-in-progress, and point out imprecisions—say in his instructions for accessing a particular program option. So halfway through his draft, Craig began working up an outline. As he developed his outline, he discovered ways to develop his draft; and when he began working on the draft once again, the tentative outline at his side provided him with clues for improving the organization of the report.

TYPES OF INTERACTIVE WRITING TASKS

Table 3.1 outlines the fundamental ways in which a writer can interact with peers (either fellow writers or nonwriter specialists) when working on any document.

Of course, actual interactive writing experiences will rarely be so orderly as the table may lead you to believe. Nonetheless, being aware of the possible ways of interacting with others as you're preparing your document can help prevent writer's block or false starts.

Table 3.1 Working Interactively at Different Stages of the Composing Process

Stage of Composing	Typical Interaction Sought
Inventing	Help in generating proposal or outline content through conversation, questioning, brainstorming (oral or written, using a blackboard or computer terminal).
Information-Gathering	Help in locating sources; help in translating sources from other languages; help in understanding highly specialized material.
First draft	Response to proposal's development, organization, persuasiveness, coherence; double-checking accuracy of factual information with appropriate specialists.
Final draft	Checking for ease of reading, conciseness, spelling, punctuation, mechanics, proper formatting.

STRATEGIES OF CO-AUTHORSHIP

Frequently in the workplace, writers are asked to work together on a single document—a form of interactive writing called collaboration or co-authorship. Several variations of co-authorship exist. For example, one or more writers may be solely responsible for the first draft, while another is solely responsible for the revision and copyediting. Or, all the writers work on different sections of the document, and all are responsible for revising what they've worked on.

Co-authoring a document needs to be planned carefully. If one of the writers on the team fails to do his or her part, the whole project can be jeopardized. Collaborators must meet frequently, and if one of them doesn't make it to a meeting, this fact alone can cause trouble because many writing projects are under tight deadlines. The following guidelines should be kept in mind before embarking on a co-authoring project:

1. Make sure that your co-author(s) and you can work together harmoniously.
2. Choose one of the writers to serve as coordinator.
3. Establish a timetable for completing each major section of the document. Be sure to include sufficient time for revisions.
4. Schedule meetings at regular intervals, and agree to attend them without fail.
5. Set up an agenda for each meeting. Members of the group will arrive better prepared to participate if they know what to expect.
6. Assign someone to keep minutes; that eliminates the possibility of brilliant ideas being lost forever because they weren't recorded.
7. Decide upon the method of collaboration to use on the project. Should only one writer be responsible for a given section? Should a revision of a given section be done by the same writer or by a different writer?
8. Agree to be candid with each other. Collaborators must be willing to give and take frank criticism.

THE ETHICS OF WRITING INTERACTIVELY

Whenever people interact, ethics becomes an important concern. Simply stated, ethical behavior refers to abiding by standards of proper conduct—conduct that helps preserve respect and fairness toward the rights of others. In group dynamics, this translates into the following:

1. Listening carefully and open-mindedly to others' views, no matter how sharply you disagree with them.

2. Asking questions before drawing conclusions, to help clarify possible misunderstandings.

3. Properly crediting the persons whose ideas you use, whether those ideas are published or not.

4. Never "hogging" the conversation during meetings. And making an effort to encourage those who haven't spoken as much as others to say more.

5. Preserving the confidentiality of ideas presented, unless everyone in your group agrees to "go public" with certain information.

It is difficult to overemphasize the fact that planning such interactive strategies in advance will greatly increase the chances of success. Discord within a group can sabotage a writing project. Harmony will enable the individual talents of each group member to be exploited to the fullest.

WRITING INTERACTIVELY IN ACADEMIC SETTINGS

Quite likely, your technical communication course will offer you the opportunity to write interactively with your classmates, and perhaps with other members of the university community as well. Drafts of assignments are "peer-critiqued" in class; that is, students get into small groups, exchange drafts, and respond critically to them. You will offer fellow writers suggestions for improvement, just as they will offer suggestions to you.

If you have never critiqued someone's writing before, you may find the experience somewhat intimidating at first. After all, a first draft is usually full of insufficiently developed points, awkward sentences, and just plain stupid mistakes. That is why it is important for everyone to realize that "first draft" is a draft *in progress,* that a writer cannot tend to everything all at once, and that one purpose of group critiquing is to develop each writer's ability to see his or her own work (as well as the work of others) in a critical light. For more discussion of revision strategies, see Chapters 5, 8, and 9.

Because it is helpful to know beforehand what sorts of critical responses are expected of you—and what you in turn ought to expect from your readers—the following guidelines may be of use.

Table 3.2 Types of Reader Responses to Drafts		
Type of Response	**Function**	**Example**
Directional	Tells writer specifically what to do to improve the problem at hand.	"Give more concrete examples here."
Judgmental	Draws an evaluative conclusion about the passage in question.	"These assertions are too general."
Facilitative	Suggests a *possible* strategy for revision.	"Can you think of any examples that would convince your readers of your views?"

GUIDELINES FOR A DRAFT-CRITIQUING WORKSHOP

Note that these guidelines are subject to modification by your instructor.

I. *Preparation for the workshop*

   A. Be sure that your draft is as finished as you can make it. Anticipate reader responses as you're working on it—you'll discover that keeping your readers in mind enhances your awareness of clarity and detail. You are expected to share a first draft, as opposed to a "rough" draft. Rough drafts typically are incomplete, inadequately organized, syntactically uneven, muddled in places, full of surface errors. A first draft reflects your best writing intended for feedback from others. It will be formatted as meticulously as a final draft.

   B. Bring as many copies to the workshop as your instructor directs. If you will work in groups of four, you will need to make three copies of your paper.

   C. Be on time to the workshop. Lateness will prevent others in your group from spending sufficient time critiquing your draft.

II. *Participating in the Workshop*

   A. If your instructor distributes critique sheets, follow his or her directions exactly.

   B. Read each draft carefully, not as "critics" per se, but as interested and involved readers.

   C. Regard the draft as a *work-in-progress;* your feedback will be offered in the spirit of what can be done to make the draft a finished piece of work.

   D. Try to respond with more facilitative than judgmental or directive remarks (see Table 3.2).

E. Be as specific in your critical responses as you can. Instead of "Can you add more details to this section?" give a more focused prompt, such as "Can you cite any examples of industrial wastes being dumped into Lake Michigan?"

F. In addition to recording your responses on a critique sheet, talk with the author. Ask the author to clarify points that seem confusing or undeveloped to you. Having such writer-reader conversations will give you insight into the problem-solving strategies of other writers, as well as the way readers respond to a document.

WRITING INTERACTIVELY: A CHECKLIST

The following list of questions should be referred to for any interactive writing assignment:

1. Whenever you interact with specialists, supervisors, or fellow writers.
    - ☐ Have all my technical questions been answered in a way that I can understand them?
    - ☐ Have I asked the specialist to clarify anything I do not fully comprehend?
    - ☐ If I disagreed with anything, is it possible I may have misunderstood a technical matter?
    - ☐ Has everyone in the group had equal opportunity to speak? to counter-respond?

2. When collaborating with one or more writers on a single document.
    - ☐ Have you set up a mutually agreed-upon timetable?
    - ☐ Has the work been divided up to everyone's satisfaction?
    - ☐ Have you allowed sufficient time for revising?

3. When peer-critiquing a first draft of an assignment in class.
    - ☐ Does my draft represent my best effort prior to receiving feedback? (If you find yourself apologizing for the draft—e.g., "This is very rough"—then the answer is no.)
    - ☐ Have I printed out a neat, well-formatted copy?
    - ☐ Am I bringing along enough copies for everyone in the group?
    - ☐ Am I prepared to give detailed, facilitative responses to the drafts I critique?
    - ☐ Am I prepared to encourage my peers to be candid in their criticism of my draft?

■ CHAPTER SUMMARY

Writing on the job, because of its complexity and the tight deadlines for completion, often requires interaction with technical professionals, managers, and other writers. Such interaction is valuable because a technical document often must reflect the expertise of many specialists (including writing specialists). A specific kind of interaction is sought for each stage of the composing process—the prewriting, first-draft, and final-draft stages. Successful collaboration (co-authoring) of technical documents should be carefully planned: the co-authors should feel comfortable working together, meet frequently, prepare agendas for what they need to cover at each meeting, consider each one's point of view carefully, and offer candid criticism.

A valuable training ground for interactive and collaborative writing is the technical writing classroom. Students should prepare carefully for a peer-critiquing workshop to reap its many benefits: be sure your draft is as finished as you can make it; anticipate reader responses as you're working on it. Come to the workshop on time, with sufficient copies of your draft, and be prepared to give and receive detailed suggestions for revision. Emphasize facilitative responses over directional or judgmental ones.

■ FOR DISCUSSION

1. What are the potential advantages of writing in a group over writing alone; of writing alone over writing in groups?

2. According to Jean Ann Lutz, in her article "Writers in Organizations and How They Learn the Image: Theory, Research, and Applications":

Most employees must integrate two kinds of knowledge when they join an organization: knowledge of their discipline (or career field) and knowledge about their profession. Writers who join an organization must also integrate a third kind of specialized knowledge, an awareness of that organization's culture and image" (118).

Discuss what might constitute an organization's "culture and image"—Management structure? Ways that employees are encouraged or required to interact? Recreational facilities?—and suggest how these things can influence the way writers work or the quality of that work. Also think of organizations other than those in business and industry:

clubs, schools, churches, sororities or fraternities, and others. How do such organizations affect the way you communicate with one another?

3. Form groups of four and role-play the following scenario in class:

You are the writing team for HOT WHEELS, INC., a firm that manufactures roller skates and skateboards. Your job is to prepare a fact sheet to describe the capabilities of one of your products and how these capabilities make the product superior to its counterparts manufactured by the competition.

Follow this procedure:

1. Choose one person to coordinate (i.e., assign specific tasks for each member of the group; decide on time allotment);
2. Begin with some brainstorming and note taking;
3. Complete the writing task assigned you by the coordinator. For example, list the top-of-the-line roller skates' most desirable features or capabilities; write a paragraph describing the kinds of maneuvers possible; prepare a fact sheet that includes the technical details about the wheels or bearings, the shape of the board, quality and comfort of the shoe, etc.;
4. Respond to each other's drafts and suggest revisions.

■ FOR YOUR NOTEBOOK

1. Keep your own "minutes" of an interactive or collaborative session in which you participate. Record the way in which the group was organized, conflicts that arose within the group (and how they were resolved), and the specific manner in which your problems with a writing task were dealt with by the technical expert or the co-author.
2. Assume the role of outside observer, and keep a detailed, running record of a collaborative writing group in action. Pay particular attention to the way the group divides the writing task among themselves, the length and frequency of its meetings, and what is being discussed at those meetings.

■ WRITING PROJECTS

1. Complete the revision of the small-group writing task you have begun in For Discussion topic 3. At your instructor's discretion, you may be credited just for your section, or you may share responsibility for the entire project.

2. Working in a group, begin a draft of a short report of a campus problem, such as lack of wheelchair access to some of the buildings and facilities. Interact with others outside of class: wheelchaired students, staff, or faculty; civil-engineering faculty; personnel in the physical plant; and anyone else who can give you more information about the problem and how it might be solved. Revise the draft accordingly, and append a second short report that discusses how interacting with those persons helped develop the report. Include any problems that occurred in the course of your consultations as well.

4

Finding and Incorporating Technical Information

Information gathering is a way of life for writers, and technical writers are no exception. Knowing how and where to locate the information you need—and how to incorporate it into your documents—are fundamental skills that can save writers hours of trial-and-error foraging through library materials and databases.

This chapter aims to guide you through these vast print and electronic resources in order to locate the statistics, experimental results, schematics, maps, surveys, definitions, analyses, formulae, and many other kinds of information necessary for substantiating technical reports. You are also introduced to the techniques of interviewing and survey taking, which are essential for acquiring timely information. Finally, you are shown how to integrate such information smoothly into your documents.

AN OVERVIEW OF THE RESEARCH PROCESS

Before you examine each information-gathering task separately, it is a good idea for you to look at the entire process as one fluid, interactive whole. Like the composing process itself (see Chapter 5), the research process is dynamic rather than linear. The basic tasks include the following:

1. *Intra-Library Activities*

   - Preparing a tentative outline for your document.

   - Doing background reading (encyclopedias, introductory texts); taking note of accompanying bibliographies.

   - Examining the works cited in these bibliographies; noting the more extensive bibliographies accompanying these works.

   - Checking the on-line catalogue for book searches via author, title, subject, or keyword(s); checking periodical indexes and databases (including government databases) for articles on the topic.

   - Locating the sources most appropriate to your topic, taking good notes.

2. *Research Activities Outside the Library*

   - Interviews with experts in the topic of your research:

     professors and research aides on campus

     professors at nearby colleges

     experts in the private sector

     friends and family members who are experts.

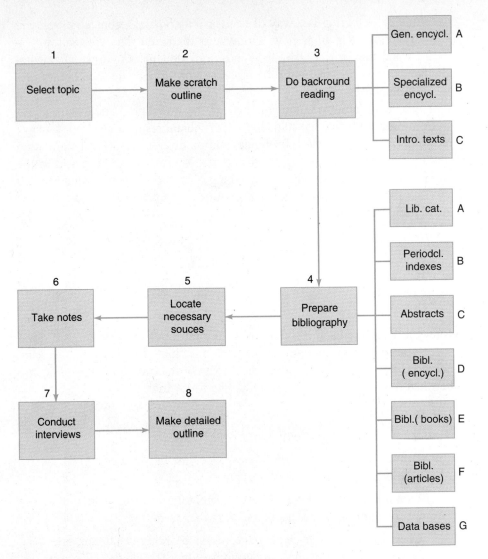

FIGURE 4.1 Research process: linear model.

- Examination of in-house company publications.
- Self-conducted surveys.
- Self-conducted laboratory or field research.

You may set out to do "first things first." That is, *first* plan your strategy, *then* do background reading in encyclopedias, *then* check databases and print indexes, *then* obtain and study the articles that seem most relevant, and so on (see Figure 4.1). But more realistically, you will find yourself working in a nonlin-

FIGURE 4.2   Research process: dynamic model.

ear manner. Discovering a new facet of your topic while reading background material, for example, could be enough to send you back to the planning stage of your document. A chance encounter with a professor who is an expert in your topic could result in an immediate interview—something you hadn't planned to do until the end of your research. Browsing through a book shelved alongside the book you intended to retrieve could add an important new dimension your thesis. Even a casual conversation with a friend during lunch could do the same thing (see Figure 4.2).

The point is simple: plan ahead, but not inflexibly. Adhering to too rigid an agenda could result in overlooked leads, unforeseen possibilities. And besides, research becomes more of an intellectual adventure, as well as a deeper learning experience, when you allow for these unknown factors.

USING BASIC REFERENCE WORKS FOR RESEARCHING TECHNICAL INFORMATION

The Reference Room of your college library is where you will find the key works for accessing the detailed information necessary for technical-writing projects. You are probably familiar with many of them: encyclopedias, unabridged dictionaries, world atlases, the *Library of Congress Subject Headings* (for helping you conduct a catalogue search by subject), and periodical indexes such as the *New York Times Index,* the *Social Sciences Index.*

But as a researcher of information relating to science and technology, it is important that you familiarize yourself with several important, somewhat more specialized reference works. These are described in the following sections; it will be worth your while to spend an hour or so in your library's Reference Room looking them over. Be sure to ask the librarian for assistance if you cannot locate a reference work or are unsure of what reference work to use for locating a particular topic. College librarians are both trained and eager to assist students—but do try to be as specific as you can in describing your needs. For example, instead of asking, "I'm writing a report about pollution; can you help me find the sources I need?" it would be far preferable to ask much more focused questions, such as, "Where would I find the most up-to-date information on mercury contamination in the Great Lakes?" The more focused your inquiry, the greater assistance the librarian can give you.

Encyclopedias

Many encyclopedias cover science and technology (or a branch thereof) exclusively, and these are ideal for obtaining a clear overview of the subject matter you are working with. Some of these encyclopedias assume that their readers are already familiar with the basics of the subject matter in question (and are so identified in the descriptive commentary that follows each title). Most article entries in these encyclopedias contain bibliographic references.

Encyclopedias for General Science and Technology

Encyclopedia of Physical Science and Technology. 15 volumes. Orlando: Academic Press, 1987. Somewhat technically oriented, but major articles open with a glossary.

The McGraw-Hill Encyclopedia of Science and Technology, 7th ed. 20 volumes. New York: McGraw-Hill, 1992. The most comprehensive of the encyclopedias devoted to science and technology; easily accessible to the layperson.

Van Nostrand's Scientific Encyclopedia, 7th ed. 2 volumes. New York: D. Van Nostrand, 1989. The concise nature of the articles makes this encyclopedia ideal for locating information quickly—although nonscience majors may find the information more difficult to assimilate than that of the *McGraw-Hill Encyclopedia of Science and Technology.*

The Way Things Work. By David Macaulay. Boston: Houghton-Mifflin Co., 1988. Colorful, fanciful illustrations accompany lucid descriptions of devices from pulleys to CD players and rocket engines.

Enyclopedias for Specific Sciences and Technologies

The Encyclopedia of Environmental Studies. By William Ashworth. New York: Facts on File, 1991. Includes biographies of environmentalists and explanations of environmental concepts from several disciplines (chemistry, biology, ecology, geology, physics), as well as expressions used by environmental engineers, meterologists, foresters, and so on. Includes a bibliography.

International Encyclopedia of Astronomy. Ed. Patrick Moore. New York: Orion Books, 1987. Visually attractive and engagingly written entries. Ideal for those lacking coursework in astronomy.

Van Nostrand Reinhold Encyclopedia of Chemistry, 4th ed. New York: Van Nostrand Reinhold, 1984. A comprehensive reference work for those who already possess a general background in chemistry.

Macmillan Encyclopedia of Computers. 2 volumes. New York: Macmillan, 1992. Targeted for nonspecialists, this work contains more than 200 articles, including use of computers in academic disciplines and professions (e.g., astronomy, literary analysis, farming, health and fitness, aviation, retailing, sports); parallel computing, printing technologies, robotics, viruses, and careers in computing.

McGraw-Hill Encyclopedia of Environmental Science & Engineering, 3rd ed. New York: McGraw-Hill, 1993. More than 200 articles, including articles on medical waste, saline water reclamation, and reforestation.

Mayo Clinic Family Health Book. New York: Morrow, 1990. A concise and comprehensive health-reference work.

The Wellness Encyclopedia: The Comprehensive Family Resource for Safeguarding Health and Preventing Illness. Boston: Houghton-Mifflin, 1991. Compiled by the University of California, Berkeley *Wellness Letter* staff, this work includes the following sections: (1) longevity, (2) nutrition (including the Wellness food guide), (3) exercise, (4) self-care, (5) environment and safety.

The Prentice-Hall Encyclopedia of Information Technology. Englewood Cliffs: Prentice-Hall, 1987. More than 150 articles on topics such as Artificial Intelligence, Assemblers & Assembly Language, Bar-Code Systems, Communications Networks, Computer Architecture, Data Base Management Systems, Electronic Mail, Fiber Optics, Integrated Software, Modems, System Design and Development, Word Processing.

Encyclopedia of Language. By David Crystal. Cambridge: Cambridge University Press, 1992. Articles on the history, theory, and structure of language.

Encyclopedia of Mathematics. 6 volumes. Dordrecht: Reidel, 1987. Covers all branches of mathematics; useful to students of beginning as well as advanced mathematics.

Miller-Keane Encyclopedia and Dictionary of Medicine, Nursing, and Allied Health, 5th ed. Philadelphia: W.B. Saunders, 1992. Contains 13,000 entries on medical equipment, drugs and therapies related to these medical fields.

Encyclopedia of Physics, 2nd ed. Ed. Rita G. Lerner and George L. Trigg. New York: VCH, 1991. A concise one-volume work. Includes bibliography.

The Encyclopedia of Psychoactive Drugs. 25 volumes. New York: Chelsea House, 1986. Several volumes treat specific drugs within a given category; for example, the category of sedative hypnotics includes volumes on barbiturates, inhalants, quaaludes, and Valium. Other categories: alcohol, stimulants, hallucinogens, narcotics, nonprescription drugs, and understanding drugs. Ideal for the lay reader.

Encyclopedia of Psychology, 2nd ed. 4 volumes. Ed. Raymond J. Corsini. New York: John Wiley & Sons, 1994. Covers all aspects of behavioral and clinical psychology.

Encyclopedia of Statistical Sciences. 8 volumes. New York: John Wiley & Sons, 1982. Covers methods of statistical analysis across the disciplines as well as in industry, business, and government.

Systems and Control Encyclopedia. 8 volumes. Oxford: Pergamon Press, 1987. Describes the structure of complex systems such as public utilities and computer networks.

Dictionaries

Good comprehensive or specialized dictionaries in science and technology can save students hours of background reading by providing clear, concise definitions of terms alone rather than article-long coverage of entire concepts or fields—although a certain amount of overlap with encyclopedias does exist.

General Science Dictionaries

Chambers Science and Technology Dictionary. Cambridge: Cambridge University Press, 1988. The most comprehensive of the science/technology dictionaries.

The Hammond Barnhardt Dictionary of Science. By Robert K. Barnhardt. Maplewood, NJ: Hammond, 1986. A handy one-volume dictionary of important scientific terms.

The Harper Dictionary of Science in Everyday Language. By Herman Schneider and Leo Schneider. New York: Harper & Row, 1988. Many entries are illustrated and written in a lively style (see sample page, Figure 4.3).

INTIMA *See* ARTERIOSCLEROSIS

INTRUSIVE FORMATION (GEOLOGY) *See* ROCK CLASSIFICATION

INVERSE SQUARE LAW

If the light on this page is too dim, you can improve matters by moving closer to the source of light, thus capitalizing on the inverse square law. The law states that energy from a point source (call the bulb a point), if unhindered by mirrors, lenses, or other impediments, spreads out, equally in all directions and that its intensity diminishes as the inverse square of the distance. At a distance of 1 foot from the bulb, the strength of the light is, let's say, x. At 2 feet the strength is $1/4\,x$. At 3 feet, it is $1/9x$. So when you move the book from 3 feet to 2 feet, you have more than doubled the brightness of the light falling on the page.

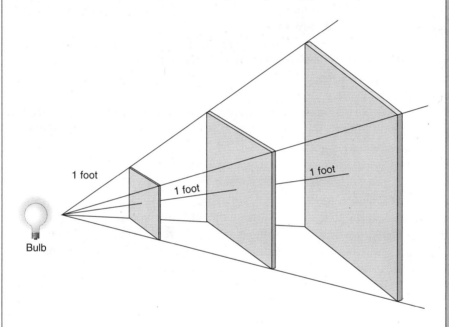

The law of inverse squares applies to any form of energy that spreads out equally in all directions. Magnetism, electromagnetism, and gravity operate under that law. Radio waves and television waves (which are electromagnetic) are strongest at the point of emission at the broadcasting station and weaken rapidly thereafter. A magnet's pull is strongest close to its poles, and so on. The moon's gravity causes the waters of the earth to heap up in tides, while Jupiter, about 26,000 times more massive than the moon, has almost no effect on tides because of its distance from the earth.

156

FIGURE 4.3   Sample page from *The Harper Dictionary of Science.*

McGraw-Hill Dictionary of Scientific and Technical Terms, 5th ed. New York: McGraw-Hill, 1994. Over 105,000 terms with 122,000 definitions from all facets of science and technology.

Dictionaries for Specific Sciences

The HarperCollins Dictionary of Biology. By W. G. Hale and J. P. Margham. New York: HarperCollins Publishers, 1991. 56,000 entries; many diagrams and charts.

Dictionary of Computing, 3rd ed. Oxford: Oxford University Press, 1990. Includes electronic data processing entries.

The Concise Oxford Dictionary of the Earth Sciences. Ed. Ailsa Allaby and Michael Allaby. Oxford: Oxford University Press, 1991. Contains 6000 cross-referenced entries on climatology, geology, oceanography, paleontology, and so on.

The HarperCollins Dictionary of Electronics. By Ian R. Sinclair. New York: HarperCollins, 1991. More than 2000 entires, with diagrams and charts.

Dictionary of the Environment, 3rd ed. New York: NYU Press, 1989. A comprehensive dictionary of environmental terms.

Dictionary of the History of Science. Ed. W. F. Bynum, E. J. Browne, and Roy Porter. Princeton: Princeton University Press, 1981. 700 entries on terms and names relating to the origins, meaning, and development of scientific ideas.

Mathematics Dictionary, 5th ed., Ed. Robert C. James and Glenn James. New York: Van Nostrand Reinhold, 1992. Covers elementary and advanced mathematical terms; lists mathematical symbols; includes French-English, German-English, and Russian-English indexes.

The Oxford Dictionary of Natural History. Ed. Michael Allaby. Oxford: Oxford University Press, 1985. The scope of natural history is steadily expanding, and this dictionary reflects that expansion, with terms from biochemistry; the atmospheric, environmental, and geological sciences; as well as from more traditional natural-science topics such as plants, animals, and their habitats.

Collective Biographies

American Men and Women of Science, 18th ed. 8 volumes. New York: R. R. Bowker, 1993. A full-scale biography of persons in the physical and biological sciences.

Asimov's Biographical Encyclopedia of Science and Technology, 2nd revised ed. By Isaac Asimov. Garden City: Doubleday, 1982. Profiles of over 1500 scientists from ancient times to the present day.

Biographical Dictionary of American Science: The Seventeenth Through the Nineteenth Centuries. By Clark A. Elliott. Westport, CT: Greenwood Press, 1979. Includes bibliography and index.

Blacks in Science and Medicine. By Vivian Ovelton Sammons. New York: Hemisphere, 1990. Biographies of African American scientists; includes bibliography and index.

Women of Mathematics: A Bibliographic Sourcebook. Ed. Louise S. Gurstein and Paul J. Campbell. New York: Greenwood, 1987. Includes bibliographic notes for each subject; cross-referenced.

Handbooks and Research Guides

No matter how much experience you have had with library resources or with information-gathering in general, you will find these guides useful.

Guide to Reference Books, 10th ed. By Eugene P. Sheehy. Chicago: American Library Association, 1986. Reference works in all disciplines: general reference works; the humanities; social and behavioral sciences; history and area studies; science, technology, and medicine.

Handbook of Research Methods: A Guide for Practitioners and Students in the Social Sciences. By Nancy Sproull. Metuchen, NJ: Scarecrow Press, 1988. A valuable guide to research for all advanced students, not just those in the social sciences. Sproull covers such topics as the research process, research ethics, how research problems originate, hypothesis testing, determining the proper sampling and data-collection methods.

How to Find Chemical Information: A Guide for Practicing Chemists, Educators, and Students, 2nd ed. By Robert E. Maizell. New York: John Wiley & Sons, 1986. Includes bibliography and indexes.

Information Needs in the Sciences: An Assessment. By Constance C. Gould and Karla Pearce. Mountain View, CA: Research Libraries Group, 1991. This work aims "to present a broad view of the shape of each discipline." Each discipline's characteristics, uses of information, serial and monograph literature, major indexing and abstracting services, are discussed.

Information Sources in the Life Sciences, 3rd ed. By. H. V. Wyatt. London: Butterworths, 1987. Includes bibliographies, indexes. This is the revised edition of *The Use of Biographical Literature,* ed. Bottle and Wyatt (1971).

Information Sources in Science and Technology. By C. Hurt. Englewood, CO: Libraries Unlimited, 1988. Describes handbooks, field guides, dictionaries, biographical dictionaries and encyclopedias for all major disciplines of science and technology.

The Official World Wildlife Fund Guide to Endangered Species of North America. 3 volumes. Washington, DC: Beacham, 1990. Each of the more than

Red Wolf
Canis rufus

Steve Maslowski/USFWS

Status	Endangered
Listed	March 11, 1967
Family	Canidae (Canine)
Description	Tawny red canine, adults averaging about 23 kg (50 lb).
Habitat	Heavily vegetated areas, coastal prairie and marsh.
Food	Carnivorous.
Reproduction	Litter of 2 to 8 pups.
Threats	Reduction in habitat, hybridization.
Region 4	North Carolina

Description

The red wolf has the size and appearance of a large dog. It is intermediate in size between the Endangered gray wolf (*Canis lupus*) and the coyote (*C. latrans*). Adults weigh between 18 and 36 kilograms (40 and 80 lbs) with males being larger. Despite its common and scientific name, this wolf shows a wide range of coloration, including red, brown, gray, black, and yellow.

Biologists believe that the species *Canis rufus* originally consisted of three subspecies: *C. r. floridanus*, an eastern subspecies that became extinct early in this century; *C. r. rufus*, a western subspecies, which was actually a red wolf-coyote hybrid and now believed extinct; and *C. r. gregoryi*, the only extant subspecies.

Behavior

Little solid scientific knowledge exists on the life history of wild red wolves. It is believed that they have a less rigid social structure than gray wolves. Red wolves usually travel in groups of two or three, but lone wolves are not uncommon.

It is unknown whether red wolves form as strong a pair bond as gray wolves. Although captive wolf pairs exhibit a fondness for each other, greeting, playing, and nuzzling, they are not faithful to a single mate. Red wolf pups are born in April or May after a gestation period of about two months. Litter sizes in captivity have ranged from two to eight pups.

The red wolf is an opportunistic predator, taking species that offer an easy capture. In

In Minnesota, where the wolf is federally listed as a Threatened species, federal agents each year have killed a few dozen wolves that have preyed on livestock. The state also has a program to compensate ranchers for animals lost to wolves. An effort in the early 1980s by the FWS to return management to the state of Minnesota would have allowed a limited wolf hunting and trapping season. Although championed by state wildlife officials as a way to control wolf populations and increase public acceptance of wolves, the attempt was blocked by a federal court after a coalition of conservation groups filed suit.

The FWS is currently working on plans to reintroduce an experimental population of gray wolves into Yellowstone National Park. In its Recovery Plan for the Northern Rocky Mountain Wolf, the FWS foresees an increase in wolves in Glacier National Park and in national forests in central Idaho. But because of the geographical isolation of Yellowstone from current wolf populations, the FWS will try to reintroduce a gray wolf population there. As a first step, the FWS expects to recommend reintroduction in late 1989, which will be followed by the drafting of an environmental impact statement. In order to deal with wolf predation on domestic livestock, the reintroduction plan will include provisions for trapping and relocating problem wolves, or, if necessary, killing them.

Defenders of Wildlife, a private conservation organization, is also working to persuade stockmen to support the reintroduction program. Defenders has brought Wyoming ranchers to Minnesota to learn from ranchers there about the effects of a local wolf population. The organization has also established a private fund to compensate Wyoming ranchers for stock losses caused by wolves.

By 1993 the FWS hopes to see the first reintroduced gray wolves back in Yellowstone National Park.

Bibliography

Clarkson, E. 1975. *Wolf Country.* E. P. Dutton, New York.

Lopez, B. 1978. *Of Wolves and Men.* Charles Scribners Sons, New York.

Mech, L. D. 1970. *The Wolf: The Ecology and Behavior of an Endangered Species.* Natural History Press, New York.

Peterson, R. O. 1986. "Gray Wolf." *Audubon Wildlife Report 1986.* Academic Press, San Diego.

U.S. Fish and Wildlife Service. 1987. "Northern Rocky Mountain Wolf Recovery Plan." U.S. Fish and Wildlife Service, Denver.

U.S. Fish and Wildlife Service. 1978. "Eastern Timber Wolf Recovery Plan." U.S. Fish and Wildlife Service, Twin Cities.

Zimen, E. 1981. *The Wolf: A Species in Danger.* Delacourt Press, New York.

Contact

Regional Office of Endangered Species
U.S. Fish and Wildlife Service
P.O. Box 25486
Denver Federal Center
Denver, Colorado 80225

Regional Office of Endangered Species
U.S. Fish and Wildlife Service
Federal Building, Fort Snelling
Twin Cities, Minnesota 55111

FIGURE 4.4 Entry from *The Official World Wildlife Fund Guide to Endangered Species of North America.*

500 entries includes a photograph of the species; its geographical location and taxonomy; descriptions of habitat, current distribution, threats to its existence; also includes bibliography and names of offices to contact for further information. (See sample entry, Figure 4.4.)

Space Almanac. By Anthony R. Curtis. Woodsboro, MD: Arcsoft, 1990. A thousand pages of information on astronauts, artificial satellites, space stations, rockets, telescopes, the solar system, stars, supernovas, galaxies. Includes calendar and index.

Telecommunications Technology Handbook. By Daniel Minoli. Boston: Artech House, 1991. Basic information on telecommunications networks, network design philosophy, digital communication systems, satellite and microwave transmission systems, fiber-optics technology, and so forth.

The Timetables of Science: A Chronology of the Most Important People and Events in the History of Science. By Alexander Hellemans and Bryan Bunch. New York: Simon and Schuster, 1988. Comprehensive chronology. The tables are designed for easy cross-referencing.

The Timetables of Technology. By Bryan Bunch and Alexander Hellemans. New York: Simon and Schuster, 1993. Major events in the history of technology, from ancient times to the present.

Specialized Technical-Writing Handbooks

Designing Technical Reports: Writing for Audiences in Organizations, 2nd ed. By J. C. Mathes and Dwight W. Stevenson. New York: Macmillan, 1991. Details the writing and formatting of reports for business and industry.

Elements of the Scientific Paper: A Step-by-Step Guide for Students and Professionals. By Michael J. Katz. New Haven: Yale University Press, 1985. Concise guide for scientists beginning their careers. Includes a sample professional paper.

The Scientist's Handbook for Writing Papers and Dissertations. By Antoinette Miele Wilkinson. Englewood Cliffs: Prentice-Hall, 1991. Includes bibliography and index.

Technical Writer's Freelancing Guide. By Peter Kent. New York: Sterling Publishing Co., Inc., 1992. Includes information about negotiating with agencies, drawing up contracts, and building relationships with clients.

Directories

These reference works are valuable in leading you to bibliographic compilations you may not have been aware of.

Gale *Directory of Databases.* Detroit: Gale Research, 1993. Includes directories for information storage and retrieval systems; also includes machine-readable bibliographic data.

The Federal Data Base Finder, 3rd ed. By Matthew Lesko. Kensington, MD: Information USA, 1984. Includes files available from the federal government

The Index and Abstract Directory: An International Guide to Services and Serials Coverage, 2nd ed. Birmingham, AL: EBSCO, 1990. A 2776-page guide to periodical- and abstract-services worldwide.

Access EPA (1991). Annual guide to Environmental Protection Agency resources, services, and products.

Periodical Indexes

The periodical literature in science and technology is vast and continues to grow at a prodigious rate. Indexes list, by category, author, and title, all the articles published in a given field for a given period. When using the indexes, keep these pointers in mind:

1. See if the index includes a cumulative index. Cumulative indexes cover all citations on a given topic for several years (the number of years will vary), and this will save you the trouble of having to look through each annual-index volume for articles on your topic.

2. After locating article citations that seem useful to you, check your library's list of periodical holdings—usually via the on-line catalogue or in bound computer printouts, several copies of which are placed throughout the library—to make sure the periodical containing the article you need is among the holdings (no college or public library subscribes to *all* existing periodicals). If it isn't, check with other libraries. It may be possible to have another library fax a copy of the article you need to your library— for a fee, of course.

3. Be attentive to citations from peripheral topics, and to cross-references. One of the delights of library research is discovering aspects of your project that you hadn't anticipated.

General Science and Technology Indexes

Applied Science and Technology Index (see sample page, Figure 4.5)
General Science Index
Index to How-to-do-it Information

Plastics—cont.

Failure
See also
Polymers—Fracture

Fillers
See also
Polymer matrix composites

Finishing
The basics on rack plating of plastics. N. Anis. *Plat Surf Finish* v81 p31 Ja '94
Electrostatic painting of plastics using metal backing. A. A. Elmoursi and D. P. Garner. bibl il diags *IEEE Trans Ind Appl* v29 p1053-7 N/D '93
Metallizing nonconductors [Guidebook and Directory, 1994] C. Davidoff. *Met Finish* v92 no1A p330-6 Ja '94
Polishing and buffing [Guidebook and Directory, 1994] A. Dickman and B. Millman. *Met Finish* v92 no1A p16+ Ja '94

Fire hazards
Extinguishment of plastics fires with plain water and wet water. S. Takahashi. bibl diag *Fire Saf J* v22 no2 p169-79 '94

Fire resistance
Flame retardants. V. Wigotsky. il *Plast Eng* v49 p20-5 O '93; Discussion. v50 p22 Ja '94

Glass reinforcement
See Polymer matrix composites—Glass fiber reinforcement

Joining to metal
Highly transient elastodynamic crack growth in a bi-material interface: higher order asymptotic analysis and optical experiments. C. Liu and others. bibl (p1947-8) il diags *J Mech Phys Solids* v41 p1887-54 D '93

Medical use
Bioabsorbable resins target metal use in surgical parts. E. Culp. *Mod Plast* v71 p31+ Ja '94
IKV has process for polymer bone screws. J. K. Rogers. *Mod Plast* v70 p11 D '93
Medical molder integrates CIM with people [MedTech] J. Ogando. il diag *Plast Technol* v39 p57-60 N '93
Rexene cites unconventional properties in flexible olefins. R. D. Leaversuch. *Mod Plast* v70 p18-19 D '93

Moisture content
Analysis of modulus and strength of dry and wet thermoset and thermoplastic composites loaded in transverse tension. F. Pomies and L. A. Carlsson. bibl diags *J Compos Mater* v28 no1 p22-35 '94
Field removal of moisture from power cables. J. A. De-Dad. diag *Electr Constr Maint* v92 p32 D '93

Optical properties
See also
Plastics, Transparent
Polystyrene—Optical properties

Preforming
Automation of preform fabrication makes SRIM viable for volume parts. K. F. Lindsay. il diag *Mod Plast* v70 p48-9+ D '93

Prices
Pricing. See issues of Plastics World

Radiation effect
See also
Polystyrene—Radiation effect
News update: UV stabilizers. L. M. Sherman. il *Plast Technol* v39 p35-9 N '93

Shrinkage
See Plastics—Expansion and contraction

Stabilizers
Cadmium pigments no hazard in molding. M. H. Naitove. *Plast Technol* v39 p89 N '93
News update: UV stabilizers. L. M. Sherman. il *Plast Technol* v39 p35-9 N '93

Stretching
Molecular orientation in plastic optical fibres. X. Llop and others. bibl diag *J Phys D* v27 p25-8 Ja '94

Swelling
Localized Raman scattering probes of molecular-scale motions in case II swelling of polystyrene in *n*-hexane. P. A. Drake and P. W. Bohn. bibl diags *Anal Chem* v66 p79-84 Ja 1 '94

Thermoforming
See Thermoforming

Uses
Plasticoncepts. See issues of Plastics Engineering
Plastics today. See issues of Plastics Engineering
Plastics, Cellular *See* Plastics, Foamed
Plastics, Effect of radiation on *See* Plastics—Radiation effect

Plastics, Foamed
Curtain rising on next generation of urethane foams. B. Miller. il diags *Plast World* v52 p28-31 Ja '94

Manufacture
Developments in open-mold processing highlight product introductions at expo [Fabrication '93, Nashville, Tenn., 1993] K. F. Lindsay. il *Mod Plast* v70 p21-2 D '93
New approach to polyester foaming. il *Plast Technol* v39 p62-4 N '93
Novel gas process makes unique foams. B. Miller. il *Plast World* v52 p10 Ja '94
PUR Congress showcases advances in chlorine-free foam technologies [Polyurethanes World Congress, Vancouver, BC, Oct. 10-13, 1993] A. Vallens. il *Mod Plast* v70 p16-17 D '93

Recycling
EPS reuse gets lift from densification. R. D. Leaversuch. il *Mod Plast* v70 p75-6 D '93
Polyurethane; formulation technologies make recycling, CFC-free foams viable. A. Vallens. il *Mod Plast* v71 p61+ Ja '94
Plastics, Irradiated *See* Plastics—Radiation effect

Plastics, Laminated

Manufacture
Developments in open-mold processing highlight product introductions at expo [Fabrication '93, Nashville, Tenn., 1993] K. F. Lindsay. il *Mod Plast* v70 p21-2 D '93

Plastics, Transparent
See also
Methacrylates
New generation of clarifiers energizes PP use in markets. R. D. Leaversuch. il *Mod Plast* v71 p19-21 Ja '94

Testing
Utility of the spread function in reflectometric applications and design. B. K. Tsai and others. bibl diags *Opt Eng* v33 p102-8 Ja '94

Plastics engineering
Equipment. See issues of Modern Plastics begining September 1990
New materials. See issues of Plastics Engineering begining with January 1989

Competitions, awards, etc.
Plastics on parade; winner's of the 1994 Society of Plastics Engineers awards. L. Brooke. il *Automot Ind* v173 p16-17 D '93
Suspension link wins Grand Award at Auto Division's 23rd annual awards program. M. W. Shortt. il *Plast Eng* v49 p18-19 D '93

Plastics engineering software
Software eases web-gauge control. J. Ogando. il *Plast Technol* v39 p21 N '93

Plastics industry
See also
Computer integrated manufacture—Plastics industry
Plasticscope. See issues of Modern Plastics
Resins 1994; plotting a course for supply [special report] il *Mod Plast* v71 p45-68C Ja '94

Acquisitions and mergers
Consolidation yields giant in N. American polyester supply. J. K. Rogers. *Mod Plast* v70 p26+ D '93

Competition
Plastics vs. metals. V. Wigotsky. il diags *Plast Eng* v49 p20-5 D '93

Exhibitions
See also
Interplas Exhibition
SPI Composites Institute Conference

Statistics
Key markets post solid gains in 1993. *Mod Plast* v71 p73-84 Ja '94

United States
Dow Plastics forecasts moderate growth in 1994. A. Thayer. *Chem Eng News* v72 p15 Ja 3 '94; Discussion. v72 p4 Ja 24 '94

Plastics machinery
See also
Plastics machinery industry
Winding machines
Equipment. See issues of Modern Plastics begining September 1990
Equipment & processing. See issues of Plastics Engineering beginning with January 1989

Plastics machinery industry

Exhibitions
In-mold powder coating highlights big European show; Interplas '93. il diag *Plast World* v52 p32-4 Ja '94

FIGURE 4.5   Sample page from the *Applied Science and Technology Index.*

Science Citation Index
Technical Book Review Index

Biological Sciences and Biotechnology

Biological and Agricultural Index

Business

Business Periodicals Index

Computers

ACM Guide to Computing Literature

Education

Education Index
Resources in Education

Engineering

Engineering Index
Index to IEEE Publications

Environmental Sciences

Environmental Index

Geological Sciences

Bibliography and Index of Geology

Medicine

Cumulated Index Medicus

Social Sciences and Psychology

Social Sciences Citation Index
Social Sciences Index

Abstracts

These journals provide abstracts (summaries) of periodical, book, and report literature in a given field. Most disciplines in science and technology have at least one journal devoted to abstracts. Here are some of them:

Air Pollution Abstracts
ASCE Publications Abstracts (Civil Engineering)
Biological Abstracts
Botanical Abstracts
Chemical Abstracts
Electrical and Electronic Abstracts
Energy Research Abstracts
Environment Abstracts Annual
Geophysical Abstracts
Information Science Abstracts
Nuclear Science Abstracts
Psychological Abstracts
Sociological Abstracts
Water Resources Abstracts

LOCATING GOVERNMENT PUBLICATIONS

The U.S. government publishes vast quantities of information through its numerous departments and agencies. Of the three million people employed by the federal government, three-quarters of a million of them are information specialists, according to Matthew Lesko (*Information USA,* rev. ed.; see listing below). Many college and public libraries are government-document repositories, and they house these documents in separate rooms. The following reference works are used for locating government documents:

> *Government Reports Announcements and Index* (for locating information on government-sponsored research projects)
> *Government Reports Annual Index*
> *Government Abstracts Annual*
> *Index to U.S. Government Periodicals*
> *Information USA,* revised ed. By Matthew Lesko. New York: Viking-Penguin, 1988.
> *Monthly Catalog of United States Government Publications*

Another useful government reference work is the *Statistical Abstract of the United States* (U.S. Dept. of Commerce, 1991). This work includes statistical tables on population, health and nutrition, education, law enforcement, federal government finances and employment, national defense, the cost-of-living index, energy consumption, and so on.

USING DATABASES

Databases are computer repositories of bibliographic citations, some of which are on CD ROM disks, which may be accessed without charge in your college library. Other databases are controlled by vendors, such as DIALOG, and may be accessed for a fee (calibrated according to time spent on-line and the number of items retrieved).

One popular and versatile CD ROM database system is InfoTrac, which contains three separate indexes for periodical citations: *Business Index, National Newspaper Index,* and *Expanded Academic Index.* The system is very user-friendly: all you need to do is sit down at an available terminal, press a key to access the database you want, press another key to locate either broad topics or material based on a keyword search, and scroll through the citations that quickly appear. NOTE: InfoTrac gives citations for recent material only—articles published within the last three or four years. You would therefore need to supplement your search by consulting the standard indexes.

Let's say you are doing research in astronomy and would like to find articles about those bizarre stars called pulsars. Wishing to begin with the most recent studies, you go to an InfoTrac terminal, access the *Expanded Academic Index*, request keyword search, then type in "pulsars." That's all there is to it! The computer will tell you how many citations are on hand, and whether your library has the periodical that the articles are in. You may also press a key to receive a printout (see Figure 4.6).

Another versatile database service is DIALOG, which includes a great many separate databases, such as *EnergyNet; Language and Language Behavior Abstracts; Food Science and Technology Abstracts;* GeoRef.; INSPEC (for physics); and many others. See your librarian for information about accessing this database and the fees involved.

A good handbook for inexperienced database users is *How to Look It Up Online,* by Alfred Glossbrenner (New York: St. Martin's, 1987).

ACQUIRING INFORMATION
FROM INTERVIEWS

Information found in books and periodicals represents the knowledge of experts that has been printed and made public (the literal meaning of "publish"). But a great many experts in any given field do not publish (or have not yet gotten around to publishing). Many professors, for example, possess a wider range of expertise than their publications may reveal, and their expertise grows as they continue to do research in their fields. Other members of the university commu-

```
Database: Expanded Academic Index
Subject:  pulsars
Library:  Orradre
Holdings: * indicates that library subscribes to this journal
------------------------------------------------------------

     Lost & found: pulsar planets.  David Bruning.
     Astronomy, June 1992 v20 n6 p36 (3).

*    Pulsed high-energy gamma-radiation from Geminga
     (1E0630+178)
     D.L. Bertsch, K.T.S. Brazier, C.E. Fichtel, R.C
     Hartman, S.D. Hunter, G. Kanbach, D.A. Kniffen, P.W.
     Kwok, Y.C. Lin, J.R. Mattox, H.A. Mayer-Hasselwander,
     C.V. Montigny, P.F. Michelson, P.L Nolan, K. Pinkau,
     H. Rothermel, E.J. Schneid, M. Sommer, P. Sreekumar
     and D. J. Thompson. Nature, May 28, 1992 v357 n6376
     p306 (2).
     Holdings: v. 173, 1954-    PERIODICALS ROOM      PAPER

*    A cosmic mystery is called solved; scientists seeking
     the source of Geminga gamma rays report it is a pulsar.
     The New York Times, May 24, 1992 v141 p15 (N) p33 (L)
     10 col in.
     Holdings: MICRO: 1851-     PERIODICALS ROOM    PAPER/MICRO
          CURRENT ISSUES IN CURRENT SECTION OF PERIOD. ROOM
```

FIGURE 4.6   Sample printout from the *Expanded Academic Index* database.

nity—technical support staff; librarians and archivists; public safety personnel; communications directors; counselors; business- and records-office staff; bookstore, food-service, career-center managers—are valuable sources of information for anyone wishing to conduct research on a topic related to these domains.

Of course, you may also want to consider interviewing individuals off campus. If so, you will probably need a wider margin of time—especially if the individual's workplace is more than walking distance from campus. Remember, too, that those who have their own practices (such as doctors, lawyers, brokers)—or those who are itinerant (investigators, salespeople, district managers) may not be easy to locate, much less interview.

As a student of technical communication, then, it is important that you develop your ability to obtain information from these first-hand, "primary" information sources. But before you embark on an interviewing crusade, keep these pointers in mind:

1. Check the published literature, so as to have a better sense of the information you need to obtain, as well as to make yourself more conversant with the field of the expert you wish to interview.

2. Call or visit the prospective interviewee to request permission for an interview, to explain the purpose of the interview, and to set up an appointment. Never "just drop in" to interview someone.

3. As soon as you have an appointment, begin drafting interview questions. Use these questions as prompts, rather than as a script. In other words, plan to *converse with* your interviewee, not interrogate him or her. The best interviews are those in which unanticipated questions or responses arise. Be ready for them!

4. Your questions should relate clearly to the purpose of the interview and be sharply focused.

5. Direct your questions to the particular expertise of the interviewee, within a specific context (e.g., in the context of an experiment-in-progress or a recently completed project). Avoid extraneous matters (except as a conversation starter). Also avoid questions that could prove to be digressive (e.g., "Did studying science as much as you did interfere with your social life?")

6. Refrain from asking "leading" questions, such as, "Don't you think that Dr. Jones's experiment was worthless?"

7. Don't hesitate to suspend your prepared questions in order to request clarification on a certain point. If you are not comprehending your interviewee's explanation fully, neither will your readers.

8. You may wish to record the interview on tape, but this really isn't necessary. Listening carefully and taking judicious notes often will suffice, unless the interview focuses on highly technical subject matter—in which case, you would want to record the responses in order to play them back several times if necessary.

9. If you have been assigned to write up the interview (see next section), it is important to begin the draft as soon after the interview as possible, ideally within the next hour, while the details of the conversation are fresh in your mind, and you can still make sense of the rapid notes you took.

10. It may be necessary to contact the interviewee for follow-up questions; ask him or her if this is permissible.

Interviewing is not only a valuable skill to cultivate; it is also an enjoyable experience that enhances your involvement both in campus affairs and with the community at large.

WRITING UP AN INTERVIEW

Most of the time, the information you obtain from interviews will be incorporated into a report or article (and you will be expected to document it as you would any other information source—see Appendix B. But interviews with scientists, engineers, medical and health-care workers, and other specialists make for engaging and informative reading in their own right. The typical format for an interview is described as follows:

1. *Introduction.* Here you describe, in one paragraph, the interviewee's background, with emphasis on the person's education, current position and duties, notable achievements, major publications. The introduction thus keeps the interview itself from getting bogged down with background information. The last sentence of the introduction should state the topic of conversation.

2. *Text of the interview.* Present the interview in a Question-Answer format (use either Q-A or surnames to designate your question and the interviewee's response. Follow a logical sequence, beginning with questions relating to an overview of the topic of conversation, and then moving to more specific questions about the topic. Remember to include one or more "spinoff" questions—those that ask for clarification or elaboration of a point made in the answer immediately preceding. Such spinoff questions will make the interview seem more like a conversation and make it more interesting and readable.

3. *Conclusion.* This may take the form of a brief summative paragraph, or a summative concluding question/answer.

Here are some additional pointers to keep in mind when writing up the interview:

1. It is not necessary to reproduce each answer verbatim unless that answer involves very precise, technical details. Always make sure, during the interview, that you have recorded such answers accurately.

2. Any time you are in doubt about the details of a response, recontact the interviewee. CAUTION: Before making such a call, be sure to make a written list covering exactly what is crucial for you to know. Pare your questions to a bare minimum, and be prepared to write the answers as you listen.

3. Often, in the throes of spontaneous conversation, anybody can make a factual or grammatical error. Be alert for these and edit them out. This is known as "silent editing." However, if you are in doubt about whether to leave in or delete a response, check with the interviewee.

Read the following interview with space scientist Dr. Harold Klein, conducted by student Daniel J. Kiely:

An Interview with Harold Klein

By Daniel J. Kiely

May 4, 1992

Harold Klein once referred to a lecture on life in outer space as "Baloney!" Now he is an active participant in SETI (Search for Extra Terrestrial Intelligence) and headed life detection experiments on the Viking Mission to Mars in the 1970's. Klein became the director of all life sciences at the NASA-Ames Research Center at Moffett Field in 1963. One of his first projects was setting up close ties with universities. Santa Clara was one of the first and today Klein is the in-resident scientist who also teaches classes in Cosmic Evolution and The Origin of Life. By the time he left NASA in 1984 his program included 250 universities. Harold Klein's expertise on life in outer space was clear in his ability to simply convey the technicalities of extraterrestrials and SETI to a curious English Major.

Q: Do you believe that there is life on other planets?

A: I have a different view than a lot of my colleagues and that is that I don't think there is much of a chance of life beyond the planet Earth. There are many stars with their own systems, so there are a great number of opportunities, but I believe that the chance of life is very small. On the planet Earth the origin of life began with simple chemicals combining to form more complicated chemicals. These chemicals somehow generated the ability to copy themselves; RNA and DNA formed randomly in prebiotic life and they kept producing. The set of conditions to form life is very complicated, and these types of circumstances happening on another planet are very slim. Life has been on the planet Earth for four billion years, and we as people have only had the technology to communicate with other planets for 100 years. A blink of an eye in the total amount of time life has existed on Earth. On another planet this time it takes to get the technology to communicate with other planets might take longer or it might take a shorter amount of time. So even if there is extraterrestrial life on other planets, coordinating communication would be incredibly hard.

Q: What is SETI's opinion?

A: SETI on the other hand has a very optimistic opinion. They look at the situation from the point that there are billions of stars with solar systems and that in all of the possibilities there has got to be life somewhere.

Q: What are SETI's basic goals?

A: SETI is not seeking communication with life on other planets. It is mainly trying to confirm the existence of life there. This is because of the problems of communication with stars that are so far away. The nearest star is four and a half light years away, and this star does not have any orbiting planets. To reach a star with planets it would take us between fifty and one hundred light years to communicate. Their big hope is that there is a more intelligent species out there and that they hear one of our random transmissions. Then this species would set up a beacon and broadcast a simple and understandable signal back. So, basically they are relying on this species' technology to make the connection. We now are very close to having this technology. In a matter of years we will be able to look through a telescope at a star and tell if planets are orbiting. We will also be able to tell which stars have the greatest possibility for life.

Q: How will we be able to tell which stars are and aren't conducive to life from so far away?

A: What they would use is a spectroscope, which analyzes atmospheres' chemical content by detecting different shades of light. And what we look for is the type of atmosphere that the Earth has. Life on Earth produces an atmosphere that shouldn't exist. Everywhere except the Earth, atmospheres are dominated by hydrogen gas, which is the result of the big bang. On this planet we have O_2, which is produced by plants. This is life. Oxygen should not exist. It should clamp on to hydrogen and constantly be creating water. The Earth is the only planet which has a disequilibrium in its atmosphere. Life exists because of this disequilibrium, which goes against the laws of thermodynamics.

Q: What is SETI doing right now?

A: SETI right now is involved in radio searches. They are trying to pick up stray radio or TV waves. They will be scanning the universe in all directions. At the moment they are doing a few minor searches. They will be doing what is called an all-sky survey where they will try to listen to the entire spectrum of the sky. They are hoping to pick up an extraterrestrial radio station or an extraterrestrial version of the Cosby Show. The only problem is that SETI has no idea what frequency to scan, so they have built some very expensive machines that will scan one star at many frequencies. Another problem with this is that they can only stay on one frequency for a certain amount of time. So they must be scanning the correct frequency at the perfect time in order to make a connection.

Q: How many different types of searches are there?

A: There are two major types of searches, the all-sky I just mentioned and then what we call a target search. Right now we have star maps of about one hundred light years distance. We know the ages of these stars, and there is more likely to be life on the older ones. So other than the all-sky survey, there is the target survey where there are a certain number of stars which are picked out because of their age and investigated more in depth. Where the all-sky might spend a half hour on a single star, a target search would spend hours. The maps are small and not extremely accurate, so the all-sky survey is the most practical. SETI hopes that at some time they can be doing both searches at the same time.

Q: Do you think this type of searching will come up with any positive results?

A: Personally I don't think they are going to find anything, but if they do, I believe the civilization will be more advanced. I don't think that we can find a means of communication with extraterrestrial life forms. It's like talking with dolphins. We know that they communicate, but we have no idea what they are saying. I think it would be the same with life in space—they wouldn't be able to understand a thing we said.

Q: Do you think that all of this research and its budget is worth it when there are so many other pressing problems in the world today?

A: The NASA budget from the government today is thirteen billion dollars, and SETI gets about one million dollars a year. So I look at it like a lottery ticket. It is such a small amount to pay for the chance for such a huge pay-off. If we were to discover life on other planets which were more technically advanced, the pay-off would be incredible in science as well as in sociology and philosophy. And even if we never find anything, we will have important information on stars. We will have radio signatures of stars, their radio intensity, and an ever-growing star map from the all-sky searches. So even if life on other planets is never discovered, we still will have valuable information from the research.

You will note that Kiely begins with a concise introduction to Klein, including his most noteworthy achievement (directing the famous Viking Mars Lander life-detection project in 1976) and his current positions (Director of life-sciences research at NASA's Ames Research Center; Scientist-in-Residence at Santa Clara University). Kiely ends the introduction with a statement of the interview topic: searching for life beyond the earth.

Kiely's opening question to Dr. Klein stimulates curiosity at once; it is also a logical place to begin the conversation: "Do you believe that there is life on other planets?" Notice also Kiely's next question: a "spinoff" that is tied to Klein's response.

Notice that Kiely's questions are clear, concise, and specific—and that they elicit substantive responses from Klein. This is an important technique to develop in interviewing. Tempting as it is for interviewers to ask clever, elaborately worded questions, the goal of any interview is to highlight the person being interviewed, not the interviewer.

CONDUCTING A SURVEY

Survey taking is an imperfect instrument at best. In a way, it is like trying to interview many people all at once, using only prefabricated questions on a form known as a questionnaire, which must be distributed—sometimes a problem in itself!—either through the mail (expensive), or through hand distribution (time-consuming). Businesses wishing to tap into the sentiments of their clientele ("Are you satisfied with our products?") may insert questionnaires with monthly billing statements; public-interest groups may mail out questionnaires using a special "nonprofit organization" postal rate, assuming they can qualify for it.

The biggest problem with survey taking, however, is that the majority of those receiving questionnaires, whether in the mail or directly from the survey taker, will decline to respond. The usual "it will only take a couple of minutes" is not persuasive. This is not to say, however, that an attractively designed questionnaire, with easy-to-answer questions, wouldn't increase the response rate.

Because survey taking can yield useful information about reactions to policy or states of awareness about important concerns (either globally, nationally, or locally), to plan a sound strategy for conducting the survey—using an attractive and well-written questionnaire—will be worth your while. Consider these guidelines:

1. Make sure that the information you need is not obtainable elsewhere. It wouldn't make much sense, for example, to survey companies nationally about the quantities of toxic waste they dispose of in a given period; such information

already exists (e.g., in EPA publications). However, if you are investigating the quantities of toxic wastes disposed of by new companies in your own community (in order, say, to compare them to statewide or national averages), then taking your own survey might prove useful.

2. Do not try to cover too much ground in one survey. Your topic should be sharply focused, the questions limited only to that one topic. Trying to survey people on their attitudes toward energy conservation, for example, will yield equally unfocused responses. However, finding out who is willing to purchase energy-efficient automobiles—why or why not; under what circumstances; where they would draw the line, etc.—could yield useful information for someone analyzing the positive and negative attributes of the most energy-efficient automobiles.

3. When asking people to fill out your questionnaire, tell them four things: who you are, the purpose of your research, how much time it would take them to fill out the questionnaire, and the date by which you would prefer to have the questionnaire returned. Provide them with a self-addressed envelope, returnable via campus mail (if the respondent is not a member of the college community, be sure to place a stamp on the return envelope). If you are contacting people by mail, include a cover letter containing the information indicated.

4. Courtesy goes a long way. Whether you contact individuals in person or by mail, communicate in a friendly tone; thank them for their time and consideration even if they choose not to respond.

DESIGNING A QUESTIONNAIRE

Effective questionnaires are models of concision, clarity, and attractive formatting. To elicit the most specific responses possible, you will need to employ a variety of question types (see Table 4.1).

Most questionnaires will make use of more than one type of question, depending on the kind of information sought. Be aware of the limitations that each type of question may possess. Binary questions, for example, may oversimplify a situation. For example, some yes-no questions could be answered Yes AND No, depending. At the opposite end, comment-type questions should be used sparingly, ideally at the end, and should be quite specific. For example: "Evaluate the governor's commitment to antipollution legislation. Can you describe a particular example of such commitment?"

The questions you include should also be impartial. Always make sure that you are not "begging the question" or providing biased options.

Table 4.1 Types of Questions for a Questionnaire

Question Type	Definition	Example
Either/or	Answerable by one of two available choices (yes or no; A or B)	Do you recycle plastic? ☐ Yes ☐ No
Open multiple	Answerable by choosing as many or as few of the options presented	Which of the following do you recycle? ☐ paper ☐ glass ☐ metal ☐ yard waste ☐ plastic
Restricted multiple	Answerable by choosing no more (or no fewer) than the number requested.	Which kinds of nonfiction books do you read most often? Check no more than three. ☐ history ☐ science ☐ biography ☐ religious ☐ philosophy ☐ sports [etc.]
Hierarchical	Answerable by ranking options in order of preference.	Rank in order of preference (1 = highest) your favorite nonfiction reading: __ hist. __ science __ biog. __ relig. __ philos. __ sports [etc.]
Fill-in	Answerable by filling in blank spaces with words or numbers	Number of hours/day you spend on the job _____.
Comment	Answerable by responding with one or more complete sentences of description or explanation	What else might we do to serve you better? _____ _____

Begging the question: "Has the uninspired energy policy of our state affected your commitment to conserve energy?" The word *uninspired* is a value judgment presented as though it were an indisputable fact—it begs the question, "*Is* the state's energy policy uninspired?" Even if the offending word were deleted, the tone of the question still suggests a bias.

Biased options: Imagine that you are asking people to estimate what percentage of logging should be permitted each year in areas where some species of wildlife are known to be endangered, and you provide the following options:

☐ 10% ☐ 50% ☐ 75% ☐ 90% ☐ 100%

Clearly, such a set of options is biased in favor of the logging industry; only the first option is less than 50%, and 0%—the negative equivalent of 100%—is not offered as an option.

Once you are satisfied with the questions you are going to use, you must then take care to format them in a way that will be pleasing to the eye as well as capable of being perused quickly. Here are three suggestions:

1. Avoid clustering the parts of a question together, especially if check boxes are included. Instead, arrange the options in columns (the number of columns you use per question will depend on available space and total number of questions needed).

2. Use indentation or boldface type to keep the question itself distinct from the options to be checked.

3. Consider dividing long questionnaires into subsections.

Study the questionnaire in Figure 4.7. This questionnaire, prepared by a public utilities company, asks consumers to provide information about their households, heating and cooling systems, and appliances, so that the company can in turn provide them with an energy analysis. Be prepared to comment on the effectiveness of the layout, the clarity of the central purpose, the variety, wording, and arrangement of the questions.

INCORPORATING YOUR INFORMATION

It's a classic example of mixed emotions: you have spent a week locating information for your research project, have taken carloads of notes, and feel like the world's foremost authority on your subject. But index cards and printouts and a file filled with tables and graphs and diagrams do not a report make. It seems as if you have scarcely even begun.

Actually, you have made dramatic progress if you've gotten this far. Information gathering is a very real part of the composing process. Subconsciously if not consciously, you are figuring out how what relates to what, and where it ought to go in your document. The time has come now to push a little harder in this direction, following these guidelines:

1. **Arrange** your notes in a tentative order: What seems most appropriate for the introduction? Which of the collected data will need to be examined first? Does any of the information seem naturally to cluster together as a coherent group? What do you want to save for last? Once you've arranged your notes, begin working on an outline (see Chapter 5).

2. When you have begun the actual drafting of your document, it may be tempting to try and incorporate everything into it; but this can be a mistake which often results in a cluttered and incoherent report. Most of the

ENERGY SAVINGS PLAN

Pacific Gas & Electric Customer
Energy Use Survey

Customer Information

NAME _____

DATE _____

ADDRESS _____

CITY _____ STATE _____ ZIP _____

PHONE: DAYS (____) _____ EVENINGS (____) _____

FOR OFFICE USE ONLY

PG&E ACCOUNT NUMBER _____

PG&E CONTROL NUMBER _____

DMC NUMBER _____

Household Information

What energy source would you like analyzed?

1. ☐ Electric   4. ☐ Propane Gas
2. ☐ Natural Gas   5. ☐ Electric &
3. ☐ Electric &   Propane Gas
   Natural Gas

What is the total square footage of your residence (excluding garage and unfinished basement)?

_____ sq. ft.

How many people live in your home?

Is someone typically home:

Summer Winter

_____ _____ Monday through Friday between
                12 noon and 6 pm

_____ _____ Three or more hours between 12 noon
                and 6 pm Monday through Friday

_____ _____ Less than three hours between 12 noon
                and 6 pm Monday through Friday

_____ _____ No one is home between 12 noon and
                6 pm Monday through Friday

How many rooms does your home have?

Is your home a:

1. ☐ Single Family   4. ☐ More Than 5 Unit
2. ☐ Duplex (two family)   Apt/Condo
3. ☐ Four-Plex   5. ☐ Mobile Home
                     6. ☐ Other

(For office use only)
Nearest primary weather station: _____

Code: Direct Mail

ENERGY SAVINGS PLAN

Please answer all questions. For those customers who use propane gas (which is marked as Box 4 or 5 in the Household Information section of the form), a four-month usage reading in gallons is necessary to analyze your propane use. Also, if you do check propane in answering this question, remember that the remainder of the survey will read all gas responses as propane gas, unless otherwise indicated.

If you need help answering any of these questions call us at 1-800-675-5499

PG&E

ENERGY SAVINGS PLAN

Energy Savings Plan Questionnaire

Find out how easy it is to analyze your home energy usage. All you have to do is fill out and return this questionnaire.

PG&E

FIGURE 4.7 Sample questionnaire.

ENERGY SAVINGS PLAN

Heating and Cooling Systems

Which fuel do you use to heat your home?
1. ☐ Electric   3. ☐ Oil
2. ☐ Gas   4. ☐ Other

What type of heating system do you have?
1. ☐ Forced Hot Air   3. ☐ Heat Pump
2. ☐ Hot Water Radiator   4. ☐ Other

What is its input rating (MBtu/hr)?
_____ *(optional question)*

What is a typical temperature setting for your primary heating system?
☐ No thermostat or cannot control thermostat

Mornings (8 am to 12 noon) _____ °F
Afternoons (12 noon to 6 pm) _____ °F
Evenings (6 pm to 10 pm) _____ °F
Nighttime (10 pm to 8 am) _____ °F
Weekend Days (8 am to 10 pm) _____ °F
Weekend Nights (10 pm to 8 am) _____ °F

Do you use a supplemental heating system?
0. ☐ No   1. ☐ Yes

Which type of fuel does this system use?
1. ☐ Electric   3. ☐ Oil
2. ☐ Gas   4. ☐ Other

What type of heating system is it?
1. ☐ Forced Hot Air   3. ☐ Heat Pump
2. ☐ Hot Water Radiator   4. ☐ Other (e.g., space heater)

Do you typically use your secondary heating system?
Monday through Friday between 12 noon and 6 pm
0. ☐ No   1. ☐ Yes

What type of cooling system do you have?
0. ☐ None   3. ☐ Window/Wall
1. ☐ Central Electric   4. ☐ Heat Pump
2. ☐ Central Gas   5. ☐ Other

What is its cooling capacity (tons)?
_____

What is a typical temperature setting for your cooling system?
☐ No thermostat or cannot control thermostat

Mornings (8 am to 12 noon) _____ °F
Afternoons (12 noon to 6 pm) _____ °F
Evenings (6 pm to 10 pm) _____ °F
Nighttime (10 pm to 8 am) _____ °F
Weekend Days (8 am to 10 pm) _____ °F
Weekend Nights (10 pm to 8 am) _____ °F

Hot Water Use

Is your water heater:
1. ☐ Electric   3. ☐ Other
2. ☐ Gas

ENERGY SAVINGS PLAN

Is it a free-standing tank?
0. ☐ No   1. ☐ Yes

Is your water heater covered with an insulating blanket or was it purchased in the last two years?
0. ☐ No   1. ☐ Yes

Which temperature is your water heater set at?
1. ☐ Low (110-120°)   3. ☐ High (141-160°)
2. ☐ Medium (121-140°)   4. ☐ Very High (161° & up)

How many showers are taken in your home each week?

	Between 12 noon and 6 pm Monday through Friday	All other times	Total

Do you use energy-saver showerheads?
0. ☐ No   1. ☐ Yes

How many baths are taken in your home each week?

	Between 12 noon and 6 pm Monday through Friday	All other times	Total

Cooking

Which is your primary cooking fuel?
1. ☐ Electric   3. ☐ Other
2. ☐ Gas

How many hours per week do you use your stove?

	Between 12 noon and 6 pm Monday through Friday	All other times	Total

Do you use a microwave oven?
0. ☐ No   1. ☐ Yes

How many minutes per week do you use your microwave oven?

	Between 12 noon and 6 pm Monday through Friday	All other times	Total

Refrigerators

How many refrigerators do you use?
Please complete the following information for each refrigerator you listed.

MAIN REFRIGERATOR
Defrost Method:
1. ☐ Frost-Free (automatic)   3. ☐ Manual
2. ☐ Switch Control

Size:
1. ☐ Dormitory (under 14 cu. ft.)   3. ☐ Full (18-21 cu. ft.)
2. ☐ Apartment (14-17 cu. ft.)   4. ☐ Large (over 21 cu. ft.)

Age: _____ (years)

ENERGY SAVINGS PLAN

SECOND REFRIGERATOR
Defrost Method:
1. ☐ Frost-Free (automatic)   3. ☐ Manual
2. ☐ Switch Control

Size:
1. ☐ Dormitory (under 14 cu. ft.)   3. ☐ Full (18-21 cu. ft.)
2. ☐ Apartment (14-17 cu. ft.)   4. ☐ Large (over 21 cu. ft.)

Age: _____ (years)

ADDITIONAL REFRIGERATOR
Defrost Method:
1. ☐ Frost-Free (automatic)   3. ☐ Manual
2. ☐ Switch Control

Size:
1. ☐ Dormitory (under 14 cu. ft.)   3. ☐ Full (18-21 cu. ft.)
2. ☐ Apartment (14-17 cu. ft.)   4. ☐ Large (over 21 cu. ft.)

Age: _____ (years)

Stand-Alone Freezers

How many stand-alone freezers do you use?
Please complete the following information for each freezer you listed.

MAIN FREEZER
Defrost Method:
1. ☐ Frost-Free (automatic)   3. ☐ Manual
2. ☐ Switch Control

Style:
1. ☐ Upright   2. ☐ Chest

Size:
1. ☐ Small (under 12 cu. ft.)   3. ☐ Large (17-20 cu. ft.)
2. ☐ Medium (12-16 cu. ft.)   4. ☐ Industrial (over 20 cu. ft.)

Age: _____ (years)

SECOND FREEZER
Defrost Method:
1. ☐ Frost-Free (automatic)   3. ☐ Manual
2. ☐ Switch Control

Style:
1. ☐ Upright   2. ☐ Chest

Size:
1. ☐ Small (under 12 cu. ft.)   3. ☐ Large (17-20 cu. ft.)
2. ☐ Medium (12-16 cu. ft.)   4. ☐ Industrial (over 20 cu. ft.)

Age: _____ (years)

FIGURE 4.7 (continued)

ENERGY SAVINGS PLAN

Clothes Washer and Dryer

Do you use a clothes washer?

0. ☐ No   1. ☐ Yes

On average, how many loads of laundry do you wash a week?

	Between 12 noon and 6 pm Monday through Friday	All other times	Total
In hot water?	_____	_____	_____
In warm water?	_____	_____	_____
In cold water?	_____	_____	_____

Which type of clothes dryer do you have?

0. ☐ None   2. ☐ Gas
1. ☐ Electric   3. ☐ Other

On average, how many loads of laundry do you dry a week?

	Summer	Winter
Between 12 noon and 6 pm Monday through Friday	_____	_____
All other times	_____	_____

Entertainment

How many color TV's are there in your home? _____

How many hours per day do you use it (them)? _____

How many black & white TV's are there in your home? _____

How many hours per day do you use it (them)? _____

How many stereos are there in your home? _____

How many hours per day do you use it (them)? _____

Waterbeds

How many heated waterbeds are there in your home? _____

How many are covered with an insulating blanket or pad? _____

Dishwasher and Well Water Pump

Do you have a dishwasher?

0. ☐ No   1. ☐ Yes

If yes, how many loads do you do each week?

Between 12 noon and 6 pm Monday through Friday	All other times	Total
_____	_____	_____

ENERGY SAVINGS PLAN

Do you have a well water pump?

0. ☐ No   1. ☐ Yes

If yes, how many hours per week do you run it?

Between 12 noon and 6 pm Monday through Friday	All other times	Total
_____	_____	_____

Large/Seasonal Appliances

Please indicate your monthly and daily usage of the following appliances.

The *start month* is the month you begin using the appliance closest to the first day of the month. The *end month* is the month you stop using the appliance closest to the last day of the month. Please be sure your entries indicate the correct amount of usage. (For example, if you use your pool heater from November 20th to April 5th, the correct entry would be December for the *start month* and March for the *end month*)

APPLIANCE	START MONTH	END MONTH	HOURS/DAY IN USE Weekdays (12 noon–6 pm)	All other times
Attic/Whole House Fan	_____	_____	_____	_____
Portable Fan	_____	_____	_____	_____
Pool Pump	_____	_____	_____	_____
Size _____ (horsepower)				
Pool Heater	_____	_____	_____	_____
Fuel Type: 1. ☐ Electric 2. ☐ Gas 3. ☐ Other				
Sump Pump	_____	_____	_____	_____
Dehumidifier	_____	_____	_____	_____
Humidifier	_____	_____	_____	_____
Electric Blanket	_____	_____	_____	_____

Other Large Appliances

Do you have any other large appliances (such as a spa or home computer) that you use more than once a year and that have not been listed elsewhere in this survey?

0. ☐ No   1. ☐ Yes

If yes, please describe the appliance(s) and its (their) use below.

Unit # 1 Description: _____

WHICH FUEL DOES IT USE?	HOURS/WEEK IN USE? Weekdays (12 noon–6 pm)	All Other Times	MONTHS IN USE? Start Month / End Month
1. ☐ Elec.	_____	_____	_____
2. ☐ Gas	_____	_____	_____
3. ☐ Other	_____	_____	_____

ENERGY SAVINGS PLAN

Unit # 2 Description: _____

	HOURS/WEEK: Weekdays (12 noon–6 pm)	All Other Times	MONTHS: Start Month / End Month
1. ☐ Elec.	_____	_____	_____
2. ☐ Gas	_____	_____	_____
3. ☐ Other	_____	_____	_____

Unit # 3 Description: _____

	HOURS/WEEK: Weekdays (12 noon–6 pm)	All Other Times	MONTHS: Start Month / End Month
1. ☐ Elec.	_____	_____	_____
2. ☐ Gas	_____	_____	_____
3. ☐ Other	_____	_____	_____

Optional Questions

Race

☐ White   ☐ Asian
☐ Black/African American   ☐ Hispanic
☐ Indian (American)   ☐ Other

Age

☐ 18-34   ☐ 55-64
☐ 35-44   ☐ 65 and over
☐ 45-54

PG&E now offers a discount rate to our low income customers and customers receiving AFDC, SSI, Food Stamps, Veterans or Survivors Benefits. If you think you may qualify for this rate, please check the box below and we will send you a rate application.

Special Considerations/Comments

Please use this space for any comments concerning this survey or PG&E.

_____
_____
_____

Have you ever participated in any free PG&E program (for example: free weatherization, TCAP, etc.)?

☐ Yes   ☐ No

Thank you for completing the PG&E Energy Savings Plan Questionnaire. Now to receive your analysis, please fold it, seal it, and drop it in the mail—no postage is necessary. Propane users, don't forget to attach at least four months of usage in gallons.

If you need help answering any of these questions call us at 1-800-675-5499

FIGURE 4.7 (continued)

time, you will have gathered much more information than is usable. To decide wisely on what to include and exclude, you must **select** what is most relevant or necessary for supporting your premise.

3. Next, you must find the best way to **integrate** the information into the report. This can be accomplished in three ways:
   a. By parenthetical reference—e.g., "Many galaxies such as Andromeda (Figure 3) are spiral-shaped."
   b. By using footnote designators: a superscript numeral at the end of a quoted passage, with the citation at the bottom of the page or at the end of the document.
   c. By using transitional phrases, such as "according to. . . "; "as so-and-so explains. . . "; "in the following table. . . "; ". . . as the following illustration reveals":

   As we approach a black hole described by the Schwarzschild solution, shown schematically in the following figure, we first come to the *photon sphere . . .*

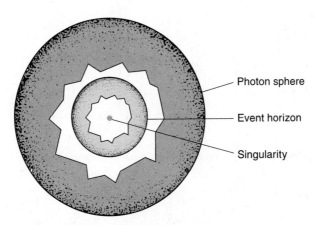

The black hole. *A schematic diagram of a black hole showing the photon sphere, the event horizon, and the singularity.*

—William J. Kaufmann, III, *Relativity and Cosmology.* New York: Harper & Row, 1973.

Remember that in documents containing frequent outside references or visual aids, clarity and coherence can sometimes be hard to maintain. Always double check to ensure that readability is not compromised.

FINDING AND INCORPORATING
TECHNICAL INFORMATION: A CHECKLIST

☐   Do I have a general idea of what I need to research and what kinds of sources I will need to consult?

☐   Have I checked Sheehy's *Guide to Reference Books* to get an idea of the reference materials available in the field I am researching?

☐   Have I checked the appropriate periodical indexes and computer databases to locate articles I might need to read?

☐   Are there government publications relevant to the subject I am researching?

☐   Have I contacted experts on campus for information?

☐   Does my topic require me to interview certain experts for in-depth information?

☐   Does my topic require me to take a survey?

☐   Have I asked for the right kind of information in my questionnaire? Are my questions clear and unbiased?

☐   Have I effectively integrated my researched data into the body of my report?

■ CHAPTER SUMMARY

Your knowing how and where to locate the information you need for your writing projects will save much time and frustration, and it will even serve as a problem-solving aid for organizing and drafting the document. As you plan your research strategies, have a general sense of the steps you will need to take to locate information, but keep your plans flexible enough to allow for unexpected discoveries. The key reference works to become familiar with include periodical indexes and abstracts; databases (especially those accessible free of charge via CD ROM); handbooks and research guides and directories. It is also worthwhile to become acquainted with the extensive range of government publications. Besides library resources, timely information is obtainable through well-prepared surveys (using questionnaires) and interviews. Once you have gathered your research notes together, putting them in tentative order can serve as a first step toward composing a draft of your document. Researched material can be easily integrated into the text of your report by using transitional phrases or simply by using parenthetical or footnote designators.

■ FOR DISCUSSION

1. Interview a fellow student on a specific topic; afterwards, use your notes to introduce the person orally to the class.

2. Which reference work(s) would be most useful for locating the following kinds of information?
   a. Technical articles on the generation of hydroelectric power.
   b. The development of non-Euclidean geometry.
   c. Differences between Freudian and Jungian psychoanalysis.
   d. Where to apply for freelance technical-writing jobs.
   e. How a thermostat works.
   f. Bibliography of articles on parasites
   g. Bibliography of field guides in geology.
   h. Reference books devoted to architecture.
   i. The clearest possible explanation of general relativity.
   j. Descriptions of endangered species in the United States.
   k. Additional topics suggested by your instructor.

3. Explore your college library's on-line and CD-ROM databases. Which ones would be most useful to you for the kinds of research you are doing in your major field? Describe these databases to the class.

4. Discuss the advantages and disadvantages of conducting surveys or interviews as a means of obtaining information.

■ FOR YOUR NOTEBOOK

1. Over the next few days, locate and examine those reference works mentioned in this chapter which seem most useful to you. Jot down brief summaries of entries that you may wish to investigate further at a later time.

2. Compare the discussion of a topic of your choice in one encyclopedia (say the *McGraw-Hill Encyclopedia of Science and Technology*) with that in another encyclopedia (e.g, *Van Nostrand's Scientific Encyclopedia*). What accounts for the differences?

3. Prepare a list of professors' specialties in the department of your major field of study. This information may be obtainable, in brief form, from your college catalogue; but find out more by arranging to interview the professors themselves.

4. Do key-word and subject searches for a specific topic, using your library's on-line catalogue. How successful were you in obtaining a useful bibliography? What pleasant or unpleasant surprises did you have?

■ WRITING PROJECTS

1. Write a 2-page evaluation of the way two or three encyclopedias or encyclopedic dictionaries cover a given topic. Consider what each work emphasizes or neglects to emphasize, apparent target audience, the clarity and readability of the prose, use of examples and analogies to help clarify concepts, illustrations, cross-references, and other notable features.

2. Prepare a questionnaire that would help you acquire necessary information for your term project. Pay particular attention to formatting, manner and variety of phrasing questions, and relevance to your central purpose. Include an introduction; concluding remarks are optional.

3. Interview an expert in your major field of study, or in the area you are researching for your term project. This expert may be any member of the college community, including seniors and graduate students involved in advanced research; he or she may also be an administrator or employee from a private company or practice. Next, write up the interview, using the question-answer format suggested in this chapter.

Dynamics of Composing Technical Documents

Writing, unlike grammar, is not rulebound, and technical writing is no exception. Teachers might impose a set of rules for beginners, such as: "First, write an introductory paragraph that ends with a thesis statement; next, write three support paragraphs; and finally, write a conclusion." That all-too-recognizable set of rules for the "five-paragraph theme" can help beginning writers understand fundamentals of organization and development, but eventually one feels cramped by it. Writing is rarely so formulaic.

It is more realistic to suggest *guidelines* for writing, with the expectation that they be modified as the writer sees fit, or even ignored altogether. This may sound more arbitrary than it actually is. Writing, especially writing in business, scientific, and technical professions, requires more than simply putting words on paper. It also involves interacting with others—with technicians, programmers, supervisors, clients—whose expertise you need to tap into for the writing to be thorough and accurate. As rhetorician Jack Selzer points out, the composing process in the workplace is "a public and social act of exploring ideas and transacting business."[1] The act of composing also involves consulting library resources and visiting labs and other sites to examine materials and activities firsthand. All the while, you are looking up information and taking notes, brainstorming and interviewing, outlining and drafting, taking more notes, revising. While predetermined steps for how to write a technical document can be helpful, they are at best an approximation of what a particular writer may need to do in order to complete a particular writing task. This chapter can give you a clearer picture of the dynamics of composing by your taking a closer look at particular writing situations.

COMPOSING AS A DYNAMIC PROCESS

Following a dynamic process of composing means that you select strategies depending upon (*a*) the nature of the subject matter; (*b*) the purpose of the document; (*c*) the complexity and length of the document, (*d*) your experience writing this kind of document; (*e*) your temperament at the time of composing. Moreover, composing dynamically means that you do not feel obliged to proceed in a linear, step-by-step manner; for example, you may wish to begin a draft before finishing (or even starting) an outline, and then to keep returning to the outline as your draft progresses; perhaps you may even revise your topic as a result of unforeseen insights that occurred to you as you generated text or as you did some background reading (see Figures 5.1 and 5.2). Guidelines will

---

[1]Jack Selzer, "Composing Processes for Technical Discourse." *Technical Writing: Theory and Practice,* ed. Bertie E. Fearing and W. Keats Sparrow. New York: Modern Language Association, 1989: 43.

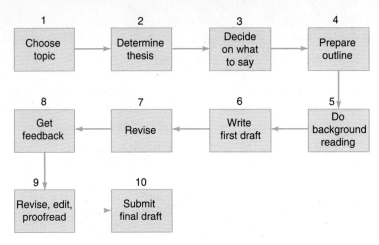

FIGURE 5.1   Composing process: linear model.

help you along, but they are not like lists of ingredients in a recipe. Ultimately, you must work out your own strategy for getting the document written. You are the final authority; that's what *author*ship is all about.

Let us consider one fairly typical example of how one writer, Tony Shimoto, composes dynamically in a work-related situation: Tony needs to send out a memo that describes a problem he encountered with his company's new laserjet printer and suggests how the problem might be resolved. Tony's usual habit is to do some freewriting in order to generate these details: to write down thoughts as they occur, spontaneously, without worrying about style; then to construct an outline; then to draft the memo; and finally to revise it as much as necessary.

This time, however, he feels that freewriting and outlining would interfere with his ideas. He knows exactly what he wants to say about the problem, so why bother freewriting or outlining? Instead, Tony decides to plunge right into a draft of the memo—but halfway through it, he realizes that he's leaving out important technical details; he then decides to do some freewriting after all. Had Tony opted for the "rule" of freewriting first, he might not have discovered what he needed to say as quickly as he did. The point to stress here is that sometimes freewriting will "dislodge" ideas; other times it won't help at all.

Let's take a different writing situation. Imagine that you have just gotten started on a paper on sedimentary-rock formation for your geology course. You find that you're stuck; your mind is a total blank. Not even freewriting is able to yield anything useful. What to do? As a focusing device, you can first make a list of the points you *think* you need to make in your paper. This is sheer stab-in-the-dark guesswork, but you may be surprised where it can lead. Such a list might look like this:

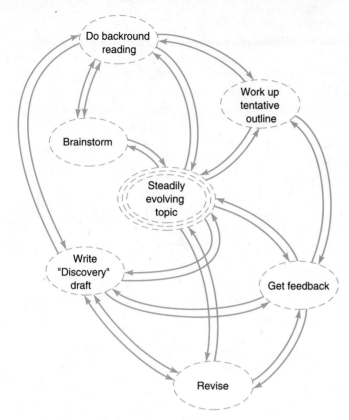

FIGURE 5.2   Composing process: dynamic model.

Original igneous rocks (those which formed out of molten lava) are
   broken down by moving water and extremes of temperature.
Disintegrated rocks settle (e.g., in riverbeds).
Pressure builds as the sediments accumulate.
Ages pass; the sediments repetrify.

Most of these points are obvious (at least from the viewpoint of a geology class),
but they have done the job of getting you to think hard along particular lines.
What's next? Plunge right into a first draft? Work up an outline? These tasks
seem premature at the moment because you feel insufficiently informed about
your subject. Outlining would only aggravate your sense of unpreparedness, so
you hike over to the college library, armed with a pack of blank index cards.

   Eventually, after having spent a few additional hours reading up on sedimentation processes and putting key bits of information on cards, you feel
ready to work up an outline. But as soon as you get into it, a sudden insight hits,

and your impulse now is to set aside the outline for the moment and write out a paragraph about how the sediments repetrify.

Should you follow your impulse, or should you stick to the tried-and-true routine? Writers who feel obliged to adhere to a standard routine will quell the impulse. And that could be a distinct disadvantage: sudden insights are not to be slighted! They arise when the mind is deeply engaged. When you feel the impulse to capture an idea, you ought to take advantage of it. You might momentarily lose sight of an overall structure, but the payoff will likely be several pages of draft in a relatively short time.

Writing, like mathematical computation, involves a problem-solving series of tasks; but unlike mathematical computation, the means of solving problems—and even the solutions themselves—have a greater range of possibilities; they are less dependent upon a strict routine, such as an established sequence of operations required to solve an algebraic equation or a geometric theorem. For writing, it is important to recognize what composition theorists refer to as its *affective* dimension. It means simply that an important part of what gets the writing going, even technical report writing—and of what sustains the momentum—is enthusiasm, intuition, the excitement of capturing an elusive thought in words.

Not everyone, of course, can be comfortable with what might seem like a helter-skelter way of writing. Perhaps you work more efficiently by following a set routine: listing key points, writing out a thesis statement, constructing an outline, then (and only then) starting the draft. Some professional writers do work this way—but it is also true that many professional writers do not. The only way to know for sure what would work best for you is to try the approach you are less familiar with at least once before deciding it is not your cup of tea.

Let us look in on professional technical writer Ramona Luzon, whose task at hand is to prepare a document describing an upgrade for a new word-processing program. Ramona must convey the information to registered users in a direct-mail announcement. In this and the following sections you will follow Ramona through her process of composing a two-page memorandum to the thousands of users of SuperWrite, to inform them of the new upgrade.

Because Ramona is able to think through the task more efficiently by writing down her *own* steps, she comes up with the following scheme:

Step 1: Determine my purpose(s) for writing the document, and what kind of document it should be.

Step 2: Decide upon the key elements to be included in the document.

Step 3: Map out a structure.

Step 4: Decide on which outside sources I'll need to consult.

Step 5: Write a first draft.

Step 6: Review all criteria.

Step 7: Get feedback from others.

Step 8: Revise.

Step 9: Proofread.

Here are the notes Ramona jotted down to herself alongside each of these self-imposed steps:

Step 1: Determine my purpose for writing the document.

a. What kind of document?

A direct-mail announcement

b. What is its primary objective?

To inform registered users of the most significant new features of the word-processor upgrade.

c. What is its secondary objective?

To inform customers that they can purchase the upgrade at a special savings if they are registered owners of an earlier version of the program.

Step 2: Decide upon the key elements to be included.

a. What specifically will the readers need to know about the upgrade?

It has a more thorough grammar checker.

It has a more powerful spell checker.

It has integrated graphics capability.

Its find-and-search capability operates twice as fast.

It has more font sizes and styles to choose from.

b. What specifically will the readers need to know about purchasing the upgrade?

If they've purchased their current program within the last year, they will get a free upgrade upon showing proof and date of purchase.

If they've purchased their current program between one and two years ago, they can receive the upgrade for just $50 upon showing proof and date of purchase.

Step 3: Map out a structure.

[Ramona prefers to begin with a brief "scratch" outline]

I. Introduction (to kindle enthusiasm for the upgrade).

II. Information about the features of the upgrade.

    III. Information about how to purchase the upgrade.

    IV. Conclusion.

Step 4: Decide which outside sources to consult.

    a. Which print sources will I need to consult?

      User manual for last upgrade.

      Galleyproofs of user manual for new upgrade.

    b. Whom shall I interview for "inside" information about the upgrade?

      The product tester (for info. regarding keyboard/mouse controls; speed of accessing capabilities).

      The promotions manager (for info. about the deadline for purchasing the upgrade at the special price).

Step 5: Write a first draft.

    [Ramona makes the following notes to herself]

    a. *Be more informational than persuasive. Readers will get suspicious if the announcement sounds too much like a come-on.*

    b. *Emphasize practicality of the new features, rather than their ingeniousness.*

Step 6: Review the draft for the following criteria (think SAS FAS).

    a. Statement of purpose: is it clear? Does it actually reflect the content of my document?

    b. Analyses and explanations: are they clear for the intended audience?

    c. Structure: strong introduction? fully-detailed body? reinforcing conclusion? logical recommendations (if relevant)?

    d. Format: attractive, uncluttered, useful subheads and graphics—text integration where warranted.

    e. Accuracy: have I double-checked facts, sources? If the information is time-dependent, are the data up-to-date?

    f. Surface correctness: grammar, spelling, punctuation, capitalization.

Step 7: Get feedback from others.

    (Here Ramona listed the names of two fellow writers, a software engineer, and the director of the upgrade team, as potential readers for her draft.)

Step 8: Revise.

    (Ramona's habit is to do at least two separate rewrites: the first concentrates only on content and organization; the second concentrates only on readability (conciseness, sentence variety, clarity, coherence from sentence to sentence and from paragraph to paragraph.)

Step 9: Proofread.

> (Ramona's habit is to check over the revised draft for errors in grammar, spelling, punctuation, capitalization; she reads each line backwards, then forwards, keeping a ruler under each line as she proofreads it, to prevent her from slipping into normal reading habits.)

Ramona has successfully followed a step-by-step approach to composing her document: not a predetermined set of steps, but her own, occasioned by the particular circumstances of the task at hand, as well as by her own temperament.

INVENTION STRATEGIES: FREEWRITING AND BRAINSTORMING

As you have noted earlier, when you peeked over Tony Shimoto's shoulder while he was composing, invention strategies are useful to know, but they don't always work; they are dependent on the writing situation at hand and on the writer's temperament at any given moment.

Ramona, like Tony, fairly consistently has had mixed success with invention strategies; at times it can help her work out a complex outline and get her to first-draft stage more quickly. "I sometimes get impatient with freewriting or outlining," she explains. "I like whipping through a draft because it gives me the best sense of what the whole document is supposed to look like. After that, I don't mind doing a zillion more drafts, if necessary." The kinds of inventing Ramona finds most useful are spontaneous scratch outlines (refer to her Step 3) to give herself a quick sense of structure. The first page of the memorandum will highlight some of SuperWrite's major features; the second page will present, in a more systematic manner, all of the major features.

Ramona also likes to scribble cryptic notes on any available piece of scrap paper while she's outlining or drafting; that is, she likes to write whatever comes into her mind, with no thought to structure or style:

> Key features of SuprWr: better spellchkr, grmr. chkr, draw. tools for on-the-spot visuals, improvd Srchr, more avail. fonts/styles/sizes; add supr/subscript capab., able to format cols. Any other desktop-publishing-like stuff to incl.?? check w/ Proj. Coord.

This kind of inventing is called *freewriting*. You simply write out thoughts as they occur, not stopping to plan or edit. Such freewriting can be done at any stage: before, during, or after the first draft or outline.

Another writer might wish to engage in a kind of idea-generating called *brainstorming*—a freewheeling jotting down (or thinking aloud, in small groups) of questions, hunches, free-associative ideas to consider about the task at hand:

> Should I open the letter with a dramatic (dazzling?) new addition to the program to get attention? "Imagine pressing one key and calling up a grammar checker in two seconds flat."
>
> And more dramatic emphasis? "That's just one of seventy-five new features added to what is already the most powerful word processor on the market."

The virtue of freewriting or brainstorming is that it shuts off the Editor inside your skull, that nagging voice tripping you every step of the way: Check your spelling! Improve your sentence construction! Watch your commas! Is that the most accurate term? These modes of invention free you of those rules. They are often messy, lacking organization; but they accomplish something crucial: to get out the substance of what you need to say, and to get it out fast.

A MORE SYSTEMATIC INVENTION STRATEGY: THE PENTAD

Devised by rhetorician Kenneth Burke, the pentad (a Greek word meaning "group of five") is a way of rapidly generating the particulars of a topic by focusing on five prompts:

Act (the event)
Scene (where the event occurs)
Agent (the person or thing performing the act)
Agency (the means by which the act is performed)
Purpose (the reason for the event)

So far, this looks like a mere variation of the journalist's "5Ws and the H" (*Why* does *Who* do *What*? *Where* and *When* does he or she do it, and *How*?). But the versatility of the pentad lies in the "dramatizations" (Act, Scene, etc.) suggested by these prompts which help the writer visualize the idea taking shape. The writer's mind, in other words, becomes a stage on which

FRANK & ERNEST reprinted by permission of NEA, Inc.

the elements of a discourse take shape. Another feature of the pentad is that it suggests what Burke calls the *ratios,* or relationships, which can be generated between any two of the five prompts. Here is how one engineer-writer, Eleanor Linsky, used the pentad to generate details for a report on seat-belt safety:

> Act: Seat-belt testing.
> Scene: Crash-Test Lab.
> Agent: Scientists concerned about the reliability of seat belts.
> Agency: Crash-testing equipment used to simulate conditions of four kinds of crashes: head-on, broadside, rear-end, and flip-over.
> Purpose: Determine whether seat belt mechanisms are sufficiently reliable.

Eleanor has quickly generated enough basic data to begin an inquiry. Next, by setting up ratios—Act to Scene, Agent to Agency, Agency to Purpose, Act to Purpose, and so on—she can generate a large quantity of data that can help guide her through several in-depth investigations she might not have considered otherwise. Table 5.1 shows a couple of examples.

Table 5.1 Ratios Used by Eleanor to Generate Content	
Ratio	**Example of Content Generated**
Act : Scene	Seat-belt testing should be done using simulations that come as close to actual car crashes as possible; the Crash-Test Lab has designed equipment to do just that.
Act : Agent	Seat-belt testing should be conducted by independent scientists to ensure maximum objectivity. The scientists at the Crash-Test Lab are all highly trained specialists who have absolutely no affiliation with the automobile industry or any of its subcontractors.

FROM IDEA TO STRUCTURE: CREATING USEFUL OUTLINES

Outlining, in a sense, is another form of invention—one with more of a concern for structuring. An outline frames the overall shape of a document. This is important because keeping the structure of the document in mind during the actual writing ensures against digressing, helps maintain overall coherence. It is easier to write about Point A if you are already aware that Points B, C, and D must follow. Far from being confining, an outline frees the writer to work out of sequence (e.g., to work on section II-C before section I-B for whatever reason). An outline also permits one to experiment with the structure—writers commonly rework their outlines several times during the course of a single draft.

Outlining is easy and even fun (it *is* fun to see a concoction of half-baked ideas suddenly take on a coherent shape). The first thing to realize, however, is that outlining should begin simply, with a scratch outline. One of Ramona's scratch outlines (she typically makes a dozen of them) for her project looks like this:

SUPERWRITE 6.0

    I. New time-saving features
       automatic dating
       pull-down palette for inserting symbols

    II. Additional conveniences
       faster "Finder"
       improved spell and grammar checkers
       improved thesaurus
       improved search/replace capability

    III. Specifications

With small documents such as the two-page memorandum Ramona is planning, experienced writers can jump from scratch outline to first draft. Writers who prefer to play it safe would probably wish to draw up a more detailed outline, a topic outline (similar to a scratch outline, except that more particulars are alluded to and a standard outline format is used), or a sentence outline, which consists entirely of complete sentences. Topic outlines are ideal for early planning stages when the writer needs to map out a general structure for the document. If Ramona had opted to prepare a topic outline, it might have resembled the following:

NEW FEATURES OF SUPERWRITE 6.0 FOR THE
GALAXY COMPUTER

I. Introductory remarks
II. Additional time-saving features
   A. Automatic dating
      1. one-key command
      2. range of formats
   B. Graphics fully integrated
      1. symbols
      2. table and graph formats
III. More interactive capabilities
   A. Insights into a given move in context
   B. Choices from list of recommended moves

This kind of outline provides a hierarchical relationship of topics to subtopics and can remind writers of where they need to work most on development.

Sentence outlines, by contrast, can be thought of as a dress rehearsal for a draft. Indeed, many writers merely remove the outline framework, add appropriate transitions, and thereby transform their sentence outline into a first draft.

NEW FEATURES OF SUPERWRITE 6.0 FOR THE
GALAXY COMPUTER

I. [Introductory] The long-awaited update of SuperWrite is here!
   A. Update was three years in the making.
   B. Thousands of 5.0 users made suggestions; many were incorporated into the new version.
   C. The new version may be ordered for $25 to 5.0 owners; for $50 to 4.0 users; for $100 to those who have versions 3.0 and earlier.
II. Version 6.0 has several major new features.
   A. Users can open files twice as fast.
   B. More menu-driven options enable the user to master the program quickly.
   C. The spell checker and grammar checker are built in and can be accessed with function-key commands.

Writers who prepare detailed sentence outlines are able to make an easy transition to a first draft.

WRITING THE FIRST DRAFT

Many writers do not draw a distinction between freewriting and first draft. Some are agile thinkers who can organize spontaneously and well inside their heads (at least for short documents), right from the start. Their freewriting comes out so structured that it can indeed function as a first draft. Others proceed much more slowly and painfully—not that the quality of their thinking isn't high, but that it has to be wrestled into shape over a series of drafts, with a good, messy freewrite to start things off. The best guideline here is, If you feel ready to begin a first draft, do it. You can always shift to outlining or freewriting if you get blocked.

STORYBOARDING TECHNIQUES

A dynamic technique borrowed from screenwriters and cartoonists, storyboarding enables writers—especially writing teams—to structure and draft a document by setting it up as a visual sequence. This can be an ideal strategy for planning such visually oriented documents as procedural manuals (see Figure 5.3), but it can work well with just about any kind of document.

Each element planned for inclusion is projected onto a storyboard, which can be a large sheet of paper or card, and taped onto a wall. When having to work out a long series of procedures, the storyboard can be invaluable: individual cards may be shuffled and reshuffled (or eliminated altogether) for proper sequencing. Written segments can be placed on separate cards and then clipped to their respective visual cards. These, likewise, can be rearranged to improve coherence.

REVISION STRATEGIES

The essayist William Gass once explained that he has to rewrite everything many times just to achieve mediocrity. Once a writer learns what it takes to make writing work in all the ways that it is supposed to work (to be clear, coherent, substantial, convincing, etc.), the first draft is only the starting point. The real shaping and enriching begins with the second draft. William Gass's assertion is humorous, but it is entirely representative: the great majority of writers revise numerous times before they allow their work to be published.

There are two kinds of substantive revision—holistic and sectional. The first kind, what F. Scott Fitzgerald referred to as "revising from spirit," is virtually starting over, when one is dissatisfied with the entire approach to the first draft. Holistic revising is both satisfying and frustrating: satisfying because a fresh start enables you to do what you feel you need to do without having to fit it into an existing framework; frustrating because you feel that you've wasted a lot of time

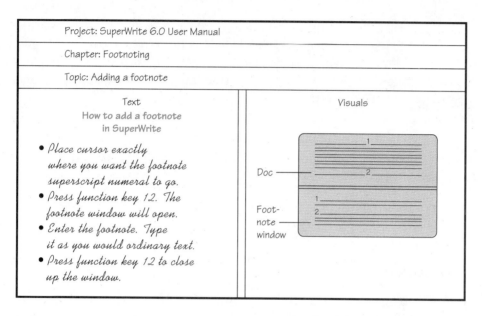

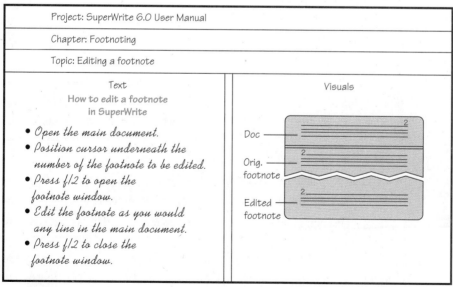

FIGURE 5.3   Storyboards for a chapter in a user manual.

writing a useless first draft. Of course, that first draft, no matter how flawed, was far from useless: it helped you sharpen your sense of purpose and style.

Sectional revision is what writers do when they rewrite paragraphs or whole chunks of a document; or rearrange paragraphs; or add more details; or get rid of extraneous details. Substitution, rearrangement, addition, deletion: these are the four kinds of sectional revision you will want to become skilled at using.

However, one does not mechanically screen a draft with revision strategies per se in mind; one screens a draft with standards of clarity, development, and coherence in mind. When a draft does not seem to meet these standards, the writer should regard the matter as a problem needing to be solved.

The following example is the first draft of Ramona's letter publicizing the latest upgrade of SuperWrite. The readers are those who have been using the earlier version of SuperWrite.

Dear SuperWrite Customer:

According to our records, you are a registered user of SuperWrite, version 5.0, as well as two earlier versions. This means you may order the latest update, 6.0, containing many important new capabilities, at a very special price. Just fill out the enclosed form and return it, with your payment.

Please contact us at 800-876-5432 for further information.

Sincerely,

Ramona Luzon, Director,
SuperWrite Project Supervisor

What would be an effective problem-solving strategy for revising this letter? Here is one two-step approach: ·

1. Review past letters for introducing a new product update to customers. Pay particular attention to

   introductory paragraph: does it clearly explain the purpose of the correspondence?

   tone: friendly and welcoming?

   background information: is it sufficient?

   the instructions for obtaining the new update: are they clear?

   the technical specifications for the updated program: are they mentioned? Are they complete?

2. Read the letter through the eyes of your primary readers. Ask yourself, What specific information do they need, and what can I add to the draft to meet that need?

Using those two guidelines, the writer revised the letter as follows.

Dear SuperWrite Customer:

SuperWrite 6.0 is here! We are pleased to tell you that, as a registered user of SuperWrite 5.0, you automatically qualify to receive the new, powerful upgrade for only $25, plus handling. We have enclosed a registration form for your convenience; merely fill it out and return it to us with your payment.

SuperWrite 6.0 has many more options than the previous version; for example:

- More versatile grammar checker.
- Larger spell checker dictionary that includes thousands of geographical and biographical names.
- Built-in thesaurus.
- Find-and-search capability that operates twice as fast.
- Three times as many type fonts, styles, and sizes to choose from.

SuperWrite 6.0 can still be operated with a minimum of 1 megabyte of RAM. However, to access the new built-in thesaurus, a minimum of 2 megabytes of RAM will be necessary.

If you have any questions about SuperWrite 6.0, or about ordering it, please call me at my toll-free number, 800-876-5432. We value you as a customer, and are here to help you in any way we can.

Sincerely yours,

Ramona Luzon, Director
SuperWrite Project Supervisor

Revising a paper need not be a kind of groping around in the dark, trusting your intuition. By approaching the task as a problem to be solved, you can revise more efficiently.

COMPOSING A TECHNICAL DOCUMENT: A CHECKLIST

1. INVENTING
   ☐ Have I tried freewriting to generate content?
   ☐ Have I participated in a brainstorming session with two or three other students as a way of generating content?

☐ Have I tried using the pentad to generate content in a more systematic manner?

2. NOTETAKING AND OUTLINING

☐ Did I take notes during my library research, and have I used these notes, arranged in a logical sequence, to work out a scratch outline?

☐ Does my topic outline cover every subtopic needed for my document?

☐ Does my sentence outline contain complete sentences for every subtopic in my document?

3. DRAFTING

☐ Did I write a "discovery draft" or "rough draft" as a way of generating content rapidly, without concern for structure or style?

☐ Have I used my outline to help develop and organize my draft? Have I used my draft to strengthen my outline?

☐ Have I used storyboarding techniques to develop my draft?

4. REVISING

☐ Have I worked out a plan for revising each section of my draft?

☐ Have I approached my revision task as a problem to be solved?

☐ In peer-critiquing groups, have I encouraged my readers to offer candid suggestions about my draft, or at least to ask questions about anything they were confused by?

■ SUMMARY

Writing, unlike grammar, is not rulebound; it is a dynamic process that can be modified to suit a writer's individual temperament or the particular task at hand. Writers should acquaint themselves with the various techniques of invention, outlining, drafting, and revising. They can then adapt these techniques in whatever manner seems most effective. Invention strategies such as freewriting and brainstorming are ideal for generating a lot of material quickly. The Burkean pentad is a way of generating content within a context of interrelationships.

Outlining, which helps writers establish a coherent framework for their document, can also help generate content. Storyboarding is another powerful means of organizing by creating a series of visual elements to which appropriate written passages may be added.

Drafting can proceed rapidly or slowly, depending upon the amount of planning, the conceptual habits of the writer, and the nature of the project. Revising strategies also vary greatly. Sometimes writers need to do a holistic revision "from spirit"; other times, writers can revise a draft section by section. What works well with one writing circumstance will not necessarily work with another. Writers should be flexible.

■ FOR DISCUSSION

1. After trying two or three of the invention strategies suggested in this chapter, discuss which of them work best for you and why.

2. Discuss the strengths and the limitations of outlining. When have you personally found outlining most useful? When did it seem otherwise?

3. Describe the manner in which you revised your last essay. What gave you the most difficulty?

4. Compare your composing habits as a freshman with your current composing habits. What do you do differently, if anything? To what do you attribute these changes?

■ FOR YOUR NOTEBOOK

1. Keep a running log of everything you do to complete a writing assignment while you're working on it. Don't overlook any detail, no matter how trivial it may seem.

2. Use your notebook to practice using the pentad to generate ideas. After several entries, evaluate the effectiveness of this mode of invention.

3. Prepare storyboards for technical writing projects; begin with a storyboard for Ramona's SuperWrite 6.0 project or Eleanor's report on seatbelt safety.

■ WRITING PROJECTS

1. Write a paper detailing the processes you employed in completing two recent writing assignments. Include everything you remember doing, and

try to reconstruct the sequence in which you undertook each task. Finally, provide a rationale for the approach you used for each paper.

2. Compare the composing habits of two classmates with regard to a common assignment. Give a detailed account of each student's approach to the assignment and his or her rationale for that approach.

6

Basic Rhetorical Techniques I: Presenting Information Clearly and Accurately

Defining and Categorizing Terms

Describing the Characteristics of Objects, Functions, Problems

Narrating Events

Establishing Relationships

Speculating

Presenting Information: A Checklist

echnical writing shares the fundamentals of all writing; and if the essence of these fundamentals had to be captured in just two words, those words would be *clarity* and *accuracy.* If your statement lacks clarity, the information in it is likely to be misunderstood or misinterpreted. Without accuracy, the information you convey, however clearly, is worthless. Or worse: it can result in rejection of proposals, loss of contracts, or even (if products are built using imprecise or inadequate information) loss of life. It's true: your words on paper not only constitute the script for what we do (the utility factor of technical writing discussed in Chapter 1), they can sometimes make a difference between life and death.

This chapter will guide you through those elements of expository writing that together yield clarity and accuracy: defining and categorizing terms, describing objects, telling about events, comparing one thing with another, examining cause–effect relationships, and speculating (hypothesizing). Probably you have studied most of these elements in earlier composition courses, but here you are studying them in the context of conveying technical information—for the general public as well as for more specialized job-related contexts.

DEFINING AND CATEGORIZING TERMS

Defining and categorizing are important rhetorical techniques in technical communication. What might pass as a sufficient definition in conventional situations is often unacceptable in situations requiring a high degree of precision. For example, a journalist may define a term like this:

> The patient began to hyperventilate (gasp for air).

In the context of a news story, such a parenthetical definition may be sufficient, if indeed necessary at all. In the context of a medical report, however, such a definition could be seriously inadequate. How long was the patient gasping for air? Were the inhalations deep or shallow? Was the hyperventilation accompanied by dizziness? by nausea? These questions would most likely need to be answered in a medical report.

The kind of definition a writer uses, then, depends on the needs of the reader, the complexity of the term to be defined, and the purpose of the communication. Basically, technical writers make use of three kinds: lexical, categorical, and operational.

Lexical (Dictionary) Definition

Lexical definition is the most common method of defining terms; it is the kind of defintion used in dictionaries and glossaries:

> **bungee:** an elasticized cord, typically with a hook at each end, used to bind a suitcase to a wheeled carrier; to secure a hold for climbing, etc.
>
> *—Random House Webster's College Dictionary.*

> **setup string:** A group of characters that sends a format command to a printer. Some application programs, such as spreadsheets, give you the option of changing format, such as character width, by entering a setup string before printing.
>
> *—ImageWriter II™ Owner's Manual* (Cupertino, CA: Apple Computer, Inc., 1985): 118.

Such definitions, as you can readily see, generally consist of three elements:

1. The term to be defined ("bungee"; "setup string").
2. The general category or genus of which it is a part (elasticized cords; a group of characters).
3. The specific characteristics (differentia) that distinguish the term from any other examples in the general category (elasticized cords that have a hook at each end; a group of characters that sends a format command to a printer).

   (Optional) Typical uses that help clarify the distinguishing characteristics ("used to bind a suitcase to a wheeled carrier"; "Some application programs, such as spreadsheets, give you the option of changing format. . . .").

When writers define terms parenthetically in a document, they are using a compressed version of lexical defining. "Gasping for air," for example, is a compressed lexical definition of *hyperventilation;* a more standard lexical definition would typically look like this:

> Extremely rapid or deep breathing that overoxygenates the blood, causing dizziness, fainting, etc.
>
> *—Webster's New World Dictionary,* 2nd ed.

Lexical definitions of objects sometimes need to be accompanied by illustrations, or else the definition would be too difficult to visualize. Illustrations can also succinctly add important subcategories to the differentia. Here is an example:

dodecahedron: a solid figure having 12 faces.

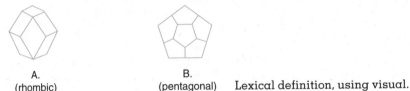

A.
(rhombic)

B.
(pentagonal)   Lexical definition, using visual.

—*Random House Webster's College Dictionary.*

Sometimes a writer may expand a lexical definition to one or more pages because the term in question is so intrinsically interesting. Typically, the author will trace the origin of the term and in so doing offer fascinating insights into the way scientific concepts evolve. One master of the expanded lexical definition is the late Isaac Asimov, the prolific multidisciplinary writer and one-time professor of biochemistry. Note how Asimov, in his following definition of "engine," concentrates on the etymology (historical origins) of the principal term, as well as the origins of related terms.

Engine

Originally, the word *engine* simply meant "an ingenious device." In fact the word *engine* is only a corruption of the word *ingenious,* which comes from the Latin "in-" (in) and "gignere" (to produce). Clever ideas, you see, are "produced in" the mind of an ingenious [person]. After James Watt invented a practical steam engine (an ingenious device run by steam power), the word was applied more and more to those devices particularly which took power from some non-living source and turned it into work by means of the to-and-fro motion of a cylinder. These days, the gasoline engines that power our automobiles, buses, trucks, and aircraft are more important than the old steam engine.

The older, more general meaning of the word still persists in *cotton gin,* a machine used to strip the cotton fibres from the seeds. *Gin* is only a slangy contraction of *engine.*

Devices which turn power into work by turning rather than by piston motion are called *turbines,* from the Latin "turbo" (a top, or other spinning object). Water wheels, which are turned by running water and which produce power as a result, are, more properly, *water turbines. . . .*

—Isaac Asimov, *Words of Science and the History Behind Them*
(Boston: Houghton-Mifflin, 1957).

New concepts are continually being developed in science and technology; consequently, new words are coined to accommodate these concepts more efficiently. Although this sort of activity can get out of hand (excessive use of neologisms can cause more confusion than clarity), it is generally helpful, and certainly a testimony to the creativeness of scientists and engineers.

Let's take a look at an especially colorful example of coinages. In March, 1992, *New York Times* writer Phil Patton explored the world of design conferences, such as the "Edge of the Millennium" conference sponsored by the Cooper-Hewitt National Museum of Design in New York. This is what he learned:

How to Talk Cyberchat

By Phil Patton

Designers have liked to get together and talk design talk at least since the Aspen Conference was established in 1951. Of late, however, there seems to be an exponential increase in the number of design conferences and a commensurate growth in design lingo.

From the Stanford Design Forum to the Design Management Institute, design talk is not only getting louder, it is getting deeper. Earlier this year, the Cooper-Hewitt National Museum of Design in Manhattan sponsored the weighty "Edge of the Millennium" conference, offering neologisms from the designers Tibor Kalman and Constantin Boym, and the behavioral psychologist Donald Norman.

Richard Saul Wurman's third annual TED Conference (Technology, Entertainment, Design) last month in Monterey featured— at $1,000 a ticket—everyone from Bill Gates of Microsoft to George Lucas, Stephen Jay Gould, Quincy Jones and the virtual-reality whiz Jaron Lanier.

Even trade shows have themes. The Westweek '92 furniture trade show this week in Los Angeles is operating under the sonorous title "Counterforce/Counterbalance: Emerging Attitudes and Esthetics in a Changing World." Moreover, its prospectus speaks of "paradigm shifts in the creative process."

Out of such conferences and other sources a new design vocabulary is emerging that reflects the schizophrenic influences of high-tech equipment and high-concept marketing, of sci-fi and Disney, of *Mondo 2000*, the magazine for virtual-reality buffs, and *Martha Stewart's Living* magazine. Here is a sampling:

(continued)

Alfa Romeo Design Coined by the industrial designer William Plumb for design that is, well, like the car: exciting, high-performance designs that are quirky and often unreliable—"giving one trouble on a foggy night on the autostrada north of Milan." (See also, Volvo Design.) **Cactus and Coyote** Used by disgruntled sellers of legitimate Southwestern art and furniture to refer to jigsawed or painted silhouettes of Arizona flora and fauna, representing the Southwestern influence in the mass market. **Cyberchat** Coined by the designer of graphic information Edward Tufte: (1) sweeping statements about the impact of computers and other high technology; (2) a popular dialect spoken at design conferences. **Cyberchat** The notion of viewing all information and its means of retrieval as a metaphorical three-dimensional landscape; popularized by the cyberpunk novels of William Gibson, Bruce Sterling and others. **Dark Deco** The gothic Gotham look created by Anton Furst, as seen in the film "Batman" and in parts of the Manhattan nightclub Planet Hollywood. The look is to have a second run later this year in "Batman 2" and in the cartoon series "Batman" that starts in September with tie-in products ranging from picture frames to bedsheets. **Demystification** "The right word to sum up what we are trying to accomplish," as Martha Stewart wrote in the February-March issue of her magazine *Martha Stewart's Living*. Her example of something needing demystification? Chocolate. **Duck and Basket** A mass-market dilution of country style. **Dude** The tendency of products of high cachet to emulate products of lower cachet emulating products of higher cachet. Ralph Lauren dungarees are a perfect example: an upscale version of a lowscale jean trying to be upscale. **Emotional** Possessing a rounded shape (said approvingly), as in "The '90s are a decade of emotional design." (See also Organic.) **Eurostyle** An ancient term redefined—now that it has reached the Middle West—to mean the replacement of simulated wood grain or chrome with matte black. A Eurostyle Chevrolet Lumina has dark trim instead of chrome; Coffee now offers Eurostyle models. **Fetishism** a.k.a. commodity fetishism, the old Marxist term, making a comeback. Psychological and fiscal overinvestment in, say, a toaster or Air Jordans. Indicates academic second thoughts about prosperity; now used almost exclusively by prosperous academics. Appearance in common conversation tends to elicit bafflement, especially among citizens of current or former Marxist states. **Florabundance** A preponderance of leafy floral patterns. A news release from Pfalzgraff, the dinnerware manufacturer that coined the term, called this style "an antidote of romance and nostalgia to the chaotic pace of living in the '90s." **Frankenstyle** Dangerous-looking design, with sharp edges, horns and metal "twigs." Associated with Philippe Starck.

(continued)

Information Rich From *The Next Economy* (Holt, 1983) by Paul Hawken, a book that touts products with a high level of information— meaning technology, design, style. A bushel of wheat is lower in information than a VCR or a Philippe Starck juicer or a pruning saw imported from Sheffield, England. **Interactive Home Systems** The future of decor as seen by Microsoft. In Bill Gates' much anticipated high-tech house, frames full of customized computer images, either moving or still, will replace paintings; if this is Monday it must be Matisse. **Organic** Having a curved shape (said approvingly). See Emotional. **Ralph** A verb meaning to decorate in the style of Ralph Lauren. "His place used to be white on white but now he's Ralphing it." **Swatchism** Coined by the Krups designer Michael Kramm. The production of many cosmetically different models of essentially identical products, on the example of the fashion-inspired Swatch watch. **Virtual** From virtual reality, a simulated world produced by computerized systems of gloves, helmets or suits. Most popular world—er, word—at current design conferences. By extension, any replication of a physical object by an electronic equivalent: A computer program of addresses is a virtual Rolodex. "Virtual reality is the ultimate design medium because there are no constraints," said Bruce Sterling, co-author with William Gibson of *The Difference Engine* (Bantam, 1991). "It will enable couch potatoes to take root." **Volvo Design** Used by the industrial designer William Plumb. Design that is sensible, sturdy, safe and dependable, well thought out and ergonomically correct, but heavier than necessary and unsleek—just like the car. With overtones of "Volvo liberal." See also Alfa Romeo Design. **Weightlessness** The fashion designer Geoffrey Beene's stated ideal for his new collection of tableware for Swid Powell—achieved by putting plastic handles on silver.

The New York Times

As useful and intriguing as lexical definition can be, it is but one of several definitional strategies that can be used to clarify concepts. We need to examine two additional strategies commonly used in technical communication: categorical definition and operational definition.

Categorical Definition

Two efficient ways of coming to understand a complex entity are (1) to place it in a classification system and (2) to isolate its component parts. This is sometimes already done, although in an abbreviated manner, in lexical definition; take, for example, a lexical definition of *scorpion:* "Any of an order (Scorpionida) of arachnids found in warm regions, with a front pair of nipping claws and a long, slender, jointed tail ending in a curved, poisonous sting" (*Webster's New World Dictionary*). The category, Scorpionida (which in turn falls within a larger category, arachnids), introduces the classification system that the term being defined is a part of, then introduces the specific category— those characteristics or component parts which differentiate the creature from other arachnids (front pair of nipping claws; long, slender, jointed tail; poisonous sting).

But if you were going to write an extended technical definition of scorpions, you would most likely want to examine the categories in more detail, as is done in the following example:

SCORPION: Any member of the arthropod order Scorpionida, class Arachnida, the class that also includes spiders and mites. About 650 species of scorpion exist; about 40 of them occur in the U.S. Usually brown in color and ranging from about 2.5 to 20 cm. (about 1 to 8 in.) in length, the scorpion has a flat, narrow body, two lobsterlike claws, eight legs, and a segmented abdominal tail. Terminating in a venomous stinger supplied by a pair of poison glands, the tail is usually curved upward and forward over the back.

Found in warm and dry tropical regions, including the southwestern U.S., the scorpion is nocturnal and feeds mainly on spiders and insects. The young are born live and remain with the mother for a short period. When capturing a victim with its claws, the scorpion inflicts a disabling sting that is painful but, in most species, not fatal to humans. The sting of the U.S. species, *Centrueides sculpturatus,* however, has proved fatal to young children and is potentially fatal to adults. Other areas of the world have more dangerous scorpion species; the poison involved is a neurotoxin, attacking the nervous system.

—*Funk & Wagnalls New Encyclopedia* (1983)

In this example, the scorpion is first *classified* within its biological class, that of Arachnida, which includes other orders, such as mites and spiders. The scorpion's distinguishing *component parts* are then itemized: flat, narrow body; two lobsterlike claws, eight legs, segmented abdominal tail, venomous stinger.

The second paragraph is, strictly speaking, descriptive (see the later section on description) rather than definitional: it concerns itself not with what the scorpion is but what it is like and where it can be found. Description, in other words, can serve to augment definition.

Operational (Functional) Definition

Some objects can only be accurately defined by what they do, how they work. Machines with moving parts are good examples. Let us return to *engine—*

specifically, to *internal combustion engine.* Of course, we can turn to a lexical definition:

> an engine of one or more working cylinders in which the process of combustion takes place within the cylinders.
>
> —*Random House Webster's Dictionary.*

But it doesn't get us very far. We do not really know what an internal combustion engine is until we know its operational principle:

> [an] engine in which combustion of fuel takes place in a confined space, producing expanding gasses that are used to provide mechanical power. The most common internal combustion engine is the four-stroke reciprocating engine used in automobiles. Here, mechanical power is supplied by a piston fitting inside a cylinder, which receives a mixture of fuel and air from the carburetor through an intake valve; the piston moves up to compress the mixture, forcing the piston down; in the exhaust stroke, an exhaust valve opens to vent the burned gas as the piston moves up. A rod connects the piston to a crankshaft. The reciprocating (up and down) movements of the crankshaft rotate the crankshaft, which is connected by gearing to the driveshaft of the automobile.
>
> —*How Things Work* (New York: Simon and Schuster, 1970).

In this passage, as in the categorical definition of the scorpion, the writer, to help readers comprehend the term being defined, complements the basic definition with a detailed *description*—in this case, of piston movement and the way it transforms heat energy (created by the burning fuel) into mechanical energy (rotation of the crankshaft and driveshaft). As is often the case with ex-

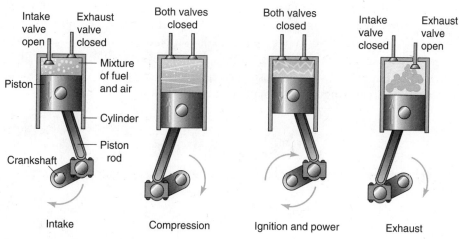

Operational definition, using visual.

tended operational definitions, the text is accompanied by a series of schematic diagrams that illustrate each "stroke" of the cycle.

Making Technical Concepts Palatable to General Readers

One of the most persistent barriers preventing the "technical" from being more readily integrated into the "everyday" is specialized terminology. Sometimes merely looking up the unfamiliar word in a dictionary will solve the problem; but more often, the strange word is only the tip of a conceptual iceberg. Lay readers need to see a connection between the technical concept and their own experience. What can a writer do to establish such a connection?

In addition to providing readers with a lexical, categorical, or operational definition of the term, a writer should tend to the following:

- Amplify the basic definition by establishing an analogy between the term in question and something out of commonplace experience. EXAMPLES: Comparing the revolution of electrons around a nucleus to the revolution of planets around the sun; comparing light waves or radio waves to water waves; comparing interplanetary distances to proportionate earthly ones ("If the sun were the size of a grapefruit, then the peanut-sized earth would be three blocks away, and the outermost planet, Pluto, would then be twenty miles away").

- Contextualize the concept; that is, give readers a sense of its larger significance. EXAMPLE: After explaining the basic definition of *ozone* (a form of oxygen—O_3), proceed to explain how ozone protects living things from harmful doses of ultraviolet radiation.

Study the extended definition on pages 119–120. First, read through it quickly; then reread it, trying to pinpoint which elements of definitional technique—categorical, descriptive, operational, comparative, parenthetical—have been incorporated into the piece. Be prepared to evaluate the organizational strategy of the piece. What purpose does each section serve? What other possible ways of organizing the information can you think of?

DESCRIBING THE CHARACTERISTICS OF OBJECTS, FUNCTIONS, PROBLEMS

Description is sensory language (mostly visual, but sometimes olfactory, auditory, tactile, or gustatory as well) that refines our sense of what an object, prod-

Plate Tectonics

DEFINITION

A unifying geophysical theory which explains phenomena as diverse as deep-sea trenches, earthquakes, volcanoes, and mountain-building. The theory has two components: *tectonics*, the geological processes pertaining to structural deformations in the earth's crust; and *plates*, the gigantic, irregularly shaped, rigid blocks comprising the earth's outer shell.

WHAT IT REALLY MEANS

The earth is a living, moving being. In the normal geological course of events, mountains are born and die in hundreds of millions of years. Continents drift to even slower tempos. Once in a while, though, one part of terra firma moves just a tiny bit more vigorously than usual, and the terrified survivors witness the most spectacular natural occurrences our planet has to offer—earthquakes, volcanoes, tidal waves.

Volcanic action and earthquakes are the most visible manifestations of plate tectonic action. Despite our belief in "solid earth," most of the inside of our globe consists of a hot, dense, liquid rock layer surrounding a denser and hotter liquid metal core. The outermost rigid crust we walk on is relatively thin, and not at all uniform. The continents themselves rest on enormous plates, each one about two hundred miles thick, which float on the liquid mantle. When two plates rub together, the resulting friction deep underground may create channels for magma to erupt through the surface, causing volcanoes. When the pressure of plates moving in opposition causes parts of the outermost crust to slip suddenly, earthquakes are created.

The scientific field of plate tectonics is still in its infancy. So far, the theory is doing what a good theory should do—it explains, simply and credibly, why and how observable geological events occur. The next step—actual prediction of those events—is awaited anxiously by those people who live near fault zones or active volcanoes.

(continued)

uct, or phenomenon is like. One could say that description is yet another elaboration of definition: by helping us perceive the specific characteristics of something, we are enhancing our sense of what it is.

Description is essential in any kind of writing. Novelists, biographers, personal-experience essayists describe their characters, settings, events in keen physical, sensory detail, to make them seem real and alive to us. Technical writers also depend on description, not so much to stimulate the senses as to establish unambiguous identifiability for practical reasons.

Describing an Object

Study the following description of pillow lava, which forms from underwater volcanic eruptions:

Tongues of lava, chilled by contact with water, develop a tough, plastic skin that is convex upward and filled with liquid. On cooling, the interior becomes crystalline, and the skin glassy and often radially cracked. Identification of pillow structure . . . is a sure indication of underwater deposition.

—Frank Press and Raymond Siever, *Earth,* 3rd ed. San Francisco: W.H. Freeman, 1982: 349.

The purpose of this description is to identify the major properties of this particular kind of formation, so that it will not get confused with other types of lava formations, which could lead to mistaken inferences.

Technical descriptions of objects serve many purposes, such as identifying components in assembly instructions, or identifying the properties of natural objects such as plants, trees, rocks and minerals, fossils, and so forth. Here is a typical example of such a description:

GOLD (Au) [Crystal structure]: Cubic—hexoctahedral.

Environment: In quartz veins and in steam deposits.

Crystal description: Most often in octahedral crystals, with or without other faces. However, clusters of parallel growths distorted into feathery leaves, wires, or thin plates are most common.

Physical properties: Rich, yellow to silvery yellow. Luster: metallic; hardness: 2½–3; gravity: 19.3. Very malleable and ductile.

Composition: Gold, usually alloyed with silver. The higher the silver content, the paler the color.

Distinguishing characteristics: Confused with metal sulfides, but distinguished from them by softness and malleability. Microscopic brown mica flakes, which may be seen in streams or in mica schist, are distinguished by the blowpipe test or by crushing the mica plates with a needle.

Occurrence: The inertness of gold and its great density make it concentrate in streambeds, either in small flakes or in larger nuggets, from which it may be recovered by panning. . . .

—Frederick H. Pough, *A Field Guide to Rocks and Minerals,* 4th ed. Boston: Houghton-Mifflin, 1976: 71–72.

The organization of this description into "environment," "crystal description," "physical properties," "composition," "distinguishing characteristics," and "occurrence"—the same categories, in the same sequence, are used for every entry—not only enables the reader to comprehend the information quickly and easily, but it permits rapid referrability as well. Once again we have an example of technical writing formatted to be utilized, not just read.

Notice also how the author uses analogy to facilitate comprehension of the various properties. Gold crystals may sometimes be distorted into *feathery leaves;* the color is often a *rich* yellow, although when alloyed (i.e., intermixed) with large amounts of silver, it tends to be *paler.*

Describing a Function

One can also describe functions as well as objects. In the earlier extended operational definition of an internal combustion engine, the writer extended the definition by describing the action of the pistons. In the following description of deep-well injection (a technique of waste disposal), note how the writer brings in sensory details to enhance comprehension:

> Deep-well injection involves pumping liquid wastes deep underground into rock formations that geologists believe will contain them permanently. . . . A high-pressure pump forces the hazardous liquids into small spaces or pores in the underground rock, where they displace the liquids (water and oil) and gases originally present. Sedimentary rock formations like sandstone are used because they are porous and allow the movement of liquids.

> —Donald G. Kaufman and Cecilia M. Franz, *Biosphere 2000: Protecting Our Global Environment.* New York: HarperCollins, 1993: 407.

Functional descriptions are also commonly used in test reports. Here is an example of such a description for the field performance of a new 35-millimeter camera:

> Lens produced sharp, contrasty slides in front-lit conditions, with slightly less resolution at wide and narrow apertures of 80 mm focal length. Backlighting showed flare at all focal lengths with little improvement from stopping down.

> —"Test: Canon EF35-80mm f/4-5.6 II," *Popular Photography,* July 1994: 53.

Integrating Definition and Description

Many technical documents will integrate definition with description and will often include a labeled diagram, so that the reader can relate the parts as they are mentioned in the text of the description with the parts that are identified on the diagram. This allows for fullest comprehension. Notice how this works in the following illustrated definition/description of a capacitor:

Capacitors

Capacitors play an important role in the building of circuits. A *capacitor* is a device that opposes any change in circuit voltage. That property of a capacitor which opposes voltage change is called *capacitance*.

Capacitors make it possible to store electric energy. Electrons are held within a capacitor. This, in effect, is stored electricity. It is also known as an *electric potential*, or an *electrostatic field*. Electrostatic fields hold electrons. When the buildup of electrons becomes great enough, the electric potential is discharged. This process takes place in nature: clouds build up electrostatic fields. Their discharge is seen as lightning.

The figure below shows a simple capacitor. Two plates of a conductor material are isolated from one another. Between the two plates is a dielectric material. The dielectric does conduct electrons easily. Electrons are stored on the plate surfaces. The larger the surface, the more area is available for stored electrons. Increasing the size of the plates therefore increases the capacitance.

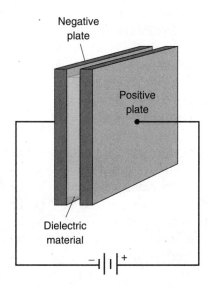

Negative plate

Positive plate

Dielectric material

Extended definition integrating description, using visual.

Rex Miller, *Electronics the Easy Way*, 2nd ed. New York: Barron's 1988: 35.

Look closely at the way the author has organized his description of the capacitor. First, he introduces it by describing its general function:

> Capacitors play an important role in the building of circuits.

Once the reader has absorbed this bit of information, the author then offers a basic lexical definition:

> A capacitor is a device that opposes any change in circuit voltage.

The document then proceeds to describe how the capacitor is able to perform its function (capacitance): it does so by storing electricity in an electrostatic field that consists of dielectric material (where the electrons are stored) sandwiched between two plates (one positive, one negative) that keep the stored electrons from escaping until they are ready to be discharged. Note how helpful the diagram is in conveying this abstract principle visually. Note also the analogy that the author draws between the capacitor's electrostatic field and that of a raincloud. Does the analogy assist readers in understanding the principle? Why or why not?

Describing a Product

Pick up almost any household product and examine its label: there you will see a description of that product's intended uses, ingredients, effectiveness, active ingredients (sometimes accompanied by a description of each ingredient's specific function), potential hazards, and what to do in case of misuse. The label for an antibacterial moisturizing soap on page 125 is typical.

Before reading on, try to pinpoint on your own what devices the author of the label used to help the consumer fully understand the nature of the moisturizing soap.

Did you find several? To begin with, notice how the use of subheads—"Effectively Kills Germs"; "Gentle to Your Skin"; "Moisturizing Formula"—enables the information to be easily and quickly read (and reviewed). Notice also the overall readability of the descriptions, unlike, say, the label on a prescription drug. Of course, a certain amount of old-fashioned advertising rhetoric is woven in ("laboratory proven to be unsurpassed at killing germs among leading antibacterial liquid soaps"; "perfect for the entire family"; and so forth)—but essential information is conveyed: "contains a . . . germ killer *Chloroxylenol*"; "rinses easily and leaves your skin feeling clean and silkened"; "contains four gentle moisturizers and conditioners to help maintain and protect the natural moisture level of your skin." That last bit of information is interestingly broken down into a table that enables the reader to match each of the four moisturizers/conditioners with its particular function.

Softsoap®
Brand
AntiBacterial
Moisturizing Soap

Effectively Kills Germs

Softsoap® AntiBacterial Moisturizing Soap is laboratory proven to be unsurpassed at killing germs among leading antibacterial liquid soaps. It contains a specially developed germ killer *Chloroxylenol* which acts directly on a broad spectrum of household germs.

Gentle to Your Skin

New Softsoap AntiBacterial Moisturizing Soap is specially formulated with mild cleansers and gentle moisturizers to help prevent drying and leave your skin feeling soft and smooth. And now Softsoap contains Silkagen,™ an exclusive blend of natural silk proteins with skin conditioners for even softer skin. Perfect for the entire family, Softsoap rinses easily and leaves your skin feeling clean and silkened.

Moisturizing Formula

Unlike other leading antibacterial liquid soaps, this Softsoap formula contains four gentle moisturizers and conditioners to help maintain and protect the natural moisture level of your skin:

Ingredient	Function
Silkagen™	Natural silk moisturizers with skin conditioners
Aloe Vera Extracts	Enhances skin suppleness and softness
Glycerin	Helps retain skin moisture
Polyquaternium 7	Makes skin feel smoother

Avoid getting liquid soap into eyes—if it does, rinse eyes with water.

ACTIVE INGREDIENT: CHLOROXYLENOL

WATER, SODIUM C14-16 OLEFIN SULFONATE, LAURAMIDE DEA, SILK PEPTIDE, HYDROLYZED SILK PROTEIN, COCAMIDOPROPYL BETAINE, POLYQUATERNIUM-7, ALOE VERA GEL, GLYCERIN, TETRASODIUM EDTA, SODIUM CHLORIDE, DMDM HYDANTOIN, CITRIC ACID, FRAGRANCE, YELLOW 5, RED 33, RED 40.

© 1991 SOFTSOAP ENTERPRISES INC., Chaska, MN 55318 Should you have questions or comments regarding this product, please contact us at 1-800-255-7552, or write: Colgate Consumer Affairs, 300 Park Avenue, New York, NY 10022-7499

Informational label.

Describing a Problem

Public agencies like the Environmental Protection Agency, the American Heart Association, and UNICEF continuously publish documents that describe the numerous critical problems that threaten people, animals, the environment. Many of these problems are complex, involving technical matters that need careful defining. Take, for example, the waterborne and water-based diseases that are debilitating or killing thousands of children in Third World regions. UNICEF publishes documents such as the one on pages 127–129 to help educate the public to the severity of the crisis.

Reread this document for its rhetorical effectiveness. For example, how well does the author define and describe such diseases as schistosomiasis? How well does the author describe the causes underlying these diseases? What might the author have clarified further?

NARRATING EVENTS

Narration is telling what happened or is happening. If the purpose of definition is to establish an entity's meaning, and the purpose of description is to establish its identity and appearance, then the purpose of narration is to present it in time and space, to help the reader see how it behaves or how it evolves.

Recall the description of pillow lava on page 121. The authors of that description follow it with an eyewitness account, by Dr. James Moore of the U.S. Geological Survey, of a pillow-lava flow in action:

> The flow was advancing under water as a wall of rubble . . . the base of which is covering over the old ocean floor to a depth of about 100 feet. . . . We [approached] one of these elongated lava tongues, in which lava was actually flowing inside and whose surface was too hot to touch on the outside. . . . There was a continuous loud mixture of noises: hissing, cracking, small explosions, and rumbling as the lava went down through these tubes.
>
> —quoted by Frank Press and Raymond Siever, *Earth,* 3rd ed., p. 349.

Notice what this narrative passage accomplishes: it enables the reader to understand what is happening moment by moment in real time. It also creates suspense—something you may normally associate with fiction writing; but suspense buildup can be equally useful in helping readers understand and appreciate a dramatic phenomenon like an underwater volcano.

Another purpose of narration in technical communication is to provide important historical background information. For example, in discussing elementary particles and the forces of nature, the distinguished astrophysicist Stephen Hawking establishes an important historical context along the way:

> Up to about twenty years ago, it was thought that protons and neutrons were "elementary" particles, but experiments in which protons were collided with other protons or electrons at high speeds indicated that they were in fact made up of

DEADLY WATER

Eighty percent of the disease in developing countries is related to poor drinking water and sanitation. The following are a few of the most serious diseases that afflict children in Africa, Asia, and Latin America when they have no alternative but to drink from contaminated ponds, lakes, streams, and rivers.

WATERBORNE DISEASES

These diseases are spread by the use of contaminated water for drinking, washing, or cooking.

Cholera - Caused by bacteria (vibrio cholerae). Cholera is contracted by drinking water contaminated by human waste. Symptoms include violent vomiting and diarrhea, abdominal pain and, often, kidney failure. Without rehydration therapy, cholera can kill quickly and is a major cause in the death of 5 million children under five who succumb to dehydration each year.

Typhoid - Caused by salmonella bacteria (salmonella typhi). Children infected with typhoid exhibit extreme lack of energy, headaches, vague pains, nosebleeds, and a high temperature of up to 104°F. At the height of the disease, children suffer delirium and deadly dehydration caused by severe diarrhea. The disease can also cause life-threatening intestinal hemorrhages.

Amoebic dysentery - This disease results from drinking water contaminated with abundant rotting organic material in which amoebas thrive. Amoebic dysentery causes dangerous dehydration from diarrhea.

Other serious waterborne diseases include polio, hepatitis A, ascariasis, and trichuriasis.

WATER-BASED DISEASES

These diseases are spread by worms and other parasites living in water used for washing or drinking. In some countries, 85 percent of five-year-olds are infested with worms.

(continued)

Schistosomiasis - Nearly 200 million people worldwide suffer from this parasitic infection. The schistosome parasite lives in the body and passes its eggs through urine or feces. If washed into ponds or slow-moving rivers, the eggs hatch and infect certain species of snails. When the larvae have developed, they leave their snail hosts and start the cycle anew by infecting people who swim or wash in the water.

Guinea worm - This is one of the most hideous and debilitating diseases to attack children in the developing world. The Guinea worm larva enters the skin of people washing or bathing in contaminated water. It grows inside the body for about a year until it matures, often at a length of more than 30 inches. The adult female worm then exits the body through a painful sore, usually in the lower leg or foot. The hapless victim must endure this agony for several weeks and carefully roll the worm up as it exits. Serious infection occurs if the worm is broken before it leaves the body on its own. Due to the pain it causes, the Guinea worm is called "the fiery serpent" by many of the 10–20 million people afflicted around the globe.

LOST TIME AND ENERGY

Two thirds of the world's households use a water source outside the home—and the water-haulers are almost invariably women and children. A person needs an average of five quarts a day for drinking and cooking and 25 more to stay clean. In some parts of Africa, women spend five hours a day hauling water in jugs that weigh 20–40 pounds. The energy needed for this task is nearly 600 calories a day, a serious problem for women on the verge of malnutrition. The time spent hauling water also detracts from mothers' abilities to care for their children.

THE BENEFITS OF CLEAN WATER

Clean water improves the health of the developing world's children in three important ways.

Disease Control - Clean water and improved sanitation have a dramatic effect on the health of children, sharply reducing diarrhea and preventing parasitic infections.

(continued)

Time and Energy Savings - Having a pump in the village center saves mothers enormous amounts of time, allowing them to spend more time feeding, educating, bathing, and caring for their children. The reduced work load for mothers and children also means less energy wasted in hauling water from distant sites.

Crop Production - Water for irrigation improves yields of family gardens that are critical for good nutrition. This improved nutrition helps children fight disease and develop normally.

WATER FOR ALL BY 2000

UNICEF brings water to more people than any other international agency. During the 1980s, UNICEF helped bring clean water to 700 million people in over 90 countries. In 1990, UNICEF helped install 104,000 wells and water systems benefiting 20 million people at an average cost of only about $4 per person.

UNICEF is committed to helping bring clean water to all the world's children by the year 2000.

*For more information, please contact the U.S. Committee for UNICEF
333 East 38th Street, New York, NY 10016.*

smaller particles. These particles were named quarks by the Caltech physicist Murray Gell-Mann, who won the Nobel prize in 1969 for his work on them. The origin of the name is an enigmatic quotation from James Joyce: "Three quarks for Muster Mark!"

—Stephen Hawking, *A Brief History of Time.* New York: Bantam Books, 1988: 65.

Narration is also important for many job-related situations. It is commonly used in progress reports, in which writers provide a detailed account of how they spent the time and money allocated to them for completing a particular project. Here is an example:

[Date] Today, the third day after feeding the Boston ferns with 200 cc. of Miracle Grow, we noted much fuller fronds and 7 cm. increased height over our control fern, which was not fed Miracle Grow.

Physicians and surgeons routinely use narration to describe the procedures and results of their treatments or surgeries. The following is an excerpt from a narrative account of an actual procedure for repairing a ventral hernia:

> After general anesthesia, the patient's abdomen was prepped and draped in the usual fashion. An oblique incision was made in the left lower quadrant of the abdomen. Subcutaneous tissue was cut. The external oblique aponeurosis was cut and separated. Underneath there was a large defect measuring 10 by 6 cm in size, through which the whole descending colon was protruding. The hernia sac edges were freed from the surrounding tissue and were closed with 0 silk purse-string sutures, 2-0 silk tie and a 2-0 Surgilon tie.
>
> —H. P. Gulesserian, M.D.

Clearly, the audience is medically trained personnel, particularly the performing surgeon himself or herself at a later time. Note how the surgeon has incorporated detailed description into the narrative. Such detailed narrative recounting of procedures is essential for future treatment of the patient.

ESTABLISHING RELATIONSHIPS

Informative writing is often analytical; that is, along with defining key concepts and describing objects and functions it also establishes relationships between concepts and objects with the use of two techniques: making comparisons and analyzing cause and effect.

Making Comparisons

What are the similarities and differences between Freudian and Jungian psychoanalysis? How energy-efficient is clothes dryer A compared to clothes dryer B? Which among the four existing brands of low-fat, low-sodium dog food on the market is the best quality for the price? How much tastier and longer-lasting is the genetically-engineered Flavr-Savr™ tomato over ordinary tomatoes? To answer questions such as these is to engage in comparison, a technique that is fundamental to many technical writing activities, particularly feasibility research (see Chapter 16).

One typically uses comparison to point out both similarities and differences; by so doing, one gains deeper insight into the natures of each object being compared. Notice how Dawn Stover, senior editor of *Popular Science* magazine, uses comparison to characterize the differences between the new Boeing 777 jets and older-model jets:

> From the outside, the 777 looks similar to today's jetliners. Inside, passengers will notice a difference. The fuselage is more than 20 feet in diameter—only the giant 747 has a wider interior. And because the side walls are not as sharply contoured as those of other wide-body planes, passengers in the window seats will have more headroom. . . . The new airline will carry 305 to 328 passengers in a three-class seating arrangement—filling the niche between the company's 218-passenger 767 model and the 420-passenger 747. . . . The new engines produce thrust of more than 80,000 pounds each. By comparison, the most powerful engines on the 747 produce about 60,000 pounds of thrust.
>
> —Dawn Stover, "The Newest Way to Fly." *Popular Science,* June 1994: 78.

Analyzing Cause and Effect

Determining the cause of a given effect is routine business in science and technology. In December 1993, the most elaborate space mission since the moon landings, the Hubble Space Telescope repair mission, took place because a flaw in the primary mirror ("cause") resulted in blurred images of distant objects ("effect"). The astronauts corrected the problem by installing corrective mirrors in the path of the light source ("cause") which cancelled out the blurriness ("effect").

Notice how the cause-effect relationship is communicated in the following passage which explains how bubonic plague was transmitted and spread:

> Rats infected with *pasteurella pestis* bacteria are bitten by rat fleas which then carry the disease to other rats. At some point the flea bites and infects humans, who in turn inflict other humans. The disease first shows as lumps (buboes, hence bubonic) in the groin or armpits. Vomiting and high temperature are followed by diarrhoea, pneumonia, and death.
>
> —James Burke, *Connections.* Boston: Little, Brown, 1978: 99.

SPECULATING

Speculating or hypothesizing might be thought of as a kind of future-oriented narration: telling what *may* happen. A more formal way of putting it is "extrapolating from available information." Weather forecasters do it all the time.

Urban planners, highway designers, school administrators, and environmentalists all engage in speculations that can determine the success or failure of coping with overcrowded freeways and schools or of preserving endangered species. Lay audiences also are eager to learn more about developments in technology that might reverse or at least stabilize such environmental crises—which is why speculative features appear in magazines and newspapers virtually every day. (See Chapter 19 for discussion of speculative feature-writing.)

The following speculation on what home-based telecommunications networks may be like early in the next century will give you a sense of how speculative writing works:

> We will use hand-held computers, watch video images combined with computer data on flat-panel screens and keep huge volumes of data on small, lightweight optical disks. Most importantly, communications technology will give workers access to huge electronic libraries, with their information scattered on computers worldwide, while intelligent computer programs dubbed "agents" will relentlessly scour those libraries for the data we need.
>
> —Rory J. O'Connor, "Wired to Work," *San Jose Mercury News,*
> August 4, 1991.

As you might guess, such speculative writing can be enormously useful for those whose business it is to plan for the future—software developers and executives for whom telecommunications systems are essential to their businesses, for example.

PRESENTING INFORMATION: A CHECKLIST

- [ ] Have I determined the purpose of the document?
- [ ] Do I know who my primary audience will be?
- [ ] Have I included all necessary elements of the definition (classification system, component parts, etymology) to suit my purpose and satisfy the reader's needs?
- [ ] Have I described all important physical and operational properties of the object so that it can be readily identified?
- [ ] Would comparing A with B, or explaining how cause A has led to effect B serve my overall purpose?
- [ ] Would a diagram of the concept I'm defining help the reader? What parts of the diagram should I identify?
- [ ] Have I discussed all the parts that I've labeled on the diagram?

☐   Is the format of the document eye-appealing and easy to refer to?

☐   Have I determined specifically what my readers will need to know about the background of the subject in order to understand its present condition fully?

☐   Have I established a clear connection between the events of the past and the present situation in my narrative?

☐   (Mainly for technical journalism contexts) Have I presented the events engagingly, generating a buildup toward a central or climactic event?

■ CHAPTER SUMMARY

Definition, description, narration, and speculation are fundamental techniques of conveying information about a given object. Definition (including categorization) establishes the meaning of the object; description establishes the object's identity; narration provides important background information that enables readers to better understand the object's nature and origins. Speculation forecasts likely situations based on present conditions. These rhetorical techniques work together to ensure the all-important goals of clarity and accuracy, without which the information cannot be put to use.

■ FOR DISCUSSION

1. What importance does description have in medicine? in engineering? in psychology? What importance does narration have for these same fields?

2. Bring a small, relatively simple object to class (e.g., a stapler, pencil sharpener, compass, nail clipper, thermometer, or ballpoint pen, or old sparkplug) and form small groups. Have each person in your group, including yourself, come up with a definition and a description of the object. Then compare each other's definitions and descriptions and discuss their similarities and differences. Which definitions and descriptions in the group are best suited for technical-communication contexts? Why?

3. Suggest some important applications for speculative writing in your major field; for cause-effect analysis.

4. What is the relationship between description and narration? What does one technique accomplish that the other technique does not? Can you suggest ways that the two techniques might overlap?

5. Bring the label of a household product to class and comment on its descriptive elements.

6. Comment on the strengths and weaknesses of the following student-written definition of a golfball:

A golfball is a sphere used to play golf. The core is usually made up of oil-filled rubber, approximately one-half inch in diameter. The core is wrapped with hundreds of feet of rubber threads, similar to rubber bands. They are used to increase the diameter of the ball to one and one-half inches. These threads are then covered with the familiar white covering. This covering is imprinted with dimples.

■ FOR YOUR NOTEBOOK

1. Write brief definitions of tools that you routinely use around the house. Add illustrations if you wish.

2. Write the most detailed description you can of a relatively simple object, such as a tennis ball or a compass (either kind). Try to include everything in the description, and accompany it with a diagram.

3. Take notes for a short narrative in which you help readers visualize what a future experience might be like. Possible topics: experiencing interactive TV; living in a lunar colony; working with a voice-activated computer. See Writing Projects, topic 3.

4. Read several speculative features in magazines such as *Omni* and *Discover*. As you do so, maintain a list of any neologisms you encounter, making sure you include their definitions. After accumulating several of these newly coined terms, try using some of them in sentences of your own.

■ WRITING PROJECTS

1. Write an extended definition of a power tool that you are familiar with: Include a lexical definition, a categorical definition (classification, component parts), and an operational definition.

2. Study the definitions for the following terms in a desk dictionary, a specialized dictionary (such as a dictionary of medical terms), and an encyclopedia; next, write your own categorical definitions for each:

a. sprain

b. amnesia

c. erosion

d. metamorphosis (biological)

e. electromagnetic spectrum

3. Write a speculative essay based on the notes you took for topic 3 under For Your Notebook. Here are additional topics you might wish to consider:
   - virtual reality games of the future
   - new uses for the Internet
   - household robots

4. Imagine that you are a writer for a public agency. Working alone or in small groups, prepare a fact sheet that will help the general public understand the nature of a health, educational, or environmental problem as well as its severity. Possible topics:

   a. inadequate schools on certain Indian reservations

   b. inadequate housing for inner city residents

   c. toxic-waste soil contamination in low-income housing areas

   d. medical-supply and hospital-facility shortages in a Third-World area recently struck by a national disaster or civil war

5. Write a one-page comparison for one of the following sets of terms that often get confused with each other:

   a. RAM and ROM

   b. serial and parallel interface

   c. centrifugal and centripetal force

   d. inductive and deductive reasoning

   e. reflecting and refracting telescopes

   f. id, ego, super-ego, libido

   g. meiosis and mitosis

   h. DNA and RNA

   i. alternating and direct current

   j. partial, annular, and total solar eclipses

6. Analyze the cause-effect relationship between one of the following pairs:

   a. sunspots and radio transmission

   b. videogame playing and reading ability among pre-teenage children.

   c. sunbathing and skin cancer

   d. unemployment rate and crime rate in a given community

   e. dietary habits and environmental problems in a given community

7

Basic Rhetorical Techniques II: Argument in Technical Writing

Elements of an Argument

Evaluating Evidence

Applying the Evidence

Structuring an Argument

Common Mistakes in Reasoning

Arguing Technical Issues: Work-Related Contexts

Arguing Technical Issues: Journalistic Contexts

The Ethics of Technical Argument

Drafting a Technical Argument: A Checklist

The purpose of any argument, in any kind of writing or speaking, is to persuade readers that your assessment of a situation, in the context of other assessments, is the most reasonable; that your recommendations, based on your assessment, should be adopted instead of the recommendations of challengers. This chapter examines the nature of argument in technical writing and offers guidelines for developing an argument that is both rhetorically effective and technically accurate.

ELEMENTS OF AN ARGUMENT

When planning the framework of an argument to be developed in a report, a writer should think in terms of seven elements that need to be included:

1. Identification of the problem or issue being debated.
2. Statement of position (thesis) regarding the problem.
3. Discussion of the problem and its importance.
4. Specific, objective evidence in support of the position statement.
5. Analysis of challenging views and their shortcomings.
6. Statement of recommendations (for enacting the thesis).
7. Summary and conclusions.

Let's examine each of these elements closely.

1. *Identification of the problem or issue being debated.* First of all, it is necessary to be clear about what a "problem" is in the context of technical communication. In everyday language, a problem arises when your expectations about something are not met: the lawn mower does not start when you pull the cord; the printer did not print when the command was entered to do so. Problems of this sort are usually solved quickly and easily. The lawn mower was out of fuel or its spark-plug wire was detached; the print command can only work when the word-processing program is out of the editing mode. Common problems such as these are often identified and directions given to solve them, usually in a "troubleshooting guide" (see Chapter 17). They are "simple" problems, not necessarily because they are easy to fix, but because there is rarely any disagreement about how to fix them.

Alas, many of the problems that confront anyone are *complex*—which is to say, much disagreement is generated by how best to solve them. Here is an example: Mono Lake, near Yosemite National Park, an ecologically sensitive lake that nurtures several unique species of waterfowl and sea life, is on the verge of drying up (and the unique species becoming extinct) because too much of its water is being diverted to Los Angeles. Debater A sees this as a major problem because the lake's ecosystem is unique in the United States, invaluable for understanding the nature of species adaptability, the symbiotic relationships among very different plant and animal species, and so on. Debater B, on the other hand, insists that this is a trivial concern compared to the need for providing water to millions of people.

2. *Statement of position (thesis) regarding the problem.* Once the problem is clearly stated, the writer should present his or her position on the issue. Is it that Mono Lake's preservation must take precedence over diverting its water to Los Angeles? Is it that L.A.'s need for that particular water supply is too great for it to be cut off? Is it that *some* of the water can be cut back without significant environmental damage, while still providing minimal water needs? In technical reports, the thesis should be as explicitly stated as possible; for example:

> In this report I will argue that the ecological damage to Mono Lake caused by steadily draining its water will soon be irreversible, whereas resorting to it for alternative water reserves would not be ecologically harmful.

3. *Discussion of the problem and its importance.* Carefully discussing the problem and its importance is the writer's next item of business. Let us continue with the example about the endangered lake. If one wishes to argue the thesis used in the example—that it is more important to save Mono Lake than to continue diverting its water to Los Angeles, then one must carefully discuss all facets of that scenario; that is, one must be able to anticipate questions that readers would need to have answered: What *kinds* of ecological damage would result from Mono Lake's dissipation? *Why* would it be irreversible? *How* can one be sure that tapping into alternative water sources would not be ecologically harmful as well?

4. *Specific, objective evidence in support of the position statement.* The heart of every argument is its evidence; but what, exactly, *is* evidence? There is no simple answer because each discipline has its own idea of what constitutes evidence. The natural sciences in general consider evidence to be the data collected from laboratory experiments—but even here it is easy to overgeneralize. The "laboratory" for geologists is often the earth itself; for astronomers it is the entire universe—and collected data from these sciences can be problematic. What an astronomer "sees" with a radio telescope—or even with the Hubble optical space telescope—is often augmented by computer enhancement and accompanying interpretive commentary. For biologists and ecologists, the great

complexity of living systems make the definition of evidence problematic as well. Evidence for Mono Lake's dissipation might consist of annual shoreline measurements over a period of ten years; but such evidence could be complicated by an eleventh year of torrential rainfall in the region, raising the water level substantially.

In the social sciences, evidence becomes even less sharply defined, for here one is dealing with highly complex behaviors and written records rather than merely physically observable subjects. A historian advancing a new argument to better explain the purpose of a military campaign may cite as evidence logbooks by commanding officers or soldiers' diaries and correspondence. (For discussion of ways to evaluate your evidence, see the next section of this chapter.)

5. *Analysis of challenging views and their shortcomings.* It is always tempting to underrepresent a view that challenges your own; but for an argument to be compelling, the writer must examine such views as fully and as objectively as his or her own. A thesis that cannot withstand challenge cannot be regarded as compelling. Back to the endangered lake scenario: for Debater B's argument to be compelling, it must present evidence of the extraordinary water requirements for the vast Los Angeles metropolitan area. Also, Debater B would want to anticipate Debater's A's counterargument that another water source can be tapped—one that would not threaten wildlife habitat, and that would be worth the moderate extra expense for pipelines. B might demonstrate, for example, that no significant alternatives to Mono Lake exist, then go on to argue that A's "moderate extra expense" would run in the billions of dollars, resulting in stiff tax hikes.

6. *Statement of recommendations (for enacting the thesis).* In technical report-writing, it is not enough merely to argue a premise convincingly. The writer must also recommend a means by which the proposed action is to be carried out. If the writer is to argue that channeling water out of Mono Lake must cease, he or she must suggest a plan for making this happen. One does not simply shut off the tap. Alternative water sources need to be identified and a program for pipeline construction—which in mountainous terrain must include expensive pumping stations—must be planned out; or, on the other hand, consequent water rationing programs must be outlined. A timeline (possibly extending over several years) also has to be determined. Finally, any foreseeable obstacles to the recommended plan for action must be discussed and resolved.

7. *Summary and conclusions.* A summary succinctly restates the problem statement and thesis, the evidence supporting the thesis, and the recommendations for carrying it out. The conclusion reemphasizes the seriousness of the issue

and the need to take action immediately. Because it provides a complete picture of the argument very concisely, the reader often turns to it first.

EVALUATING EVIDENCE

Evidence used to support a position on a debatable issue must be carefully evaluated. A merely objective tone or style is not enough. As technical communication specialist Scott P. Sanders warns, "[A] plain style and an objective tone can promote false rhetoric as surely as the most ornate rhetorical amplification."[1] Specifically, the argument should be evaluated in terms of two criteria: reliability and accuracy.

Let us apply these criteria. In 1988 two Canadian psychologists released a study arguing that a woman's ability to perform certain tasks varies in direct proportion to her estrogen level; that when her estrogen level is low (as is the case just before menstruation), a woman's performance level will be correspondingly low; and when it is high (as it is before ovulation), she is able to perform that same task more efficiently.[2]

1. What kind of evidence would you consider to be *reliable* in this argument? Would you consider individual testimony by women ("I can never think straight before I menstruate")? Not reliable enough, you say, because it is too subjective. A more scientifically legitimate approach would be to obtain performance results of women all performing certain tasks at designated times of the month, then to check performance levels in relation to the menstrual periods. Even so, this could not result in sufficiently reliable evidence in itself. What *kinds* of tasks are the women being asked to perform? Simple? Complex? Are these tasks *representative* of tasks the women in the group would normally engage in? If the experiment tested a woman's ability to perform certain word games such as tongue twisters (as was the case in the original experiment), would you consider that to be reliable evidence for denying a woman employment, say, as a computer programmer? One can easily imagine ways in which the "results" of a seemingly scientific investigation may be misconstrued to deny a woman employment for hidden, unsupportable reasons.

2. Is the evidence *accurate?* This may seem like a relatively easy matter to check out, but it can be tricky, especially in situations involving the biological or social sciences. To determine the accuracy of the evidence, one needs to know *(a)* how was the experiment structured; *(b)* whether external influences

---

[1]Scott P. Sanders, "How Can Technical Writers Be Persuasive?" in *Solving Problems in Technical Writing,* ed. Lynn Beene and Peter White. New York: Oxford University Press, 1988: 55–78.
[2]Based on an article in the *New York Times,* November 17, 1988.

such as test anxiety or illnesses unrelated to menstruation might have interfered with test peformance. The number of subjects used in the experiments, as well as the number of experiments, can also influence the accuracy of the findings. If you wanted to investigate this matter, you would need to locate the documents that the investigators had originally published; relying on journalistic pieces would be inadequate: too much is left out or simplified because of the target audience of laypersons.

APPLYING THE EVIDENCE

Another set of concerns comes into play once you begin structuring your argument. Let's say that you are a technical writer for a manufacturer of water-filtration systems and have been asked to prepare a report recommending an effective strategy for marketing water filters for home use. What would be your strategy, and how would you convince your readers (the advertising director, the vice president for consumer products, and marketing representatives) that your strategy would indeed be effective?

Consider the following three-step procedure:

1. Practice explaining, orally or in writing, the function and benefits of the water filter to those who are likely to be completely unfamiliar with a water filter's parts, function, or benefits. Your explanation should be neither condescendingly simple nor dependent on technical jargon.
2. Research every facet of the filter: how it works, how it is different from (and presumably better than) competing filters.
3. Acquire the hard evidence you need to support any claims you make.

Unlike commercials, which appeal mainly to emotions or to fantasies in order to sell products (think of automobile commercials), marketing a product such as a water filter requires emphasis on hard evidence: for example, the types and quantities of potentially hazardous substances that are known to exist in tapwater through testing, or that the filter can indeed remove these harmful substances. But as a report writer you realize that mentioning such facts alone will not constitute an effective promotion. Consumers need to be convinced that filter A can filter out those substances more safely and effectively than filters B and C. This means (the writer might argue) that the promotional strategy must include a clear, easily understandable explanation of how the filter does what it does—and how it does it better than competing filters on the market.

While it is true that facts can sometimes speak for themselves (no one needs to convince you to fill your gasoline tank when the gauge registers

Criterion	Method of Application
	Table 7.1 Methods for Applying the Criteria Used in Promoting a Water Filter
Selection of facts	Mention that hundreds of "impure" substances exist in the water supply, but most are harmless; easy to adopt an "alarmist" strategy by mentioning many substances; the best approach, however, would be to call attention to three or four substances that cause the greatest concern.
Interpretation of facts	Explain how these substances (known carcinogens?) could be harmful to fetuses. Also, explain how the filter system operates, why it can be relied on—perhaps more so than competing filters.
Prioritization	First emphasize the filter's effectiveness in filtering out the most dangerous substances. Then, when explaining how the filter works, writers should emphasize the quality of the filtering charcoal that is used: higher quality than that of competing filter systems.

"empty"), the facts you need to incorporate into your recommendation report must be *(a)* selected, *(b)* interpreted, *(c)* prioritized relative to your thesis. Let us consider one possible set of methods for applying the criteria used in promoting a water filter (see Table 7.1).

As the information in Table 7.1 suggests, you need to select, interpret, and rank the facts underlying the performance of the product to be promoted, and do so in view of the target reader's needs and level of understanding. By carefully attending to these strategies, you will produce a recommendation report that will enable managers to promote the product successfully.

STRUCTURING AN ARGUMENT

The way a writer structures an argument is as important as the substance of the argument. The world's most compelling idea for an argument will seem like gibberish if it is not presented in a logical and coherent manner.

What makes for a solid argumentative structure? You have already considered the seven elements in a technical-writing argument; for structuring an argument, you may convert these into four steps:

1. an introduction to the problem underlying the argument, which leads to . . .
2. a clear, direct statement of the thesis to be defended or challenged;
3. evidence that you, the writer, have collected to convince readers of the validity of the thesis;

4. conclusions and recommendations derived logically from the thesis.

Read the two reports on the issue of irradiation hazards on pages 144–147. Colby and Epstein argue against the practice; Cox, on the other hand, argues that irradiating can actually be beneficial. How convincingly does each report embody the four-step argumentative structure in the preceding outline?

These two documents reveal some interesting elements of argument strategy. Let's begin with their respective introductions.

In their argument against food irradiation, Colby and Epstein begin with the fact that irradiated food *will* go on sale, then quickly explain what "irradiation" means: "This food was treated with massive doses of ionizing radiation (100,000 rads, roughly equivalent to 10 million X-rays) . . . " The tone is intentionally alarmist—particularly the analogy to X rays, which most people already realize are dangerous even in small doses (hence the thick aprons technicians make you wear when they take dental X rays). The authors begin to persuade their readers even before they go into their explicit reasons against the procedure.

Cox takes a completely opposite tactic in his introduction; he cites possible evidence that exposure to radiation is beneficial rather than harmful to workers. Of course, one must look closely at Cox's wording here: the "Nuclear Shipyard Workers Study seems to show. . . " Why "seems"? Because other factors may be contributing to the workers' longevity. Also, there might have been major limitations in or objections to the reliability of the study itself which could have been raised at the time of its release.

How compelling is the evidence in each case? Colby and Epstein cite facts such as chromosomal damage in children who have ingested irradiated products and the reduction of nutrient levels by irradiation; they also debunk the FDA's claim that irradiated food has been thoroughly tested, citing as counterevidence the prohibition by three states of irradiated food from sale.

Cox's principal strategy is to cite the testimony of experts—not just the NSWS, but people like Leonard Sagan, a member of the National Council on Radiation Protection's board of directors, who testify to the benefits of "radiation hormesis."

Readers who have turned to these documents so as to make up their minds about radiation's benefits or hazards will have learned more than they'd previously suspected about the matter. Each side raises an important concern that the other side does not address. Cox, for example, does not address the matter of insufficient testing; Colby and Epstein do not address the matter of radiation hormesis. Further inquiry into the issue is necessary. (See, for example, the articles by Nielson and Mayell in For Discussion, topic 2, at the end of this chapter.) Look at the affiliations (i.e., vested interests) both of Colby and Epstein and of Cox given in the author notes at the end of each article.

Michael Colby and Samuel Epstein Say We Should Halt the Rush to Irradiate Our Food.

The nation's first irradiated food, fresh fruit and vegetables, is soon to go on sale at a small Miami supermarket. This food was treated with massive doses of ionizing radiation (100,000 rads, roughly equivalent to 10 million medical X-rays) at a large cobalt-60 facility, Vindicator Inc., which plans to treat 800 million tons of food a year for nationwide sale.

Food irradiation was the brainchild of the Atomic Energy Commission's efforts in the Eisenhower administration to find practical uses for the flood of radioactive wastes from nuclear weapons.

Atomic Energy of Canada (Nordion Ltd.), with its virtual monopoly on cobalt-60 and with strong backing from the International Atomic Energy Agency, hopes to operate a chain of U.S. plants with U.S. irradiation companies.

Industry and the Food and Drug Administration insist that irradiated food has been thoroughly tested and is absolutely safe.

However, New York, New Jersey and Maine have prohibited the sale and distribution of irradiated food, as have foreign governments, including Germany, Denmark, Sweden, Australia and New Zealand.

Claims of safety are unproven at best.

High-energy irradiation produces complex chemical changes in food with the formation of poorly characterized radiolytic products, including benzene, organic peroxides and carbonyls.

Radiolytic products kill bacteria, molds and larvae and thus ensure spoilage-free food, a major attraction to the purveyors of marginal produce and contaminated poultry.

However, concentrated extracts of these products have never been tested for cancer and other delayed adverse effects.

The overdue need for such studies is further emphasized by numerous reports of chronic toxic effects in insensitive studies on test animals fed unextracted whole irradiated food.

These include reproductive damage in rodents and chromosomal damage in rodents, monkeys and children.

Besides food safety, irradiation poses serious occupational and environmental hazards due to the transport and handling of radioactive materials.

Accidents have already been reported in facilities sterilizing

(continued)

medical supplies by irradiation.

Irradiation also reduces levels of essential nutrients in food, especially vitamins A, C, E and the B complex. Cooking irradiated food reduces these levels still further.

The industry reluctantly admits this but suggests that the problem could be taken care of by vitamin supplements!

In spite of this substantial evidence, the Food and Drug Administration approved food irradiation in 1986.

The FDA based its decision on five questionable or allegedly negative tests and on theoretical estimates on cancer risk, which was claimed to be insignificant and "acceptable."

This position is consistent with the administration's revocation of the Delaney law, which banned the deliberate contamination of food with any amount of cancer-causing chemicals, and its substitution by rubber number standards based on "acceptable" cancer risk.

Cancer rates have now reached epidemic proportions, striking one in three and killing one in four, with 500,000 deaths last year.

Further risks to the entire nation of cancer, besides other health effects, hardly seem justified by the narrow economic interest of a small industry supported by a highly politicized federal bureaucracy.

Michael Colby is national director of Food and Water Inc. in New York; Dr. Samuel S. Epstein is professor of occupational and environmental medicine at the University of Illinois Medical Center, Chicago, and author of *The Politics of Cancer*.

Media Ignore Radiation's Benefits

Patrick Cox Says the Media Ignore Anything About Radiation That Isn't a Scare Story.

More than three years ago, a study comparing workers in nuclear shipyards to workers in non-nuclear shipyards, with over 70,000 participants tracked from the '60s to the '80s, ended. The Nuclear Shipyard Workers Study revealed a shocking 24% difference in mortality rates between non-nuclear workers and those exposed to high, by Environmental Protection Agency standards, levels of gamma radiation.

Why, many have subsequently asked, have the media ignored this sensational story? The answer is simple:

(continued)

Those with the highest chronic radiation exposure had the lowest mortality from all causes, including cancers, and lived significantly longer than the non-nuclear workers—whose health was typical of the general population.

In short, the Nuclear Shipyard Workers Study seems to show that certain doses of ionizing radiation are actually beneficial, contradicting the politically correct view of those who would have us believe that human enterprise is incompatible with environmental health.

Furthermore, the study bolsters decades of experiments on plants, invertebrates and lower mammals indicating that chronic radiation—like iodine, iron, sunshine and water—is healthful and necessary at low levels but fatal at high levels.

This phenomenon, known as radiation hormesis, refutes the popular notion that all radiation is bad, an idea based on obviously dangerous exposures to brief, extreme doses from nuclear blasts. Some researchers believe the mechanism of radiation hormesis is a stimulation of the immune system.

Physician and epidemiologist Leonard Sagan, member of the congressionally mandated National Council on Radiation Protection's board of directors and editor of the Health Physics Society's May 1987 journal issue on radiation hormesis, says, "Because of the political nature of society, the media's neglect of radiation hormesis has real health implications, indirectly abetting billions in unnecessary spending."

Just one example of this is the Department of Energy's estimate that it will spend $200 billion over 10 years to reduce radiation levels, now less than one millirem per year, near the department's nuclear activities.

But 30-millirem variations in natural radiation are typical in the USA, and the Nuclear Shipyard Workers Study links over 500 millirems with improved health.

T. D. Luckey, author of *Radiation Hormesis*, places the optimal dose much higher.

Sagan adds:

"As we lack the resources even to inoculate children for measles, we must question the responsibility of those who incite needless spending by countenancing only one side of environmental issues."

In the same vein, Robert Parks, head of the American Physical Society's Washington office, says, "I think it is clear that if the NSWS had found an increase in mortality among exposed workers, it would have been trumpeted by the opponents of nuclear power."

John Cameron of the University of Wisconsin and a member of the scientific advisory panel for the Energy Department-funded study, put

(continued)

it succinctly when he wrote:

> "If the NSWS results had shown an increase of mortality of 24% for the nuclear workers, it is likely that the results of the study would have made the national TV news over three years ago when the study was completed."

How very true.

Patrick Cox is a Menlo Park, Calif., writer and economic and political analyst for the Competitive Enterprise Institute.

COMMON MISTAKES IN REASONING

Reasoning is the process of forming conclusions based on facts; it is the means by which a writer or debater shapes an argument. It is very unlike everyday conversations, which seldom reflect the systematic and accurate reasoning process that one expects to encounter in debating or argumentative writing. Because of the highly precise nature of technical argumentation, you need to be alert for reasoning errors that commonly occur not because your thinking process is flawed but because a link in the chain of reasoning can be easily overlooked when dealing with complex matters.

There are five common mistakes in reasoning:

1. *Inaccurate comparison.* A is compared to B, but B is not really comparable to A, except perhaps in a metaphorical or exaggerated way. People bustling about on city streets, for example, are often metaphorically compared to ants bustling about an anthill—but the comparison is essentially inaccurate: The behavior of human beings is far more complex than that of insects.

2. *Incomplete comparison.* A is compared to B on just one level, but there are other levels to either A or B that need to be called to attention before the comparison is complete. Light bulb A is said to be more economical than light bulb B (same wattage) because bulb A has a life expectancy of 1500 hours while bulb B has an expectancy of only 800 hours. The comparison is incomplete, however, unless the writer also takes two other factors into consideration: (1) each bulb's luminosity factor (the long-life bulb generates fewer lumens even though it has the same wattage); and (2) each bulb's energy efficiency: because bulb A generates fewer lumens per watt, it is not utilizing the potential energy as efficiently as bulb B.

3. *Effects not clearly derived from stated causes.* This is a very common pitfall to be aware of when trying to locate blame for a social ill. If you wished

to find the cause of poor reading habits among grade-school children, you might be tempted to hit on what seems most obvious: too much time watching television, or too much time playing videogames. But how do you then explain those children who have superior reading habits *and* watch a substantial amount of television? Clearly, other factors are involved, and they must be uncovered and analyzed if the inquiry is to be scientifically valid.

4. *Assumptions mistaken for facts.* This can be a serious form of deception: If gubernatorial candidate A asserts that candidate B's unwillingness to support public school reform could lead to a significant rise in juvenile delinquency—especially drug-related crimes—candidate A is presenting an assumption—that B is unwilling to support education—as if it were an incontestable fact. This is also known as "begging the question." If A believes B is unwilling to support public education, his or her first responsibility is to cite evidence supporting this assertion, and *then* go on to discuss the long-term consequences it could have.

5. *Hiding facts behind facts.* This is perhaps the commonest pitfall of them all: using positive facts to hide the existence of negative ones. Politicians may promise a moratorium on taxes, but that might mean continued traffic gridlock on the freeway because the revenue requested for road-widening is now denied.

Errors in reasoning are fairly easy to catch if you are alert for them. Can you find any of these errors in the articles on irradiation? You may also wish to consult Table 7.2 for a quick overview when structuring an argument.

ARGUING TECHNICAL ISSUES: WORK-RELATED CONTEXTS

Argument is important in all facets of industry, not just in those potentially controversial areas like environmental policy or public health. For example, in designing space station Freedom, expected to be operational for at least thirty years, NASA is concerned about possible collisions with space debris—tiny meteoroids as well as particles from artificial satellites (or even the satellites themselves). According to some researchers, the risk factor is slight, and only minimal precautions would be necessary: merely orient the station so that its narrowest dimension is in the direction of flight, thus reducing the amount of surface area that could be struck by debris. Other researchers, however, insist that the risk factor is substantial, and that more money has to be allocated to cover additional protective devices for the station.

If you were a member of the subcommittee established to resolve this issue, what information would you need to consider to inform your recommendation? As you may surmise, facts alone would not bring you to an unambigu-

Table 7.2 Common Mistakes in Reasoning

Problem	Example	Commentary
Inaccurate Comparison	Cows behave like machines.	The comparison raises more problems than it attempts to solve. Machines do not struggle to survive, have no nervous systems, etc.
Incomplete Comparison	We should purchase modem A because it transmits data more than 7 times faster than modem B (9600 baud instead of a mere 1200 baud).	Other points of comparison besides speed are necessary in selecting a good modem: durability, cost, reliability.
Effects Not Clearly Derived from Stated Causes	Because of Einstein's discovery of mass-energy equivalence ($E = mc^2$), nuclear annihilation is a constant threat.	Einstein's discovery did not in itself lead to nuclear weaponry. This would be akin to blaming the prehistoric discovery of how to make fire for last week's arson blaze.
Mistaking Assumptions for Facts (Begging the Question)	Because of the logging industry's disdain for environment, important wildlife species are doomed to extinction.	(1) The writer treats a personal assumption—that the logging industry is disdainful of nature—as if it were an established fact; (2) that the logging industry is contributing in any way to species extinction (or even endangerment) is strictly conjectural.
Hiding Facts Behind Facts	We should increase our nuclear-energy use—it does not pollute the atmosphere.	Assuming this claim were factual,* it would still be deceptive because of the "hidden" fact that the *waste* from nuclear power plants would cause other, equally serious environmental hazards.

*It is not; the refining of uranium generates "greenhouse effect" pollutants.

ous conclusion. You might want to argue, for example, that even if the risk factor were "slight," the consequences of a collision could be catastrophic to crew members. Therefore, maximum shielding, along with the safest directional vector, should be recommended—despite the major additional expenses required. But then, how would you counterargue the charge that such additional costs might shut down the entire program?

Debatable issues arise in virtually every area of industry and technology. Most may not be as far-reaching as protective shielding for a space station, but they are important nonetheless. Here are three examples:

- A newly formed graphics-design company needs high-speed, versatile laser printers—as few as possible without compromising the projected output for being competitive in the field. What type of printer should it purchase, and how many?

- As manager of an apartment complex, you have been receiving many complaints from your renters that the appliances which came with the apartment—refrigerators, stoves, air conditioners, microwaves, dishwashers—have been malfunctioning. The cost for repairs or replacements, however, greatly exceeds the amount allotted by the owner of the complex. What strategy would you use to convince the owner to come up with the additional funds?

- The CEO of your firm is concerned about the physical fitness of her employees. It isn't all altruism either: she would like to reduce the frequency of sick leaves, increase productivity, and reduce on-the-job fatigue, which could undermine quality of work and cause accidents. She has asked you, a technical writer, to draw up a feasibility study for establishing an exercise room on the premises. What should that exercise room include or not include? More fundamentally, would such a facility reduce worker inefficiency and sick leaves?

ARGUING TECHNICAL ISSUES: JOURNALISTIC CONTEXTS

In 1989, a seventh-grader entered, in her school's science fair, an exhibit depicting the development of human life from 6-week embryo to 6-month fetus. During the preexhibit judging, she received a blue ribbon for the entry. However, when the fair opened, the school's principal disqualified it. Why? Instead of using illustrations, the girl used actual unborn fetuses, sealed in jars, obtained from her uncle (a pathologist), to depict the stages. All of the fetuses had been miscarried naturally—but no matter. The principal ruled that preteen children would be too emotionally disturbed by such an exhibit.[3]

Do you agree or disagree with the principal's ruling?

Two rare species of birds are on the verge of extinction in the rapidly diminishing forests of several regions in the United States. Environmentalists file lawsuits against the logging industry so that new logging restrictions would keep the birds from going extinct. Action against the loggers would result in the loss of many jobs.[4]

Would adopting or rejecting the new logging restrictions be in the best interests of the communities involved?

---

[3]*San Jose Mercury News,* April 30, 1989: 3C.
[4]*Time,* December 9, 1991: 70–75.

The drug, Ritalin, has been shown to effectively treat hyperactive children suffering from Attention Deficit Disorder—inability to concentrate. Not all children who exhibit lack of concentration, however, have this disorder; yet the Drug Enforcement Administration discovered that unusually high quantities of the drug—which can be addictive—were being ordered, suggesting that schools were dispensing it indiscriminately.[5]

How should such a drug be controlled?

The preceding examples are but a small sampling of the scientific and technical *issues* that surround us—issues which must be understood before we can pass judgment upon them. But to make such concepts understandable to nonspecialists, writers need to find some experiential link—some analogue—to the "everyday" world.

Technical journalism is that facet of technical writing which does exactly that. In so doing, it emphasizes the humanistic concerns at the heart of science and technology: A technical journalist examines what is going on in a research lab, a manufacturing plant, a rain forest, and places it in the context of larger social issues. A technical journalist is concerned with *implications*, say, for the 1992 discovery that plastic can be organically grown, using a genetically altered weed. How will such a discovery contribute to—or detract from—human welfare in the long run? What immediate influence might it have on our daily lives?

THE ETHICS OF TECHNICAL ARGUMENT

The word *ethics* is derived from *ethos*, the Greek word for "character" or "custom." Ethics is concerned with establishing standards for the proper conduct of individuals or groups. Unlike laws, which are designed to protect people and property from physical or financial harm, ethical standards cannot be *publicly* enforced, for that would infringe on individual freedom. Sometimes, though, an unethical practice will become illegal if it can be demonstrated that physical or financial harm can result. A good example is the ban on smoking in many public places. Once, it was merely unethical to smoke beside someone who was bothered by cigarette smoke; now it is illegal, if the area has been designated "nonsmoking."

Ethics is important to argumentation—and this is especially so in technical argumentation. Every thesis one wishes to persuade an audience to accept presupposes an antithesis. One could argue that the thesis is more logical than the antithesis—but if logic were the only criterion (as in geometry), there would be no reason to develop an argument.

---

[5]F. Roberts, "Is Ritalin Necessary?" *Parents* 64 (May 1989): 52.

A system of ethical principles—or values—is also at work; and as all of us know, one system of values may be less esteemed than another. When a manager of a dry cleaners argues in a memo to the owners that she wishes to have a more powerful computer system because it would enable her to get her billing done in half the time, the owners may not see that as a compelling reason unless she could argue that the time saved could be perceived as an advantage to the business—for example, by decreasing the cleaning time for customers. If the manager did state that as her reason for wanting the upgrade but subsequently spent her extra time secretly knitting sweaters for herself, you might conclude that her conduct was unethical.

Presentation of technical information can often look convincing on the surface. Have all the facts been disclosed? Are some facts much more important, or relevant, than others? One might argue for building a sports arena downtown, on grounds that revenues generated the first year alone would get the city out of debt; but the individual taking that stance may conveniently ignore the substantial tax hike the arena would cost the citizens. It is a breach of ethics, in other words, to be aware of problems with the argument that you urge your readers to accept, and not to call attention to them—or to create the sense that the problems are trivial when an inner voice tells you they are serious.

DRAFTING A TECHNICAL ARGUMENT: A CHECKLIST

- [ ] Did I state the problem clearly and fairly?
- [ ] Have I presented a compelling, explicitly worded thesis?
- [ ] Have I marshalled specific and appropriate evidence to support my thesis convincingly?
- [ ] Have I represented challenging views fully and objectively?
- [ ] Are my counterarguments to the challenges convincing?
- [ ] Have I organized my argument in light of my readers' needs?
- [ ] Have I defined technical concepts that may be unfamiliar to my readers?

■ CHAPTER SUMMARY

Argumentation plays a major role in technical communication; arguments persuade their readers to take some sort of action, such as to improve a product or service, purchase one brand of machine over another, alter a manufacturing process, change or adopt a public policy. For such documents to be effective, they must be ethically sound (not intending to deceive or to distort or to compromise standards of proper conduct). They must also be well-structured—that is, must introduce the problem and provide sufficient background information; state the position (thesis) clearly and present sound and thorough evidence to support the position; acknowledge and objectively examine the nature and validity of challenging views; and finally, draw logical conclusions and/or make appropriate recommendations.

Facts in themselves do little; they require careful arrangement and analysis before they can qualify as proper evidence in support of an argument.

When constructing an argument, writers must be careful to avoid such pitfalls in reasoning as faulty comparisons and cause–effect relationships, mistaking assumptions for facts, and misrepresenting of facts.

■ FOR DISCUSSION

1a. Comment on Colby/Epstein's and Cox's respective uses of factual evidence to support their views on food irradiation. What are the strengths and shortcomings of each?

1b. Do Colby/Epstein's or Cox's arguments reveal any errors in reasoning, such as those discussed in this chapter? What are they?

2a. Read the arguments for and against food irradiation on pages 156–169. What new insights (or complications) are added to the debate that Colby/Epstein or Cox have overlooked or slighted? Comment on Nielson's and on Mayell's use of evidence.

2b. Point out and discuss any errors in reasoning you detect in these articles.

2c. Visit the library and locate information on the current status of food irradiation policy. Does this new information affect either the Colby/Epstein or the Cox argument?

3. Locate technical features that argue different sides of a controversial issue and check both features for errors in reasoning. Discuss your findings in class.

4. Suggest effective ways of comparing two or three different makes of the following products for the purpose of choosing one make over the other. Decide on your own criteria for selection.
   a. camcorders

b. racing bicycles
c. fax machines
d. CD players
e. microwave ovens
f. interior latex paints
g. cordless telephones
h. answering machines
i. sunglasses

5. Discuss possible ways of resolving each of the cases described under Arguing Technical Issues: Work-Related Contexts.

■ FOR YOUR NOTEBOOK

1. Over the next few weeks, scan the editorial pages of the metro or campus newspaper, or periodicals (*Popular Mechanics, Discover, Technology Review,* etc.) for controversial topics such as those mentioned in this chapter. Summarize these pieces in your notebook, adding your own reactions to the issues, if you wish.

2. Maintain a "running debate" with yourself on a single controversial topic, ideally one that you've already been wrestling with for a while. Try to look for valid assertions from both sides of the issue. You may want to use these notes later on as the basis for an argumentative feature or report.

3. Keep a record of any unethical uses of argument you encounter, whether in conversation, written reports, or publicity material.

■ WRITING PROJECTS

1. Using the notes you have accumulated on ways of supporting and refuting a debatable topic in science or technology, prepare a "pro-con" paper in which you examine each side of the issue as impartially and as thoroughly as possible. Then, in a "conclusions" section, take a stance (pro, con, or some compromise), and justify it.

2. Using the following diagram as a prompt, and the articles in this chapter as further sources, write a clear description of the food-irradiation process for lay readers.

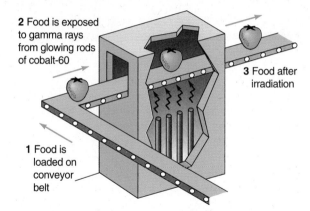

2 Food is exposed to gamma rays from glowing rods of cobalt-60

3 Food after irradiation

1 Food is loaded on conveyor belt

Diagram depicting food-irradiation process.

3. Write an argumentative paper on one of the following topics, committing yourself to an explicit position.
   a. Benefits vs. hazards of videogames.
   b. A college education: specialized training vs. "well-roundedness."
   c. Monkeywrenching (damaging) harpoons to protest whaling: ethical or unethical?
   d. Saving an endangered species vs. shutting down a 3000-job lumber operation.

4. Write up a comparison of two brands of a single product, taken from the list in For Discussion, topic 4. Be sure that your points of comparison are as comprehensive as possible, and that you end with a recommendation.

5. According to Jeremy Rifkin, in his book, *Beyond Beef: The Rise and Fall of the Cattle Culture,* the cattle industry in the United States, in Central and South America, and elsewhere, is the result of widespread environmental destruction. Read Rifkin's book; locate challenging views, such as the one in the article "The Beef Against Beef" (pages 170–173), and argue, in a 5–6-page paper, what should be done to meet the alleged crisis, if anything. Be sure to accurately represent views that would challenge your own, and to explain why these views are shortsighted.

Food Irradiation Is Beneficial

Niel E. Nielson

Most people would prefer their food to be free from disease-causing microorganisms and insect material, and would appreciate a technology that prolongs the shelf-life of fresh foods. Yet, food irradiation to ensure the wholesomeness of fresh and processed foods is under attack from a small group of antinuclear advocates who insist, most unscientifically, that they are protecting people from the alleged unknown effects low-level irradiation may have on food products.

FOOD IRRADIATION

These antinukes fail to inform their followers and the public that the established, organized scientific community worldwide encourages use of "food irradiation" done in accordance with published U.S. and international guidelines and regulations.

The focus of those who use the often misleading, all-inclusive term "food irradiation" is the processing of foods with electromagnetic energy at picometer wavelengths (trillionths of meters). It is well recognized in the scientific literature that low doses—that is, extremely small amounts of energy being absorbed   of picowave processing have the ability to disable the reproductive capabilities of essentially all insects, bacteria, parasites, and viruses. Because of this, the expanded, widespread use of this processing technique could be immediately life saving and could immediately, demonstrably, improve public health.

The basis for this assertion has been documented in studies published by the Centers for Disease Control and the U.S. Department of Agriculture, in which it is shown that thousands of lives are lost in the United States alone, annually, from such food-borne infections as salmonellosis and complications arising from such infections. In addition, there is a large amount of information on the subject that is published in highly respected scientific literature, as well as first-hand experimental and research-based knowledge of the capabilities of low doses of picowaves to kill disease-causing microbes.

Niel E. Nielson, "Food Irradiation Means Better Health," *21st Century Science and Technology*, July/August 1988. Reprinted with permission.

(continued)

Routine applications of doses of picowaves of less than 100 kilorads (this is an amount of energy equivalent to that necessary to raise the temperature of the same amount of water by less than one half a degree Fahrenheit) to foods could reduce total bacterial content in foods by more than 99 percent and could completely disable the reproduction capabilities of 100 percent of the parasites. Since the immune systems of healthy human beings can overcome smaller quantities of disease-causing organisms when ingested, this 99 percent reduction would greatly reduce the numbers of human illnesses resulting from ingesting foods containing disease-causing microorganisms.

Further, based on the findings of the Centers for Disease Control and the Agriculture Department, widespread expanded use of picowave processing of foods could save hundreds of thousands of hospitalizations annually from such food-borne infections.

THE ILLNESS QUESTION

The salmonella problem is very real and very complex, and it highlights the importance of the routine use of picowave processing. About 40 percent of the poultry in the United States is infected with salmonella. The problem becomes obscured by the obvious fact that when you properly cook the poultry, you kill the salmonella, so you should not get it from the poultry you eat. You get it from the raw poultry contaminating things that aren't going to get cooked—salads, for example, picking it up from a cutting board, or the kitchen help's hands.

This is a two-step affair, but the result is the same: People get sick. And that says that you have to prevent the salmonella from getting into the kitchen in the first place. This is the focus of the Department of Agriculture's Food Safety Inspection Service's petition to the Food and Drug Administration to expand use of this technology to include all poultry, and at higher doses.

This is a very important issue: The antinuclear people use the argument that "unknown" illnesses are possible because 50 million people have not been tested for 50 years. The antinukes are saying that they think there is a very remote possibility of food irradiation causing a problem and therefore the technology should not be used. Yet, they completely overlook the fact that people are dying right now who could be prevented from dying if we could use the technology. That means that anyone who gives credibility to the idea that 40 years of testing in the United States and elsewhere is not enough research has lost perspective.

(continued)

DOUBLING THE FOOD SUPPLY

Let's look at some of the advantages of this technology, especially for the developing sector.

There is now more food lost to insects and spoilage than would be required to overcome all the malnourishment problems worldwide, if the rest of the infrastructure for transportation, storage, and distribution were also established. If you take the whole world's production, 50 percent is lost before it ever gets on its way to market. In the developing nations, 70 to 90 percent of their own production is lost, thus losing people food and income.

For example, a significant percentage of the fish and seafood harvested in the tropics contains enough bacteria that, if it isn't frozen almost immediately as it is caught and cleaned, it is not going to meet the U.S. Food and Drug Administration standards. As a result, and in some cases as a result of poor handling, the food that could be exported from those countries to the United States is rejected in the range of 3 to 15 percent.

We could prevent a large part of the rejection of good foods—not decomposed foods—by use of routine picowave processing, which would drop down the bacteria count by 90 to 99 percent at only 100 kilorads.

This is a big number—90 percent—if you put it in perspective. In terms of sterility, it is not a big number. But you don't need sterility if you are going to eat the foods right away. All you need to do is knock the bacteria count down by 90 percent.

In the case of fish and seafood, it would be processed, frozen, and then picowaved. In the case of fruits and vegetables, it would be precooled and then picowave processed. This process would prevent insect problems that exist with a lot of these fresh fruits and currently prevent them from being brought to the United States. In many cases, the effective chemical fumigants, like EDB, are banned in the United States; in other cases, the alternative fumigant/disinfestation procedures are ineffective or damage the fruit.

One of the scare stories of the antinukes is that there are unique radiolytic products (URPs) that appear in irradiated food. Now, there has never been a chemical identified in the processing of food by picowaves that is not already in our diets. Every food group has been thoroughly examined looking for evidence of truly unique chemical products and none has ever been found. This is a myth that the opponents of food irradiation like to scare people with, that unique radiolytic products appear in food that has been treated with low-dose radiation. These unique radiolytic products are like UFOs—frequently

(continued)

sighted, but not there.

To date, in the established scientific literature, there has been *no* chemical discovered to result from processing foods with electromagnetic energy at picometer wavelengths that is not already being ingested by mankind routinely in much larger quantities. All such chemicals identified to date have been found to exist in foods naturally, or as a result of several types of widely employed cooking or preserving processes, whether in commercial or domestic kitchens or food production plants.

Obviously, there are proven toxins, carcinogens, mutagens, and so on, and the Food and Drug Administration (FDA) issues standards for their content in foods, regulates the use of such chemicals and the foods that contain them, and takes enforcement actions against food processors and suppliers who do not work within the law.

Also obviously, there are chemicals such as vitamins and minerals about which we are still learning a great deal and the FDA has published (and will publish) recommendations concerning them also. But these chemicals about which we do have knowledge are a very small fraction of the total number of chemicals we ingest each day, and because there are so many different chemicals in foods that have not been thoroughly studied for their impacts upon mankind when ingested, food scientists increasingly employ such techniques as chemiclearance for evaluation of *relative* safety.

In this chemiclearance process, any resulting chemical that is identified as not having been in the foods prior to the processing is compared with all of the known toxins, carcinogens, and so on, and if not found to be among the known problems, a search is made to determine if such a chemical is already in mankind's diet.

Many highly qualified researchers focused for many years, internationally, upon chemicals that appeared to result from picowave processing and did so in every significant food category. They found all of the so-called unique chemical species that were discovered to be already existent in mankind's diet, in one way or another. As a result of all of this, those in this field of study feel very confident of the accuracy of the statement that there is *no* evidence that there will be *any* increase in risk for consumers of picowave processed foods.

The antinukes who conveniently use the URP argument to further their purposes typically do not have anything to do with public or environmental health. This is an indictment of a lot of people, and there may be some followers to whom I should apologize for making this all-encompassing indictment. But certainly those technologists who encourage those people or who let themselves be used as references for those people, need to be indicted because they are misleading people.

(continued)

They are just not playing it straight.

The established scientific community, as represented by organizations ranging from the National Academy of Sciences to the American Medical Association and the Institute of Food Technologists—the people who really know food science—have all endorsed picowave processing by organization. This means that the majority of the people in these organizations—responsible, recognized scientists—realize that there is no increase in risk and that there are potentially very high benefits. The people who are objecting to the use of this technology have no such accreditation, and yet they would pretend to have a large high-technology following in the field, which in truth they don't.

What they are doing is attacking the established institutions, which the public sponsors and which have gotten us so far advanced in the last 50 to 100 years, in so many ways, including just plain quality of life. There is the underlying theme that the establishment is bad. This permeates all these anti-food-irradiation activities.

In reality, the largest number of the leaders of these antinuclear organizations are doing it for profit for themselves. They are doing it because of the popularity they get, the number of lines of press, the exposure on radio and television. So they are feeding their egos, feeding their pockets from it. However, here the stakes are so high that they can be traced to deaths and needless illnesses, needless misery that is not speculative—it is real.

There is one favorite example that the antinukes like to use—quoted completely out of context. There is an Indian feeding study carried out by the National Institute of Nutrition. This study has been thoroughly repudiated by very well respected scientists worldwide and in India. The work done by India's Nutrition Institute could not be duplicated, which is one of the cardinal requirements of accepting scientific research. Also, their peers in India would not support this study and, in fact, after an investigation of how the study was conducted, came out with a policy statement against it.

A subsequent, much larger and more comprehensive Chinese study reached totally opposite conclusions to those reached by India's National Institute of Nutrition. But when the antinukes quote this study, they do not tell anybody that it's been discredited.

Another thing the antinukes do is distort information. One of these distortions is that irradiation creates peroxides in the food. Well, that is true. Food irradiation creates peroxides. But so does the body—and without them we would not live. These people know that the public did not pay attention in high school when these sorts of topics were discussed. So they play on it.

(continued)

MACHINE-GENERATED PICOWAVES

The idea of using machine-generated picowaves instead of radioisotope-generated picowaves is gaining in popularity rather quickly around the world. This will cause people to refocus, for so many of the antinuclear factions are focused on the handling and storage of the *radioactive* cobalt-60 or cesium-137. Now we can tell them, "We've eliminated radioactive cobalt; we are using an accelerator to produce picowaves, now what are your concerns?" This throws out the most sensational part of the antinuclear argument.

The bottom line is that once we eliminate the radioisotope question, no argument brought up by the antinukes has merit. This is a bold statement from somebody with a scientific background, but it is true. The antinukes bring up arguments against food irradiation that have been authoritatively defeated, that are irrational, and in the face of overwhelming, established scientific-community rejection of their arguments, they continue to bring them up.

What becomes unquestionably apparent to anyone who believes in the qualifications of the established scientific organizations and has read their endorsements, or has sufficient technical training to do a thorough and objective investigation of the use of this technology, are the following points:

(1) No other food processing technique, whether employing radiation or not, has been as thoroughly studied by so many scientists worldwide.

(2) There is agreement among scientists expert in the field that there is *no* absolutely safe food.

(3) All of the arguments put forth by those who would deprive mankind of use of this technology have proven to be without substance.

(4) The motivations of those who would oppose use of this technology are often for their own personal benefit and have little or nothing to do with food safety.

(5) Thirty years of delays in use of this technology—which was originally recommended for widespread use in 1958 by the food industry, the USDA, FDA, and others—have been costly in terms of the quality of life, or even the continuation of life for millions of victims of many food-borne diseases.

(6) Any further delays in widespread, expanded use of this technology are going to result in a needless continuation of thousands of deaths and hundreds of thousands of illnesses annually from food-borne illnesses in the United States alone.

Niel E. Nielson, a high-technology entrepreneur, was the president of Emergent Technologies, Inc., a San Jose, California food irradiation company.

Food Irradiation Is Harmful

Mark Mayell

An armed guard watches as three-foot-square containers are loaded onto a moving conveyor belt. "Caution—High Radiation Area" warn signs that adorn the walls of the small industrial plant. Off to the side is a high-tech control room, complete with consoles arrayed with lights, toggle switches, dials and video screens.

Containers slide by the control room into a maze of concrete walls. At the center of the maze is a room where the walls are over six feet thick. In the middle of this room is a storage tank holding rods of highly radioactive cesium-137, a by-product of nuclear weapons production. The conveyor belt comes to a halt as the containers surround the storage tank. An operator in the control room throws some switches. In the center of the maze, radiation from the cesium-137 swarms into the air, irradiating the containers and their contents. The radiation reaches a level 200 times greater than that which would quickly kill a human being.

Ten minutes later, the conveyor belt starts up and the containers move into a separate storage area. The contents are loaded onto a truck and shipped off for further processing and packaging. Eventually, what was once exposed to cesium-137 will appear as food on the shelves of your local supermarket.

Supermarket? Food? Why should what you eat have anything at all to do with nuclear-weapons by-products and lethal levels of radioactivity? It may seem a strange idea, but scenarios similar to the hypothetical one just described do take place today, and they'll become increasingly commonplace if irradiators, food processors, the nuclear industry and the federal government have their way. Touting food irradiation as an alternative to toxic preservatives, as "a new way to end waste and hunger" and as "the microwave oven of the 1980s," they hope to see up to half of all foods in the American diet irradiated within the next decade.

FOOD FASCISM

A growing legion of food-irradiation critics disagree. The whole idea of irradiating food is "perfectly ridiculous," says Dr. Sidney Wolfe of the

(continued)

consumer activist Health Research Group in Washington, D.C. It's "food fascism," charges Jeffrey Reinhardt, a co-founder of the San Francisco-based National Coalition to Stop Food Irradiation (NCSFI). It's as absurd as the Defense Department's plans for a nuclear-powered airplane, say others. The problems with food irradiation, critics charge, range from the creation of unsafe, possibly carcinogenic by-products in the food to increased environmental and worker hazards to a further warping of an already dangerously centralized and chemical-dependent agricultural system.

Today, battle lines are forming quickly in what may become one of the most charged public debates of the decade. As Congress and the Food and Drug Administration (FDA) stand poised to allow much greater use of food irradiation in the U.S., consumers and public-health advocates are joining together to try to stop this new technology. Scientists and public-relations campaigns are being marshalled in support of both sides. Already there's a fierce battle raging over the question of whether—and how—irradiated foods should be labeled, and that may be just the first skirmish of a long political conflict.

"Food irradiation is one of the most important issues of the '80s," says Denis Mosgofian, current director of NCSFI, "because it brings together those concerned with natural foods, better nutrition, organic farming, the environment, nuclear power and even the arms race and peace. The issue is not just about food—it's about the by-product use of nuclear technology."

Food can be irradiated in any form—packaged, fresh, loose or frozen. If food is sealed in an airtight container and then given a dose high enough to kill all microorganisms, it may have a shelf life of years. For packaged foods, irradiation can eliminate the need for harmful additives, such as nitrates in meat. Since at high enough levels irradiation can kill insects and their larvae, processors may not need to use toxic insecticides and fumigants.

Promoters of the process point to its greatest potential as a means to alleviate world hunger. Up to one quarter of the world's crops spoil before they can be eaten. If a large proportion of these foods could be irradiated, proponents say, starvation in developing countries could be virtually eliminated.

When ionizing radiation in the form of gamma rays strikes food, the result is a sort of biological and chemical mayhem. Cell division is disrupted and slowed down, which will slow the ripening process of fruit, for instance. Thus, by disturbing their metabolism, irradiation can kill bacteria, and at still higher levels can also kill viruses. Ionizing radiation can also disable organisms by destroying their genetic material. It is not necessary to make food radioactive for all this to happen.

(continued)

What are left behind in the irradiation process are new biological substances. Some of these biological substances, or *radiolytic products* (RPs), are common in other food processes, such as cooking or canning. These include carbon dioxide, methane and fatty acids. Other substances left behind are *unique radiolytic products* (URPs), and have not been observed outside of the process of food irradiation.

The FDA has already approved a limited number of foods for irradiation treatment. In 1963, it said from 20,000 to 50,000 rads could be used to treat wheat and wheat flour, and in 1964 it approved 5,000 to 15,000 rads for potatoes. Since the standard means of preserving these foods are relatively cheap and safe, there has been no reason for producers to irradiate them. Also, a series of rulings that the FDA made in 1966 and 1967 meant that if these foods were irradiated, they would have to be labeled at the retail level. Since no producer wants to admit on a label that their food has been "Treated with Ionizing Radiation" or "Treated with Gamma Radiation," no one is irradiating wheat or potatoes for U.S. consumers today.

But since 1983, the FDA has made a number of rulings that could greatly expand the availability of irradiated food. In 1983, the FDA approved the use of up to one million rads to irradiate spices and seasonings. Spices are often used as ingredients, and as such don't have to be labeled when irradiated. Today, perhaps one half of one percent of all spices produced in the U.S. are irradiated. McCormick, a major spice producer, may be the only company openly admitting to selling irradiated food in the U.S. today.

THE LABELING CONTROVERSY

The big breakthrough for food irradiators began on February 14, 1984, when the FDA published in the *Federal Register* proposed guidelines that would allow the irradiation of fruits, vegetables and grains at a level of up to 100,000 rads. The FDA also proposed to eliminate the retail labeling requirement that had been in force since the 1960s. The public was invited to comment on these proposed changes.

The public did comment, too, for almost the next two years, to the tune of some 5,000 letters to the FDA, the overwhelming majority of them opposing food irradiation and favoring honest labeling. Nevertheless, on December 12, 1985, the day before her last day as head of the Department of Health and Human Services, Secretary Margaret Heckler approved for publication an FDA final rule that answers the prayers of food irradiators. The rule both broadens the ap-

(continued)

proved uses of irradiation (allowing fruits and vegetables to be exposed to up to 100,000 rads, and increasing the limit for spices from one to three million rads) and proposes a new labeling requirement that is significantly weaker than what has been in effect for the past 30 years.

Under the new labeling requirement, irradiated foods must still be labeled at the wholesale and retail levels, but no longer is the straightforward "treated with ionizing radiation" the standard. Instead, producers need only put a "radura," a tulip-shaped international symbol, and the word "picowaved" on labels. Moreover, Heckler said that after two years the FDA will consider whether the symbol is well enough known on its own, and "picowaved" can be dropped.

Opponents of food irradiation were livid at what they saw as a gutting of irradiation labeling. The innocent-looking radura could almost be a health food symbol, they pointed out. And picowaved may sound scientific, but don't look for it in any dictionaries. It is a word that was recently coined by Niel Nielson, the president of Emergent Technologies, Inc., a San Jose irradiator. "It's meant to be cute and unthreatening, like 'microwaved,'" says Denis Mosgofian.

The new FDA regulations must be okayed by the Office of Management and Budget before they can be published in the *Federal Register* and become law.

Another food that it is currently legal to treat with irradiation in the U.S. is pork. In July of 1985, the FDA granted approval to a petition submitted by Radiation Technology of Rockaway, New Jersey, for the use of up to 100,000 rads to destroy the trichinosis parasite in pork. During the short, one-month public comment period that the FDA allowed, critics of the ruling pointed out that pork's high fat content makes it more susceptible to radiation-induced damage, and that there were only 60 reported cases of trichinosis in 1984 anyway.

The Department of Agriculture, which has jurisdiction over meat, needed to put its stamp of approval on the pork irradiation proposal, and on January 15, 1986, they too came through for the irradiators, amending federal meat inspection regulations to permit the use of gamma radiation on pork. Like the FDA, they did some serious waffling on the labeling question, declaring that "decisions with respect to the labeling of such products . . . will be made on a case-by-case basis."

In order to understand why labeling is such a crucial issue in the food-irradiation debate, we must keep one salient fact in mind. That is, given the choice, most consumers would rather not eat food that has been exposed to nuclear radiation.

Overall, the food-irradiation industry is relatively small in the U.S. Analysts put industry revenues at about $3 million per year, mostly due

(continued)

to spices. Radiation Technology, the New Jersey firm, probably does the most business in other foods, such as strawberries and shrimp. These are irradiated for export only. There are a dozen or so companies already formed who hope to do food irradiation, and they're currently waiting for favorable rulings from the FDA. In lieu of food, irradiators are treating disposable medical supplies, baby powder, cosmetics, plastics and Teflon.

In addition to the growing irradiation industry, the U.S. government is another major actor in this debate with a vested political and economic interest. Both the Department of Defense (DOD) and the Department of Energy (DOE) see food irradiation as the crown jewel in their ongoing "atoms for peace" campaign. The two main sources of gamma radiation, cobalt-60 and cesium-137, are by-products of either nuclear-power processes or nuclear-weapons production.

Figuring out what to do with cesium-137 is a primary function of the DOE's Nuclear By-products Utilization Program. Most cesium-137 in the U.S. comes from the Hanford Nuclear Reservation in Washington state, where they produce plutonium for various weapons programs.

The connection between nuclear weapons production and food irradiation could be a major liability to the fledgling industry, and some irradiators are hoping cobalt-60 will be a publicly accepted alternative. Over four-fifths of the cobalt-60 used in the world's irradiators comes from Atomic Energy of Canada, a Canadian government-backed company. According to a recent Canadian business report, "Acceptance of gamma processing would mean a great deal to the troubled nuclear industry, which is aggressively marketing the process as a means of selling the by-products of nuclear research."

IS IRRADIATED FOOD SAFE TO EAT?

Perhaps the single most important question with regard to irradiated foods is simply, are they safe to eat? The federal government has spent somewhere between $50 million and $80 million over the past 40 years in an effort to answer that question.

Based on this research, the FDA says that at doses below the 100,000-rad limit being proposed, "the difference between an irradiated food and a comparable non-irradiated food is so small as to make the foods indistinguishable with respect to safety." In effect, the FDA is saying, yes, irradiation causes radiolytic products to form. And some 10 percent of these are likely to be unique radiolytic products. (So far, at

(continued)

least 42 URPs have been identified.) Moreover, foods irradiated at levels above 100,000 rads "may contain enough [URPs] to warrant toxicological evaluation." More study is needed only to determine the safety of, for instance, irradiated meat, fish and poultry. . . .

The government's reading of the scientific findings is hotly disputed by many researchers. One series of objections has to do with the reliability of the studies. Many of the early studies were commissioned by the Army or the Atomic Energy Commission, whose plainly stated support of food irradiation may have caused inconclusive or adverse findings to be discounted.

Offsetting the studies supportive of food irradiation safety are numerous ones demonstrating a variety of ill effects:

- Rats and other test animals experienced an increase in testicular tumors and kidney disease, and a shortened lifespan, while being fed irradiated chicken in tests sponsored by the USDA between 1976 and 1980;

- A recent pair of Russian studies has also found evidence of testicular damage and kidney disease;

- Studies of food irradiation in India have linked it with leukemia and abnormal development of white blood cells (children and monkeys were fed wheat three weeks after it had been exposed to 74,000 rads of radiation);

- Some animals fed irradiated wheat have been known to develop cells that contain more than the usual number of chromosomes.

Safety and labeling questions aside, opponents of food irradiation point to a number of important reasons why the process will never live up to promoters' expectations:

- There are radiation-resistant bacteria, such as the bacteria that cause botulism (botulism-affected foods would in turn be harder for consumers to spot, since other signs of spoilage might have been prevented by the irradiation);

- Irradiation causes what are known as "organoleptic" changes in food, that is, changes in foods' taste, smell and texture. Dairy products and some fruits are especially prone to such changes and are thus inappropriate for irradiation;

- Unpackaged food is not protected from bacterial contamination *after* irradiation;

- Some microorganisms can mutate under radiation, possibly creating new species even more dangerous than the original;

(continued)

- According to a 1984 EPA [Environmental Protection Agency] report, food irradiation has been tied to increased production of aflatoxins, a deadly carcinogen;

- Even as a sprout inhibitor in potatoes, its most popular use, irradiation is not ideal since it increases potatoes' sensitivity to fungal attack and therefore to rotting;

- As a high-tech, capital-intensive technology, it is particularly unsuited for use in the Third World.

Other important drawbacks to food irradiation include its harmful effect on nutrients, potential environmental hazards and worker safety.

Like canning and freezing, exposing food to a radioactive source is harmful to certain beneficial constituents of food. Studies have shown that irradiation affects vitamins, proteins and amino acids, carbohydrates, nucleic acids and enzymes.

If the food-irradiation industry were ever to treat a significant portion of this country's food supply, the potential for radioactive leaks, spills or even disasters would be significant. Since the process is not economical if foods have to be shipped long distances for treatment, what are being proposed are many hundreds, even thousands, of irradiation facilities spread over the land, concentrated in food-producing areas. All of these would require individual shipments of highly radioactive substances. Residual radioactive wastes would also have to be shipped out or otherwise disposed of.

CONSUMER INVOLVEMENT

Faced with the combined efforts of elements in Congress, nuclear weapons producers, the radiation industry and food processors in support of food irradiation, consumers have recently begun to organize on the grassroots level. One common strategy is passing local labeling laws. In April 1985, an Oregon state legislator introduced the nation's first state irradiated food-labeling law. It would have required all irradiated food sold in the state to bear a label reading: "Warning: This product has been radiated with radioactive isotopes for purposes of preservation; the health effects are unknown." Legislators later changed the focus of the bill from food irradiation to sulfites, but other states have since followed Oregon's lead.

"We've collected over 5,000 signatures on a food-irradiation labeling petition here in Vermont," says Ken Hannington of the Vermont

(continued)

Alliance to Protect Our Food. Hannington points out that 5,000 signatures are a respectable number for a small state, and he says it will be followed up by a legislative proposal by State Senator Sally Conrad. Vermont's labeling requirements, if passed, would extend even to restaurants. At least one other state, Minnesota, has introduced similar legislation. . . .

As many problems as there are with food irradiation, it might deserve further consideration if there were no alternatives. Fortunately, there are many alternatives—humanity has been preserving its food with various techniques for millennia, and such traditional techniques as cold storage, drying and fermentation work well for certain foods and climes. . . .

A discussion of alternatives brings us to the core of the issue of food irradiation and preservation in general: For what purpose? So that New Englanders can eat Hawaiian mangoes in January? So that any food can be stored indefinitely and then shipped anywhere in the world? So companies can increase their profits? If these reflect our values, food irradiation may be a plausible answer. If, however, we seek our values in more and better fresh food, greater individual and public health, regional and local self-sufficiency in the production and consumption of food, a return to sustainable, ecological agriculture and an environment free of long-term toxic hazards, then there is certainly no need to climb aboard the food irradiation express. Like the nuclear airplane, we must hope that it never gets off the ground.

Mark Mayell is the editor of *East West Journal*, a magazine that focuses on natural health and living.

Do Cows Cause Global Warming and Human Hunger? The Fault, Dear Jeremy, Lies Not in Our Cattle But in Ourselves . . .

By J. Madeleine Nash

Vermin. The word reminds most people of cockroaches scuttling across kitchen floors and rats skulking in dark basement corners. But to Jeremy Rifkin, the environmental movement's most prominent polemicist, vermin are big, brown-eyed ungulates that graze the rolling countryside, chew their cud and moo. In his controversial new book, *Beyond Beef: The Rise and Fall of the Cattle Culture*, Rifkin manages to blame the world's burgeoning population of bovines for a staggering spectrum of ecological ills. In the U.S., he charges, runoff from mammoth feedlots is despoiling streams and underground aquifers. In sub-Saharan Africa, cattle are contributing to desertification by denuding arid lands of fragile vegetation. In Central and South America, ranchers are felling tropical rain forests and turning them into pastures for their voracious herds. "The average cow," claims Rifkin, "eats its way through 900 lbs. of vegetation every month. It is literally a hoofed locust."

According to Rifkin, civilization began a long slide downhill when 18th century British gentry acquired a taste for fat-marbled beef and proceeded to spread that proclivity, like a plague, throughout the Western world. Rifkin's real argument, of course, is not with the 1.3 billion bovines that roam the planet but with modern methods of mass-producing beef that include plumping animals with hormones and stuffing them with "enough grain to feed hundreds of millions of people." Although he did not personally visit a ranch or a meat-packing plant, his stomach-churning descriptions of how cattle are treated from birth to slaughter brim with righteous indignation. (A reformed carnivore, Rifkin says he swore off beef 15 years ago after taking three bites of a revolting blue-gray hamburger, then throwing the rest away.)

Such inflammatory rhetoric sends shudders through the U.S. beef industry, which is already reeling from a nearly one-third drop in per capita consumption since 1976—the result of popular concern about fat in the diet. Now Rifkin hungers for a more decisive blow. This week he

J. Madeleine Nash, "The Beef Against Beef," *Time* magazine, April 20, 1992.

(continued)

is leading a coalition of environmental, food-policy and animal-rights groups in launching a well-financed advertising campaign aimed at slashing worldwide beef consumption by 50% over the coming decade. Members of the coalition range from the Rainforest Action Network, which blames cattle for "killing the Amazon," to the Fund for Animals, which criticizes the use of poisons and traps to control coyotes that prey on calves. The International Rivers Network blames cattle for wasting scarce water resources, while Food First denounces the feed-lot system for wasting grain that could otherwise be used for human consumption.

Not since he took on the biotechnology industry over the safety of genetic engineering has Rifkin been embroiled in a higher-profile controversy, or one with the potential for greater economic consequences. With so much at stake, it is hardly surprising that environmentalists and meat-industry advocates have locked horns over Rifkin's charges. Among the most notable areas of dispute:

Cattle Ranching Is Destroying Tropical Forests.

Without question, ranching is a factor in tropical deforestation, and a major one at that. But University of Pennsylvania biologist Daniel Janzen, for one, believes that this unfortunate epoch in the history of Latin America is rapidly drawing to a close. In Costa Rica, he says, "most of the pastureland that was easily cleared of forest has already been cleared." At the same time, the remaining forest has begun to rise in value. "Two decades ago," explains Janzen, "the choice was simple. Either the forest stood there, or someone tore it down to plant a crop." Now leaders of countries like Costa Rica are beginning to view forests as valuable assets that can help control erosion, protect watersheds and generate income from New Age industries like biotechnology and ecotourism.

Cows Are Contributing to Global Warming.

To a measurable extent, they are. The symbiotic bacteria that dwell in every cow's gut enable grazers to break down the cellulose in grass. As a by-product, these bacteria produce considerable amounts of methane, which, like carbon dioxide, is a heat-trapping greenhouse gas. The methane periodically gusts forth from grazing herds in the form of rumbling postprandial belches. But if cattle contribute to the global methane load, they are hardly alone. Swamps, termite mounds and rice paddies are all hosts to similar sorts of bacterial methane factories.

(continued)

Overgrazing By Cattle Has Destroyed Grasslands.

The "cowburnt" ranges of the American West testify to the damage wrought by decades of uncontrolled grazing, which transformed once verdant land into desert. Of more than 50 million acres of U.S. Forest Service land that is open to grazing, half remains in poor condition. Lands under control of the Bureau of Land Management are in equally bad shape. Driving the cattle off, however, as some radical environmentalists would like, is not necessarily the solution. Properly managed grazing, range ecologists agree, serves to enrich rather than impoverish grasslands. In exchange for forage, hoofed beasts deposit tons of that old-fashioned organic fertilizer known as manure.

Grain Fed to Cattle Could Feed the Hungry.

"Hunger isn't about actual scarcity," declares Stephanie Rosenfeld, a researcher for San Francisco–based Food First. "It's about the maldistribution of resources. People are hungry for different reasons at different times, but quite often the reasons have to do with beef." The link is often very subtle: in countries like Egypt and Mexico, for instance, farmland that formerly grew staples for human consumption is being switched to grow grain for beef that only the wealthy can afford. Indirectly, then, a growing cattle population threatens humans on the low end of the economic scale with hunger. D. Gale Johnson, an agricultural economist at the University of Chicago, questions this assumption. He notes that in China, beef consumption has risen in tandem with overall improvements in diet.

Rifkin's critics—and there are many—regularly accuse him of taking a nugget of truth and enlarging it beyond reason in ways calculated to raise public fears. "*Beyond Beef* is about the worst book I've ever read," exclaims Dennis Avery, director of Global Food Issues for the Hudson Institute, a think tank in Indianapolis. "It establishes Rifkin as the Stephen King of food horror stories." Among other things, Rifkin raises the specter of beef contaminated with viruses, including a bovine immunodeficiency virus that he provocatively labels "COW AIDS," though there is no evidence that the virus can infect humans. Rifkin also charges that inspection of carcasses is shoddy, which the U.S. Department of Agriculture flatly denies. However, even the American Meat Institute allows that the inspection system, which still relies on visually examining and touching meat, hasn't changed much since 1906 and needs more up-to-date techniques to detect invisible contaminants like microbes. Ironically, the primary tools for improve-

(continued)

ment could well come from biotechnology, an industry that Rifkin loves to bash.

Rifkin is using beef as a metaphor for all that has gone rotten in the modern world, wrongs that he attributes to a metaphysical loss of humans' sacred relationship to nature. And cattle, because of their prominent role in ancient mythology and their haunting presence in prehistoric pictographs, lend themselves well to this moralistic exercise.

But how much blame for environmental degradation should the cattle industry rightly shoulder? In the Netherlands, for instance, manure from pigs poses a major ecological threat, defiling water supplies with excessive nitrates and acidifying local soils. Sheep have permanently scarred the landscape in Spain and Portugal, while in India—a country that Rifkin praises for its kindness to cows—bovines are ravenous wraiths whose constant quest for food drives them to ravage standing forests. Holy or not, most of India's 200 million cows go hungry much of the time.

Cutting down on beef consumption in protein-sated countries like the U.S. is a prudent prescription that would go a long way toward enhancing general health. Red meat is the primary source of saturated fat in the American diet, and too much dietary fat has been linked to the development of both heart disease and certain types of cancer. But trimming beef in the American diet, emphasizes Felicia Busch of the American Dietetic Association, "will not solve world hunger, and it isn't going to save our planet." The environmental cost of beef is just one aspect of the multiplying burdens of producing food for an exploding human population. The real threat to the carrying capacity of planet Earth, dear Jeremy, comes not from our cattle but from ourselves.

—With reporting by Janice M. Horowitz/New York and David S. Jackson/San Francisco

CHAPTER

8

Editing to Improve Style I: Sentences

Constructing Concise Sentences

Achieving Sentence Coherence

Improving Emphasis

Using Active and Passive Voice

Maintaining Parallelism

Strengthening Sentences: A Checklist

174

The purpose of this chapter and the next is to assist you with copyediting—that all-important "polish" every writer wants to give a piece of writing that took a lot of time and hard work to create. Just as a furniture maker devotes time to carefully and patiently smoothing every centimeter of a table's surface with microfine sandpaper, then giving the wood a quality woodstain finish, writers should likewise place finishing touches on their products. Because technical documents are certainly "products" in every sense of the word, a technical writer needs to assure optimum clarity, coherence, and accuracy. Of course, when deeply engaged in a first draft, writers are most concerned with content: facts, figures, lines of reasoning that best support assertions; there's hardly any time at this stage to worry about whether the sentences are verbose or awkward, or if the paragraphs do not link up as logically as they should. That is why writers need to allow themselves sufficient time to take care of these essentials; and to know what to look for. This chapter focuses on the following elements:

- Conciseness: using no more words than are necessary.
- Coherence: checking for smooth, logical relationships from one element to the next.
- Emphasis: placing the information to be emphasized in the most strategic position in a sentence.
- Active and passive voice: deciding when to use one or the other.
- Parallelism: using symmetrical syntactic structures to improve clarity and readability.

CONSTRUCTING CONCISE SENTENCES

Verbose sentences interfere with clarity and comprehension because the reader must dig through unnecessary words in order to uncover the buried meaning. Consider this sentence:

One definition of radon is that it is a gas without a scent or any color and is produced by radioactive uranium as it decays.

Can you spot the unnecessary words? First note the redundancy: "One definition . . . is that." Redundant because it is self-evident that a definition is being presented, and that it is but one definition. Let's see the result after eliminating the redundancy:

> Radon is a gas without any color or scent and is produced by radioactive uranium as it decays.

Now another kind of wordiness sticks out—fat: an excess of relatively weak words instead of fewer, stronger words that could convey the idea even better. "The gas that was produced was without any color or scent" is an example of fat; the idea can be expressed more concisely:

> A colorless, odorless gas was produced.

With this change, you can now make a clearer, more concise sentence:

> Radon is a colorless, odorless gas and is produced by radioactive uranium as it decays.

However, it still is not as concise as it could be. The "and is" doesn't really contribute anything; and "radioactive uranium *as it* decays" is both wordy and misleading: it seems to attribute radon more to the uranium itself than to its decay. "Radioactive decay of uranium" would be more precise.

The completely revised sentence now reads:

> Radon is a colorless, odorless gas produced by radioactive decay of uranium.

The original 25 words have been trimmed down to 12 with no loss of meaning—and with a substantial gain in clarity.

Eliminating Redundant Expressions

The redundant element in the preceding example may have been rather obvious to you, but not all redundancies are as easy to spot. Which expressions in the following example would you consider redundant?

> During the month of February, employees who had been working at aviation assembly plants in the city of Los Angeles were required to undergo mandatory testing for drug use; but unlike their counterparts in the cities of Pasadena and Burbank, they were not given any prior forewarning of such a test.

Since a redundant expression is one whose meaning is already evident in another expression, you are correct in tagging "the month of" as redundant; February is obviously a month. Similarly, employees "who had been working" is redundant: "employee" by definition is one who works. And "prior forewarning" is most likely redundant because "forewarning" means "prior warning." However, if it ever would become necessary to stipulate, "We gave him a prior forewarning before today's forewarning," the expression would not be redundant.

But what about "city of Los Angeles"? Isn't it obvious that Los Angeles is a city? Yes, but Los Angeles is also a county. The passage can be tricky because not everyone is aware of this, nor of the fact that Pasadena and Burbank are cities within Los Angeles County. Here it is a question in audience targeting: if the writer is addressing non-L.A. County residents, it would make sense to explicitly identify Pasadena and Burbank as cities in L.A. County in order to emphasize the inconsistent drug-testing policies between city and county aviation employees.

Redundant	Preferable
The science of chemistry	Chemistry
The deadline for when the paper was due	The paper was due
Small in size	Small (however, "Small in stature as well as in size" is acceptable)
Sour tasting	Sour
Computer software program	Computer program (or simply "program," if it is clear from the context that the reference is to computers)
Hard-copy printout	Hard copy (or printout)
In my opinion the part should be redesigned	The part should be redesigned (such a statement is obviously an opinion)
True facts	Facts (however, "The invented facts were easily distinguished from the true facts" is acceptable)
A very shocking disclosure	A shocking disclosure
A major and significant event	A major event (select one)

Referring often to the following list of common redundant expressions can help you avoid them; however, keep in mind that in certain contexts some of them might be acceptable (e.g., for dramatic emphasis).

Eliminating Extraneous Verbiage (Fat)

Wordy sentences are most often caused by using a combination of cumbersome noun phrases and weak verbs, such as forms of "to be."

"Now can we have an eating experience?"

Drawing by M. Twohy; © 1993 The New Yorker Magazine, Inc.

Her preparation of the experiment was done with great care.

The noun phrase, "Her preparation of the experiment" is cumbersome, having absorbed what should be the verb.

She prepared the experiment with great care.

This is a stronger sentence, not simply because it is concise, but because the stronger verb "prepared" conveys the meaning more effectively than the auxiliary "was" in the original sentence.

Fat can also result from fad expressions, often spawned by promotional slogans. Examples include "digital sound system" for CD player, "driving machine" for car, and so on.

Review the list of common lard expressions and their leaner equivalents on page 179.

ACHIEVING SENTENCE COHERENCE

In any written document, the relationship among the elements of a sentence, as well as the relationship among all the sentences in a paragraph, must be immediately apparent to readers or confusion will result.

Consider the following groups of sentences:

Fat	Lean
At the present time	Now
During the period when	When
She came to the conclusion	She concluded
They reached a consensus of unanimous agreement	They all agreed
Due to the fact that	Because
The reason for this being that	Because
In the near future	Soon
Conducted in a sensitive manner	Sensitively conducted
In the event that	If
Should the situation arise that	If
A need exists for us	We need
There is an urgent need for them to receive medical aid	They urgently require medical aid
Make an attempt to reach them by phone	Try phoning them

I. There are two principal kinds of nuclear particles. One kind is the proton. The other kind is the neutron. Protons and neutrons are locked tightly into the structure of the nucleus.

II. The simplest electromagnet is a loop of wire carrying an electric current. The current produces a magnetic field. So the loop acts as a magnet.

Although the sentences in each of these clusters are grammatical, collectively they do not make for acceptable writing. They are not *coherent*. The writer has neglected to combine the bits of information in a way that would reveal how the sentences interrelate. Before reading on you may want to try your hand at improving them your own way.

I. The two principal kinds of nuclear particles, protons and neutrons, are locked tightly into the structure of the nucleus.

II. The simplest electromagnet is a loop of wire carrying an electric current. Because the current produces a magnetic field, the loop acts as a magnet.

—Robert M. Hazen and James Trefil, *Science Matters*.
New York: Doubleday, 1990: 112.

In the first example, coherence is heightened by embedding the second sentence into the first; that enables the reader to concentrate on the heart of the cluster, the idea that protons and neutrons are locked tightly inside the nucleus.

In the second cluster, the word "Because" tells the reader that a cause-effect relationship is being set up; the reader instantly learns that the loop of wire acts as a magnet because the electric current running through it produces a magnetic field.

Other words (subordinating conjunctions) that signal relationships include:

after, during, until	(temporal)
unless, if, although	(conditional)
wherever	(spatial)
who, which	(relational)

Inexperienced writers often assume that the reader will understand the intended relationship between two sentences simply because they're next to each other:

> It's raining heavily. It would be a good idea to conduct the meteorological experiment indoors.

Clearly there's a relationship to be drawn between the two sentences, but which one?

> Because it's raining heavily, it would be a good idea to conduct the meteorological experiment indoors.

> —or—

> Although it's raining heavily, it would be a good idea to conduct the experiment indoors.

Judicious use of transitional expressions is another way of reinforcing the relationship between sentences, as well as contributing to a smooth writing style. Some transitions are external and obvious: "First . . . second . . . third . . . finally"; "For example"; "On the other hand"; "Also"; "According to" are common examples. In many instances, however, transitions can almost appear intrusive or redundant. Moving smoothly from a general statement at the beginning of a paragraph to specific support statements is often enough to ensure coherence, as in the following example:

> The motor vehicle industry is a cornerstone of the modern industrial economy. It contributes almost 7 percent of global economic output and employs more than 4 million people. Automobiles account for about 4 percent of all industrial employment in the United States and Spain, 6 percent in Britain and Canada, 7 percent in Japan, and 8–9 percent in France, Italy, Sweden, and the former West Germany. Vehicle sales continue to grow but, due to automation, employment has expanded much less. For example, sales revenues for West German auto companies more than doubled from 1977 to 1987, but the number of jobs grew by less than one quarter.

> —Michael Renner, "Creating Sustainable Jobs in Industrial Countries," in *State of the World 1992*, ed. Lester Brown et al. New York: Norton, 1992: 146.

Although there is but one "overt" transitional expression—"for example"— the passage is highly coherent throughout. The first sentence states the generalization that the following sentences illustrate. Readers have a clear sense of how the argument is advancing.

IMPROVING EMPHASIS

How efficiently readers will comprehend the information in a sentence has a lot to do with the way in which that sentence is constructed. Which of the following sentence pairs, for example, seems easiest to comprehend?

1. The rate at which work is done, which is the definition of *power,* can be expressed in terms of watts and joules. Watts (W) is the metric measurement term for power; joules (J) is the term for energy or work.

2. *Power* is defined as the rate at which work is done. It is expressed in metric measurement terms of watts (W) for power and joules (J) for energy or work.

—Rex Miller, *Electronics the Easy Way,* 2nd ed.
New York: Barron's, 1988: 14.

Clearly, 2 is more readily understandable (and is what the author actually wrote)—but why? See if you can determine the reasons before reading on.

Although the sentences in 1 are, on the surface, well constructed, they do not give proper emphasis to the central idea, the definition of power. Instead, the term is virtually buried in the middle of the first sentence, and the audience is not immediately aware, as it is in 2, that it is reading a definition of power. The reader has to hop back and forth across the first sentence to piece meaning together.

USING ACTIVE AND PASSIVE VOICE

First of all, the terms "active voice" and "passive voice" are old metaphors —so old that we rarely think of them as metaphoric. "Active" carries a positive connotation of strength, energy, determinism; "passive" carries just the opposite connotation. Not surprisingly, active-voice constructions ("The seismologists carefully studied the long-dormant fault") are often favored over passive-voice constructions ("The long-dormant fault was carefully studied"), without regard to context.

The choice of voice should depend on what is being emphasized. In the sentence "The seismologists carefully studied the long-dormant fault," emphasis is given to the seismologists; whereas in the sentence, "The long dormant fault was carefully studied," emphasis is given to the fault; who did the studying is not the concern—only the fact that the fault is being studied is relevant. Here is another example of passive construction used (and properly so) in the context of a scientific procedure:

> The presence of nitrogen, sulfur, and the halogens in an organic compound **may be detected** through the sodium fusion test, in which the organic sample **is dropped** over red-hot sodium. These elements, if present, **are converted** into easily identifiable inorganic ions.
>
> —C. N. Wu, *Modern Organic Chemistry,* vol. 1.
> New York: Barnes & Noble, 1979: 3.

Passive constructions can interfere with clarity when they are used inappropriately. Consider these sentences:

1a Mind-altering drugs were heavily experimented with by psychologists as well as crackpots in the sixties.

The passive voice construction makes the sentence awkward; the true subject is buried in a prepositional phrase.

1b. Psychologists as well as crackpots experimented heavily with mind-altering drugs in the sixties.

1c. Mind-altering drugs provided psychologists with new insights into the nature of consciousnes.

Shifting to active voice strengthens the predication ("experimented" or "provided" instead of "were"), and the reader instantly understands who did what (or what caused what) and when.

MAINTAINING PARALLELISM

In a sentence, when two or more groups of words share the same pattern, they are said to be parallel:

> It is important to study math, not only because it teaches logical relationships, but also because it teaches design.

We can better discern the parallelism in this sentence by aligning the parallel elements as follows:

It is important to study math

not only

because it teaches

logical relationships

but also

because it teaches

design

Parallel constructions are very useful in reinforcing comprehension. On the other hand, phrasing that breaches parallelism sounds awkward and causes confusion:

It is important to study math, not only because it teaches logical reasoning; design is another quality of math that can be taught to students.

Not surprisingly, parallelism is one of the most conspicuous characteristics of aphorisms; their grammatical symmetry helps us to remember them easily:

We do not want to live in a world where the machine has mastered the man; we want to live in a world where man has mastered the machine.

—Frank Lloyd Wright

Parallelism is also important for maintaining clarity in lists and headings. Consider the following list:

Prerequisites for the technical communication major:

First Year

• One introductory course in technical writing
• Technical illustration: two courses
• TC 140 Introduction to Document Design

Second-Year Course Offerings

• Procedural Writing (TC 151)
• Hypermedia
• One course in oral communication

As you can see, neither the lists nor their headings are parallel with respect to one another. One can easily become confused about the program as a result. For example, students might wonder why the first heading does not include the phrase "course offerings"; are the courses listed below it something else? The listings themselves are not parallel; some items are titles of actual courses while others are areas from which courses should be taken. One course number is given parenthetically, another not.

Here is one possible revision of the preceding list:

First-Year Course Offerings

- TC 140 Introduction to Document Design

One introductory course in technical writing from the following:
- TC 100 Introduction to Technical Communication
- TC 110 Technical Writing for Engineers
- TC 115 Writing in Industry

Two courses in technical illustration from the following:
- TC 130 Introduction to Technical Illustration
- TC 119 Graphic Design for Engineers
- TC 129 Presentation Graphics

Second-Year Course Offerings

- TC 151 Procedural Writing
- TC 163 Hypermedia

One course in oral communication from the following:
- TC 170 Introduction to Oral Presentations in Industry
- TC 172 Electronic Oral Presentation Techniques

STRENGTHENING SENTENCES: A CHECKLIST

- ☐ Have I trimmed away the fat from my sentences?
- ☐ Have I checked for redundant expressions?
- ☐ Do my sentences connect smoothly and logically?
- ☐ Have I neither avoided nor overused transitional expressions?
- ☐ Have I used active or passive voice appropriately?
- ☐ Have I checked for breakdowns in parallelism?

■ CHAPTER SUMMARY

The way sentences are constructed and worded influences the way they are comprehended. When copyediting a draft to improve style at the sentence level, writers should look specifically for *redundancy* (repetition of an idea that is captured by another expression or is clearly implied); *verbosity* (more words than are needed to convey the idea); *incoherence* (lack of clear continuity among the elements of a sentence, or among groups of sentences); inappropriate use of *voice* (active or passive); and breakdowns of parallelism (two or more groups of words patterned the same way).

■ FOR DISCUSSION

1. Revise the following redundant or verbose sentences:
   a. Of the nine known planets in our sun's solar system, only two do not have satellites in orbit around them.
   b. We received from Intel Corporation a hard-copy printout of a purchase order via our fax machine.
   c. The coral snake and the king snake are both black and yellow in color, but whereas the yellow bands are wider in size on the king snake, they are narrower in size on the coral snake.
   d. Being an environmentalist who is concerned about the effects of pollution upon our planet, as well as the fact that precious natural resources were being depleted, Craig decided to begin recycling his trash.
   e. In the event that an earthquake strikes, the best place to be in order to be safe would be inside a doorway in your house.
   f. Individual cells all have the capability of responding to such stimuli as light, hot and cold temperatures, and pressure arising from their surroundings.
   g. The inhabitants who lived in Egypt during ancient times possessed a sophisticated engineering technology, evidence for which is plainly seen in the pyramids that are still standing.

2. Combine the following sentence clusters into single sentences:
   a. We removed the old wiring. We replaced the old outlets. Then we installed the floodlights.
   b. Not enough insulation had been installed in the attic. The heating bill was twice as high as it should have been.
   c. A home remodeling show was being held downtown. I decided to attend. The tickets were very expensive, however.
   d. The medical researchers designed a program that could simulate live-animal responses. Laboratory rats were not needed.
   e. Tarantula spiders are venomous. The venom is rarely toxic to humans.
   f. Tarantulas are frightening. They are gentle spiders, however. They attack only if provoked.

3. Locate a document (perhaps a draft of a report you wrote) that contains problems with parallel construction. Reproduce these examples on a transparency and discuss possible revisions with the rest of the class.

4. Improve the parallel elements in the following passages:
   a. To prepare wild rice, first place a cup of rice in a quart of chicken broth. Then sliced mushrooms and grated onions should be added. About ninety minutes' cooking time on low heat is recommended.
   b. The height of your birdhouse should be about seven feet, the diameter of the house opening about four inches, and also be sure to include a wide-enough platform for two or three birds to land on at a time.
   c. Technical writers at this company are expected to write and edit newsletters, design user manuals, and in some cases it may be necessary to have some skill in programming.

5. Discuss the effectiveness—or ineffectiveness—of passive voice used in a document that you will have located and reproduced (ideally on an overhead transparency) to share with the rest of the class.

■ FOR YOUR NOTEBOOK

1. Dig up a draft of a previous assignment from this course or another and check it for problems with conciseness, coherence, voice, or parallelism. Record the original sentences and your revisions of them in your notebook.

2. Experiment with sentences. For example, write a sentence; then rewrite it several different ways, using essentially the same content but changing the emphasis.

3. Rewrite sentences by other writers—your classmates, the authors of the books you're reading (including this one!), even your instructor's comments (if you dare). Try to make these sentences (*a*) more concise; (*b*) more emphatic (or reflect a different emphasis).

■ WRITING PROJECTS

1. Using a paper you have written for another class, write a detailed critique of your use of passive and active voice, emphasis, parallelism.

2. Locate three technical articles on a similar topic and write a paper in which you analyze each author's sentences in his or her introductory paragraph. Pay special attention to use of active and passive voice and methods for maintaining coherence (both within each sentence and from sentence to sentence).

CHAPTER

9

Editing to Improve Style II: Diction

The Importance of Verbal Precision in Technical Writing

Common-Sense Pointers Regarding General and Specific Diction

Editing Ambiguous Language

Jargon: When Needed, When Not

Word Origins and Their Value

Commonly Misused Expressions

Using Inclusive Language

Editing for Diction: A Checklist

Words, and the sentences they build, function interactively. Wordy or redundant expressions, for example, cause problems with sentence construction, as was demonstrated in the preceding chapter. Words misused or imprecisely used can also interfere with comprehension, which can lead to a serious misreading not only of the sentence involved but of the entire document.

THE IMPORTANCE OF VERBAL PRECISION IN TECHNICAL WRITING

Precise diction (word choice) is important in any kind of writing, of course; but in scientific and technical writing, it is crucial. Imprecise diction, say, in a set of directions can result in incorrect assembly. Like the components of an equation or a schematic drawing, words must be precise enough to convey one meaning only, not several possible meanings; otherwise, the decisions those words call for may be unreliable. To give an obvious example: if a promotional writer for a watch manufacturer assumed that the terms *water resistant* and *waterproof* were synonymous, he or she would cost the company a lot of money in refunds for damaged watches!

Much of the imprecise diction that crops up in technical documents comes from a simple laziness, the failure to consult a dictionary. And if a standard dictionary is inadequate for a certain field, then one should consult a specialized dictionary. (See Chapter 4, pages 60–62, for information about specialized dictionaries.)

COMMON-SENSE POINTERS REGARDING GENERAL AND SPECIFIC DICTION

Ever since your freshman composition days or earlier, you have been admonished, "Be specific!" But "general" and "specific" are relative terms; what may be quite specific in one context may be considered a gross generality in another. "Beetle" may well be a specific example of insect; to a beetle expert, however, it would be a vague generalization, since there are thousands of species of beetles. Also, it is not necessarily true that specifics are preferable to generalities. Both are essential. Without specific examples, the generality is not rendered convincing or understandable; without generalities, readers will wonder, What's the point? Also, just as importantly, both are context dependent, as Table 9.1 demonstrates. The table shows that any level can be sufficiently spe-

Table 9.1 Levels of Generality, Specificity

Level	Topic	Possible Aim at This Level	Possible Format at This Level
Ia	Astronomy	Define astronomy for general audience.	Encyclopedia article
Ib	Astronomy in the Renaissance	Show impact of astronomy on Renaissance culture.	Chapter in a book
Ic	Kepler's method for determining planetary motion	Show how the most gifted of Renaissance astronomers practiced his craft.	Scholarly article for historians of science
IIa	The solar system	Introduce the planets and their satellites; other solar system members (comets, asteroids) to young readers.	Encyclopedia article; chapter in introductory textbook; book (one chapter devoted to each planet)
IIb	Theories of solar system's origin	Examine theories in light of available evidence to determine which theory seems most plausible.	General article for amateur astronomers; scholarly book examining each theory in depth
IIc	Newest evidence for solar system's origin	Examine in depth a recent hypothesis.	Scholarly article for astronomers; also could work, in a simplified version, as a feature in a popular science magazine
IIIa	Mars	Introduce the red planet: evolution; history of exploration; major discoveries.	Video; article for young readers; book for older readers
IIIb	Composition of Martian soil	Infer planetary dynamics from soil chemistry.	Scholarly article for nonspecialist scientists
IIIc	Oxidizing agents in the Martian soil	Analyze the phenomenon of Martian soil oxidation.	Scholarly article for specialists in Martian geochemistry

cific for a given aim. It also demonstrates that any parallel level (e.g., Ia, IIa, IIIa) can be general or specific relative to each other. Because specificity and generality are relative gauges, it makes no sense to assume that the more specific ("narrowed down") one's topic, the better, unless a context is given.

Technical writing often demands specific information. In a discussion of fishing line, readers expect particular test strength numbers, not general categories. Not "moderately durable" or "highly durable"; not even "sea bass strength" or "marlin strength"—but "20-pound test"; "50-pound test"; "200-pound test." One could then add examples of the kinds of fish one could catch with line of a particular gauge or test number.

Although generalities are as important to a document as specifics, it is true that specifics are harder for apprentice writers to come up with. Hence, it is a good idea to practice being specific as much as possible until it becomes second nature. A good way to start is to practice the habit of breaking down generalities into different levels of specificity. Here are four ways to break down a generality:

1. *By example:* "For example, you can catch marlins and tuna with a 200-pound test line."

2. *By case in point:* "Consider this scenario: You are deep-sea fishing in the Gulf of Mexico and suddenly you snag a marlin with your 50-pound test line. What do you suppose will happen?"

3. *By descriptive details:* "When I snagged the marlin he fought with incredible strength; alas, my line was not strong enough to hold him and it snapped like thread."

4. *By explanatory details:* "Tensile strength, such as that of fishing line, is the resistance to lengthwise (longitudinal) stress, which is determined by the greatest amount of pull without breaking the material."

EDITING AMBIGUOUS LANGUAGE

An ambiguous word or phrase is one that can have more than one meaning or referent. Ambiguity generally arises from one or a combination of the following problems:

- improper word choice
- unclear pronoun reference
- squinting modification
- ambiguous relationship

Improper word choice: It is important to avoid using a word that can have more than one meaning in the context you have established. Consider:

Erika was **surprised** to hear the price of the new laser printer.

Since a surprise can be either pleasant or unpleasant, a more precise modifier or a hint at the cost is needed.

Revision 1: Erika was surprised to hear how little the laser printer cost. (or: how much)

Revision 2: Erika was upset by the price of the new laser printer.

Unclear pronoun reference: A pronoun must refer only to the noun for which it is intended. That is not the case in the following sentence:

> We tried to get new parts from Ajax, but **they** were not there.

Does **they** refer to the new parts or to the people at Ajax?

> Revision 1: We tried to get new parts from Ajax, but they did not have them in stock.
>
> Revision 2: We telephoned Ajax to order the new parts, but they did not answer.

Squinting modification: Once in a while, a word can modify the phrase preceding it, meaning one thing, as well as the phrase following it, meaning something else.

> Waxing the car **thoroughly** pleases Bobby.
>
> (You can't tell whether Bobby waxed the car thoroughly or was thoroughly pleased.)
>
> Revision 1: Thoroughly waxing the car pleases Bobby.
>
> Revision 2: Waxing the car pleases Bobby thoroughly.

Ambiguous Relationship: Subordinating conjunctions (because, after, during, wherever, etc.) help to establish causal, temporal, or spatial relationships in discourse. If you use a coordinating conjunction, such as "and," when you actually intend a subordinate relationship, ambiguity could result:

> The rain was heavy and the concert was canceled.

Was the concert canceled because of the rain, or were the two events unrelated? The ambiguity can be removed in two ways:

> Revision 1: The concert was canceled because of heavy rain.
>
> Revision 2: Although the rain was heavy, the concert was canceled because of a musician's strike.

JARGON: WHEN NEEDED, WHEN NOT

All disciplines have their specialized vocabularies, commonly referred to as *jargon.* The purpose of specialized language, remember, is to avoid ambiguity and to get an idea across quickly. The medical term for *feverish,* for example, is *febrile.* Why not simply say *feverish?* The answer is that the latter word has nonmedical meanings, such as "frenzied state of excitement." When medical

Table 9.2 Jargon Frequently Misused	
Jargon	**Preferable Expression for General Contexts**
accessorize	equip
ergonomic	practical, functional
high correlation factor	reliable
input (noncomputer)	response, reply
interface with	meet
liquidate	sell
methodology	method, approach
open-architecture system	upgradable
prioritize	do first, rank
somnolent	sleepy
sphygmomanometer	blood pressure cuff
systematize	arrange, order

professionals communicate information about a patient's symptoms, the ideal is to eliminate any possible misinterpretation, hence the use of jargon. A specialized term gets a bad reputation when it is misused, not because it is inherently a specialized term. Tell someone in casual conversation that you've been febrile the past few days, and you might get a lot of blank looks, as well as be accused of speaking pretentiously. In such a context, the specialized term is inappropriate—"jargony" in the negative sense.

Sometimes technical expressions do become part of mainstream usage as a result of the popularity of the technology to which they refer. One no longer has to be a specialist to understand terms like megabyte, interactive software, CD-ROM, hypertext, or turbomouse. However, these and similar terms can still be misused. Make sure your audience is familiar with laser printers before waxing eloquent about a certain model's font bank, 600 dpi resolution and 8 ppm performance. For a list of jargon expressions frequently misused, see Table 9.2.

WORD ORIGINS AND THEIR VALUE

Words are tools in more ways than one: like mechanical tools, the better you understand their nature and the more experience you have with them, the more use you can get out of them. Words are living fossils of sorts. Embedded in the word *radish* is the Latin word for root, *radix;* in mathematics, a *radical* pertains to a root number. The word *disaster* once had astrological connotations (*dis* = bad; *astro* = star). Words also have interchangeable parts: prefixes, suffixes, roots that come together to form compounds (laser + disc); roots that compress into acronyms (like *laser:* Light Amplification by Stimulated Emission of Radiation) or into parts of words (like *modem:* modulation-demodulation); and some parts wear out or become obsolete. The more familiar you become with the anatomy of words and their origins, the more control you will have over them.

Take the word *coronary,* a general reference to the heart (coronary arteries; coronary bypass, coronary thrombosis, etc.); the word comes from Latin: *corona* (crown), so named because of the arteries that surround the heart like a crown, or so it seemed to the ancient Romans. The word, which still exists in words like *coronation,* and [solar] *corona* (which forms at the climax of a solar eclipse), is thus an ancient metaphor, once bringing to mind the structure of the arteries leading to and from the heart.

Other words whose origins are instructive:

epidemic	Gr. *epi* (upon) + *demos* (people, population)
galaxy	Gr. *gala* (milk); as in Milky Way
helium	Gr. *helios* (sun); the element was first detected on the sun before it was detected on earth
hysteria	Gr. *hystero-* (of the womb); it used to be thought that hysterical behavior (in women) was caused by agitation in the uterus. The root meaning still exists in the term *hysterectomy*

Prefixes and Suffixes

Learning the following Greek or Latin prefixes and suffixes can save you much time consulting a dictionary when reading technical documents.

Prefix	Meaning	Examples
a-	not; without	asymmetrical, asymptomatic
dys-	ill; bad	dyslexia, dysfunctional
ec- or **ex-**	out(side)	eccentric (literally "outside the circle") exoskeleton, excavate
en- or **endo-**	in, within	enclose, endocardial
litho-	stone	lithograph; paleolithic
macro-	large	macro-economics; macrocosm
micro-	tiny, small	microbiology, microcosm
photo-	light	photosynthesis, photography
post-	after, back	post-hypnotic, posterior
proto-	first, primal	protoplasm, protein, proton, prototype
re- or **retro-**	back	regress, retrofire; retrograde

super-	above; greater	supernova, superovulate
sym- or **syn-**	with	symmetry, symphony, synchronicity
topo-	place	topography, topology

Suffix	Meaning	Examples
-able or **-ible**	capable of being	inflatable, immersible
-al	having the character of	filial, celestial
-ate	designating function, or one who exercises a function	surrogate; prelate
-ion	condition	inflation, immersion
-ism	indicating a practice or system or belief	plagiarism, optimism
-ment	resulting state	agreement, increment
-osis	designating a specific condition	psychosis, osteoporosis
-ous	being full of	superfluous (lit. overflowing), luminous
-ped **-pod**	reference to feet	biped, cephalopod
-y	characterized by	steamy, fiery

COMMONLY MISUSED EXPRESSIONS

The following errors are among the most common for writers in business and industry. It is recommended that you commit as many of these to memory as you can. For additional usage guidelines, see Appendix A.

affect, effect
Except in psychology, **affect** is nearly always a verb; **effect** is usually a noun. "Caffeine **affects** my pulse rate"; "Caffeine has a strong **effect** on my pulse rate"; "Those are the inmate's personal **effects** (belongings)." **Effect** is also a verb meaning "to bring about": "The new policy **effected** a dramatic change."

as, like
In formal usage, use **as** or **as if** rather than **like** to introduce a clause: "We arrived on time, **as** we promised"; "The parrot spoke **as if** it understood what it was saying." Use **like** for simple comparisons: "It seems like rain;" "That cloud looks **like** a rabbit."

etc.
It is best to avoid this abbreviation in formal writing, since it is often used improperly or redundantly:

	Improper: "We need to take along a can opener, coffeepot, etc." ("Need" suggests that *every* item should be mentioned.)
	Redundant: "Some of the groceries we purchased include milk, eggs, bread, juice, etc." ("Some" *means* that you aren't going to list everything, so "etc." is redundant.)
	NOTE: **etc.** is incorrect. Remember that **etc.** is the abbreviation for *et cetera,* Latin for "and the rest."
evidence, proof	**Evidence** refers to that which *contributes* to **proof,** but may not establish **proof.**
fewer, less	**Fewer** is used with quantifiable units: **fewer** coins; **less** is used with a nonquantifiable substance: **less** money.
inter-, intra-	Between/within: An **inter**departmental communication takes place *between* one department and another; an **intra**departmental communication occurs *within* a single department.
it's, its	**It's** = "it is" or "it has": "**It's** good to see you"; "**It's** been fun." **Its** is a neuter possessive pronoun, equivalent to masculine *his* and feminine *hers.* "The spider spins **its** web at night."
	NOTE 1: Using **it's** for **its** (or vice versa) is an easy typographical error to make; be careful.
	NOTE 2: There is no such construction as **its'.**
lay, lie	These may be the most frequently misused words in English. The problem is that the past tense of one word is spelled the same as the present tense of the other word. **Lie** means to recline: "The dog **lies** (not **lays**) in the shade"; "The dog **lay** in the shade all day yesterday." (Participles: **has lain/had lain.**) **Lay** means to put down: "Please **lay** the slides beside the microscope"; "I was sure that I **laid** (not **layed**) my book on the table." (Participles: **has laid/had laid.**)
media, medium	plural/singular. "The **media** always *seem* (not *seems*) to emphasize the sensationalistic"; "TV is the most influential **medium.**"
principle, principal	The first word is always a noun meaning "basis": "Please explain the **principles** of relativity to me." **Principal** is a noun meaning "main or chief (person or object)": "The school **principal** said the income would cover the interest on the loan, but not the **principal.**" As an adjective, it also means "main or chief": "What are the **principal** duties of your new job?"

unique   Means literally one of a kind; hence, most qualifiers (**very unique, somewhat unique**) do not make sense. It does make sense to say **possibly unique** or **almost unique,** however.

USING INCLUSIVE LANGUAGE

A fascinating thing about language is that it can change to accommodate social needs. The growth of inclusive language—that is, language that dissociates activity or occupation or grammatical structure from gender—has been taking place over the last quarter century (an amazingly short time for fundamental linguistic change); but the need was urgent, long overdue. Inclusive language is now part of standard usage, no longer considered extremist or associated with radical feminism. It makes solid good sense.

Thus, one violates standard usage by using gender-specific terms like fire*man,* wait*ress,* un*manned* probe. The proper expressions—firefighter, server, robot (or robotic) probe—make more sense because the work one performs is not inherently related to whether one is male or female. Even when, say, a firefighter is a man, it is not logical to link his gender with his activity.

Some expressions are difficult to overcome because the alternatives sound funny. *Freshman* is an example. To say *freshperson* sounds somewhat comical (although that could change; and besides, *freshman* has always sounded vaguely comical too); *first-year student* sounds verbose; *frosh* sounds slangy.

Use Table 9.3 to remind you of acceptable alternatives to old-style, noninclusive expressions.

EDITING FOR DICTION: A CHECKLIST

☐   Have I used the most suitable word or expression for the context?

☐   Did I check for ambiguity?

☐   Is my level of specificity appropriate for my aim and audience?

☐   Have I used examples, cases in point, descriptive and explanatory details where appropriate?

☐   If I have used jargon, will my readers understand it?

☐   Have I checked the dictionary for words whose meanings I am unsure of?

☐   Did I adhere to nonsexist usage?

Table 9.3 Sexist Expressions and Their Nonsexist Equivalents

Sexist	Nonsexist Equivalent(s)
businessman [-woman]	businessperson, executive, marketing [sales] representative
chairman [-woman]	chair, head, CEO (Chief Executive Officer)
girl [when referring formally to an adult female]	woman
his [her] (referring to both sexes)	his or her (or rephrase in plural)
male nurse	nurse
mankind	humanity, humankind
manpower	workers, workforce, staff
Miss or Mrs.	Ms. (a title, like Mr., should not disclose one's marital status)
paperboy	paper carrier
policeman [-woman]	police officer
repairman [-woman]	mechanic, technician, repairperson
salesman [-woman]	salesperson, sales representative
seamstress	tailor
steward [stewardess]	flight attendant
waiter [waitress]	server
woman doctor	doctor, physician
workmen	workers

CHAPTER SUMMARY

Precise word choice is important enough to give it separate attention when editing a manuscript, especially in technical communication, where a misused or vague expression could have serious consequences. It is necessary to check for possible ambiguity, insufficient use of generalities or specifics, inappropriate jargon, and noninclusive language. Acquiring familiarity with the origins of words, the meanings of common prefixes and suffixes, can help writers deepen their command of language.

FOR DISCUSSION

1. Discuss how generality and specificity interact.
2. Look for errors in diction in an essay you wrote earlier. In class, form

small groups to discuss these errors with each other and collaborate on revising the troublesome passages.

3. What contributes to ambiguity in language?

4. Explain the rationale behind the use of inclusive language. Do you agree or disagree with it, and why?

◼ FOR YOUR NOTEBOOK

1. Maintain a list of prefixes and suffixes and their meanings, similar to the lists presented in this chapter.

2. Revise passages from published articles whose diction, you feel, could be improved or simplified in some way.

3. Over the next few weeks, collect ambiguous phrases and sentences from your reading. Later on, practice revising them.

◼ WRITING PROJECTS

1. Write a detailed self-critique of your diction in one or two papers you have written in this or another class. Optional: Revise one of the papers.

2. Write a detailed critique of the diction in another student's draft of an assignment. Optional: Revise the paper.

3. Examine the etymology of several terms from your major field of study. Write a paper discussing the evolution of those terms.

10

Principles of Document Design

As you may have surmised by paging through this textbook, technical writing is a highly visual form of communication, both in its use of graphs and diagrams and in its textual layout and design. Because technical writers aim to transmit information as unambiguously as possible, they carefully manipulate the physical appearance of the text so that it reinforces coherence, overall readability, and ease of reference (as in conferences, when copies of a document are distributed and sections of it referred to frequently). The basic principles of design you will study in this chapter are easy to learn, fun to practice, and they will enable you to create documents that are professional in appearance.

ELEMENTS OF DOCUMENT DESIGN

The term *document design*, refers to the total page-by-page arrangement of text and visuals in the most practical and eye-pleasing manner possible. Document design, in some rudimentary form, has clearly been around since the invention of the printing press in the fifteenth century; the design principles are probably centuries older. In a fascinating article, "Visual Language: The Development of Format and Page Design in English Renaissance Technical Writing," Elizabeth Tebeaux traces the growth of pragmatic writing, with its "visual language" reflected in page layout just as today, as the reproductions from a sixteenth-century herbal text and a medical first-aid text (see Figures 10.1 and 10.2). Manipulating textual elements by themselves is called "formatting"; integrating text with visuals on a given page is called "layout." Before the days of personal computers, document design was not considered part of writers' work. Writers typed out the text of their documents, perhaps penciling in rough tables, graphs, or drawings. A layout artist or graphic designer would then prepare the finished visuals; create a "pasteup," or layout, of each page; and indicate to the printer which typefaces and sizes to use and where. Today, desktop publishing programs, and even many word processing programs, enable the technical writer to perform all of these functions. So effective is this new technology that it is changing the way we read and process information from texts. As document-design specialist Stephen A. Bernhardt states, "Influenced especially by the growth of electronic media, strategies of rhetorical organization will move increasingly toward visual patterns presented on screens and interpreted through visual as well as verbal syntax."[1]

Effective document design accomplishes several things that contribute toward the efficient processing of information:

---

[1]Stephen A. Bernhardt, "Seeing the Text." *College Composition and Communication* 37 (Feb. 1986): 77.

FIGURE 10.1 Page from the *Boke of Distyllacyon of Herbes* (1527) showing an effective balance of text and illustration.

- *Ease of reading and reference.* It is important to remember that technical documents are referred to, often in group contexts, not just "read").
- *Quick recognition of categories or hierarchical relationships.* Readers need to understand quickly how a document is organized. Are they looking at steps to a procedure? At the results of an experiment or investigation? The relationship between the whole and its constituent parts, or between generalizations and the specific reasons or examples used to support those generalizations, should be visually reinforced. Outlines, of course, are the basic tools for determining the organizational scheme of a document (see pages 100–102 for discussion of outlining strategies).
- *Ease of connecting abstract discussion with concrete pictorial representation.* While it is true that the mind processes visual images more readily than descriptions of such images, it is also true that the mind processes the *interaction* of visuals with their corresponding verbal descriptions even more readily.

THE FYRSTE

Thynges good for the heed

℟ Cububes.
Galyngale.
Lignum aloes.
Maioram.
Baulme myntes.
Gladen.
Nutmegges.
Muske.
Rosemarye,
Roses.
Pionye.
Hisope.
Spyke.
Camompll,
Meilplote.
Fewe.
Frankyncense.

Thynges good for the harte

℟ Cynamome.
Saffron.
Corall.
Cloues.
Lignum aloes.
Perles.
Macis.
Baulme myntes.
Myrobolanes.
Muske,
Nutmigges.
Rosemarye.

The bone of the harte of a redde dere.

Maioram.
Buglosse.
Borage.
Setuall.

Thynges good for the liuer

℟ wormewode.
withwynde.
Agrymonye.
Saffron.
Cloues.
Endyue.
Lyuerworte.
Lykorie.
Plantayne.
Dragons.
Rapsons greate.
Saunders.
Fenelle.
Violettes.
Rose water.
Letyse.

Thynges good for the lunges

℟ Elycampane.
Hysope.
Scabiose.
Lykorise.
Rapsons.
Maydenheare.

FIGURE 10.2   Page from *The Castel of Helth* (ca. 1539) showing tabulated text.

- *Aesthetic appeal and psychological impact.* The effect of well-packaged information goes beyond the utilitarian ideals of clarity and referrability. Attractive design helps stimulate a receptive attitude toward the material. A cluttered or unappealing layout can make the reading task seem much more difficult than it is; it can actually interfere with reader motivation.

DETERMINING A DOCUMENT'S APPEARANCE

Any writer in industry faces several questions regarding the consequences of the appearance of the document he or she is preparing; that is, of applying the general principles to a specific document. Should diagrams be placed alongside the portion of text that refers to them, interspersed within the text, or placed by themselves, on separate pages, so as not to "interrupt" the text? How large should the diagrams be? How wide should the columns of text be? What type fonts and sizes should be selected? These questions of balance, harmony, and functionality are influenced by at least four variables: (1) the target audience; (2) the purpose of the document; (3) the complexity of the material; and (4) the type of publication (manual, pamphlet, on-screen document, fact sheet, etc.). The writer-designer's own tastes matter, of course, but they are conditioned by his or her appreciation of those four external variables.

Document-specific problems are most effectively solved when one is aware of the following general principles of document design:

1. *Selecting typography.* This includes (1) **type font,** the general type design used throughout the document; (2) **font style,** the variations within a particular font, such as boldface or italics; (3) **font size,** measured in "points," usually from very small (5-point) to headline-size (72-point), with 10- or 12-point the typical size for standard text.

2. *Formatting textual elements.* Choosing type fonts and sizes must be done in the context of each document page. What is the best way to arrange the textual elements to promote reading efficiency as well as eye appeal?

3. *Arranging text and graphics ("layout").* Integrating text with visuals follows the same rationale as integrating different textual elements: what arrangement contributes best to ease of reference and balance?

4. *Manipulating white space, visuals, and text.* A particular layout of text and visuals can satisfy the criteria in item 3 yet still be unsatisfactory because of poor eye-appeal.

Two categories of fonts exist: serif (arm or base extensions of certain letters, such as, E, M, L, T,) and those without extensions (sans-serif).

SERIF FONTS		SANS-SERIF FONTS	
NEW YORK	New York	CHICAGO	Chicago
PALATINO	Palatino	GENEVA	Geneva
TIMES	Times	MONACO	Monaco
GARAMOND	Garamond	UNIVERSE	Universe
BODONI	Bodoni	OPTIMA	Optima
NEW CALEDONIA	New Caledonia	FUTURA	Futura
BENGUIAT	Benguiat	HELVETICA	Helvetica

FIGURE 10.3A   Common type fonts used in technical writing.

Let's examine each of these considerations in detail.

SELECTING TYPOGRAPHY

A document's typography is just as much a design element as the graphs and charts or drawings you will be covering in the next chapter. Dozens of fonts and styles exist, which can be set in many different sizes. For a sampling of fonts, styles, and sizes, see Figures 10.3A, 10.3B, and 10.4.

Choosing one font over another is sometimes merely a matter of taste; other times, it is a matter of convention. The first criterion, however, should always be ease of readability, unless of course the aim is to create a mood with just a few words. Gothic script, for example, is ideal for announcing the following:

$$\text{Costume Party Tonight}$$

As you might guess, Gothic or other "aesthetic" fonts are inappropriate for most technical documents; one exception would be magazine ads or promotional documents. This is not to say that typographical variety doesn't exist in mainstream technical writing; it does—but variety must always serve readability.

Some of the possible type styles for just one font--Helvetica

Helvetica Thin	*Helvetica Condensed Italic*	**Helvetica Black Condensed**
Helvetica Thin Italic	Helvetica Extended	***Helvetica Black Condensed Italic***
Helvetica Light	**Helvetica Bold**	**Helvetica Black Extended**
Helvetica Light Italic	***Helvetica Bold Italic***	**Helvetica Compressed**
Helvetica Light Condensed	**Helvetica Bold Condensed**	Helvetica Extra Compressed
Helvetica Light Condensed Italic		**Helvetica Ultra Compressed**
Helvetica Light Extended	**Helvetica Bold Extended**	Helvetica Bold Outline
Helvetica	**Helvetica Heavy**	Helvetica Rounded Bold
Helvetica Italic	***Helvetica Heavy Italic***	Helvetica Rounded Bold Italic
Helvetica Condensed	**Helvetica Black**	Helvetica Rounded Bold Outline

FIGURE 10.3B Some of the possible type styles for just one font—Helvetica.

Too much variety, like the **typographical** variations in *this* sentence, would seem quite amateurish!

It is helpful to acquire a basic familiarity with typographical options, since many word processing programs and versatile printers accommodate them. For most purposes, relatively few options can go a very long way.

FORMATTING TEXTUAL ELEMENTS

To gain a sense of the role that document design elements can play in readability, let's begin with a draft of a page from a brochure on the economics of clean water, when the writer was concentrating only on generating content, not on text layout:

> The money which we have all invested in clean water pays for many things, such as new plant construction, old plant upgrades, sewer lines, the hiring of qualified and certified professionals to operate treatment facilities efficiently, laboratory costs, equipment and maintenance costs (such as pumps, lines, trucks), collection system (sewer lines), and sludge disposal.

abcdefghijklmnopqrstuvwxyz
6 ABCDEFGHIJKLMNOPQR

abcdefghijklmnopqrstuv
7 ABCDEFGHIJKLMNOP

abcdefghijklmnopqrst
8 ABCDEFGHIJKLMN

abcdefghijklmnopq
9 ABCDEFGHIJKL

abcdefghijklmnop
10 ABCDEFGHIJK

abcdefghijklmn
11 ABCDEFGHIJ

abcdefghijklm
12 ABCDEFGHIJ

abcdefghijkl
14 ABCDEFG

abcdefghi
18 ABCDEF

abcdef
24 ABCD

abcde
30 ABC

abcd
36 ABC

FIGURE 10.4 Sizes of type compared.

Jan V. White: 50. *Mastering Graphics: Design and Production Made Easy.* New York: R. R. Bowker Co., 1983.

Even though the writer has conveyed the intended information in this initial version, the text layout interferes with rapid and efficient absorption of that information. This "nondesign" layout is usually adequate for nonutilitarian documents, but it is not suited for documents that are typically consulted for data.

But now let's look in on what this writer has done with the document in the final drafting of it (see below). You will probably agree that the visual appeal has increased considerably. First note that the text has been arranged into discrete units according to topic, and that each unit is signalled by a **subheading** (often called "subhead"). Subheads operate on different levels, each level represented by a particular typestyle. Note how one typestyle is used to emphasize the separate categories of activity ("Construction"; "Operation and Maintenance"), while another typestyle is used to indicate the particular subdivision of each category (e.g., "New plant construction," "Old plant upgrades," and "Sewer lines." The particular typestyle selections—in this case, 14-point New York, underlined; 12-point New York boldface; 12-point New York, plain—work well in calling immediate attention to the hierarchical relationship that exists between the categories and the tasks or services that exemplify them.

What Does My Community Get for Its Money?

The money which we have all invested in clean water pays for many things, including:

Construction

New plant construction. New facilities are needed to handle increased wastewater flows in growing areas.

Old plant upgrades. Outmoded equipment and plants must be replaced. Generally, treatment equipment lasts about 10–20 years; structures last at least 30 years.

Sewer lines. New sewer construction extends lines to developing areas. Repair and rehabilitation projects keep excess water from getting into the sewers.

Operation and Maintenance

Personnel. Qualified, certified professionals are hired to operate treatment facilities efficiently. They must be paid competitive wages and benefits.

Laboratory costs. Tests must be run to show that the wastewater treatment plant is operating safely and efficiently.

Equipment maintenance costs. Pumps, lines, trucks, and other equipment must be repaired and replaced as needed.

Collection system maintenance. Sewer lines must be maintained, cleaned, and repaired.

Sludge disposal. The cost of processing, hauling, and disposing of the solids removed from the treated wastewater must be paid.

Water Pollution Control Federation [Alexandria, VA], "Clean Water: A Bargain at Any Cost" (1987).

When choosing which typestyles to use, your primary consideration should be ease of reference: does the layout assist the reader in grasping the informational content as quickly as possible?

ARRANGING TEXT AND GRAPHICS

"A major problem in technical communication," state M. Jimmie Killingsworth and M. Gilbertson, "involves how graphics should relate to the written text." As with writing in general, arranging textual and visual elements are rhetorical tasks because "The technical writer must find an approach to the audience that will at once accommodate verbal and visual material."[2]

As mentioned earlier, visual aids such as tables, graphs, charts, drawings, and photographs can heighten the clarity and comprehension of a technical presentation. As Figure 10.5 suggests, technical writers now often need to decide such variables as:

- size of the visual;
- proportion of textual material to visual material on a given page;
- placement of the visual(s) with respect to the text.

There are no hard and fast rules for applying one arrangement over another in a given case. Generally, relatively complex visuals that could disrupt the continuity of the text should be offset (top or bottom of the page) rather than interspersed with the text. Visuals for step-by-step instructions in which each step needs illustrating ought to be aligned with the respective steps. It is important, as Killingsworth and Gilbertson remark, to ensure that technical documents (e.g., manuals) are task oriented rather than product oriented; that is, product-oriented documents tend to "neglect the reader's needs and focus instead on the attributes of the system" (138).

MANIPULATING WHITE SPACE, VISUALS, AND TEXT

When arranging text and graphics, you also need to be aware of a common oversight: The underuse or overuse of white space. Space is an important element in document design because it contributes to the often delicate balance

---

[2]M. Jimmie Killingsworth and M. Gilbertson, "How Can Text and Graphics Be Integrated Effectively?" *Solving Problems in Technical Writing,* ed. Lynn Beene and Peter White. New York, Oxford University Press, 1988: 131.

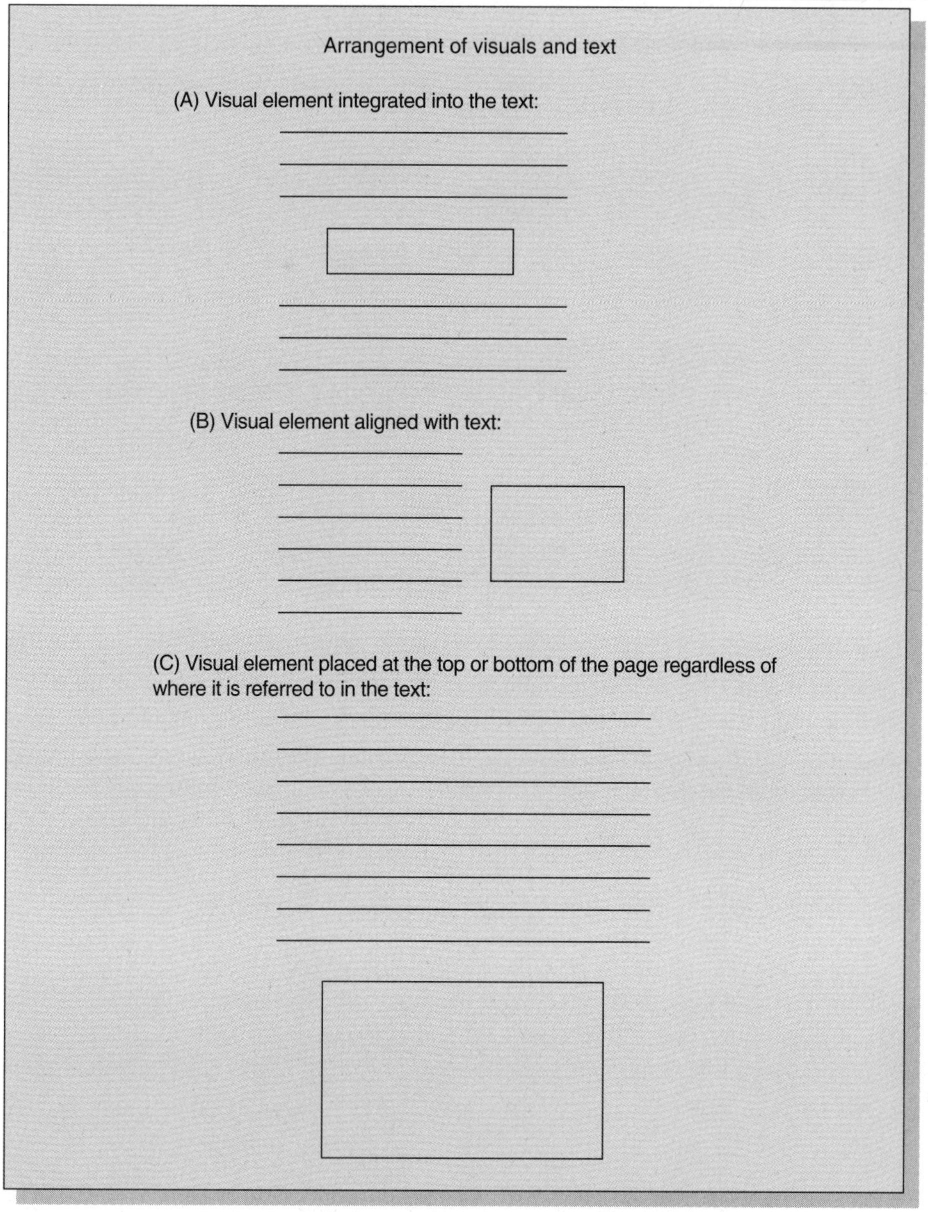

FIGURE 10.5   Three variables in the arrangement of visuals and text.

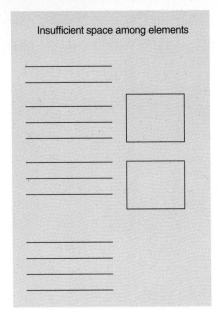

Insufficient space among elements

FIGURE 10.6

between distinction and interconnection among the elements on a given page. Include too little space (Figure 10.6) and the reader feels that the elements lose their distinctive qualities; include too much space (Figure10.7) and the reader feels that the elements are isolated from each other, sharing nothing in common.

USING TEXTUAL MARKERS

Along with selecting different type sizes and fonts, technical writers will make use of *textual markers* to highlight text and help create an overall attractive layout. The seven most frequently used types of textual markers in technical documents are as follows:

1. numbers
2. bullets (round or square)
3. item or ballot boxes
4. leader dots
5. caution or emphasis boxes
6. background or field shading
7. icons

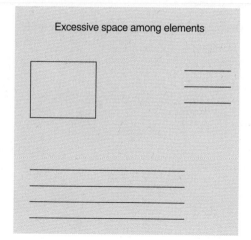

Excessive space among elements

FIGURE 10.7

Numbers

Items in a list are enumerated whenever it is useful to indicate a sequence (as in a hierarchical series of steps), or a fixed quantity (often stated explicitly, as in the preceding list).

Some writers like to use what is known as a decimal numbering system, which is similar to that of an outline. Each level of specificity in a discussion is given a decimal number, so that if the opening paragraph is identified as "1," a more specific discussion of the topic in the following paragraph would be identified as 1.1, and so on.

Bullets

Bullets [•] make each item in a list stand out conspicuously; they are an effective alternative to numbers when there is no clear reason for presenting the list as a sequence or a previously quantified set. Bullets may be square [■] as well as spherical, but the two types should not be mixed within a single document.

Item or Ballot Boxes

An item box [□] is most often used for checklists (as at the end of chapters, except Chapter 1, in this book). They are sometimes used like bullets—simply to highlight items in a list.

```
Proper use of leader dots:

Sound Technician . . . . . . . . . . Sam Jenkins
Computer Animation . . . . . . . . Diane Keller
Cinematography . . . . . . . . . . . Frances E. Gray

Improper use of leader dots:

Sound Technician . . . . Sam Jenkins
Computer Animation . . . Diane Keller
Cinematography . . . . Frances E. Gray

          --or--

Sound Technician . . . . . . . . Sam Jenkins
Computer Animation . . . . . . . Diane Keller
Cinematography . . . . . . . . . Frances E. Gray
```

FIGURE 10.8

Leader Dots

To keep the eye aligned from an item to its referent a good many spaces away, leader lines are used. Dots may be separated by one, two, or no spaces; the rows of dots, however, must be in vertical alignment. Also, the referents (right side) must be in vertical alignment. Finally, leader dots should be limited to connecting items separated by more than five spaces (see Figure 10.8).

Caution or Emphasis Boxes

Whenever a statement is so important that overlooking it could result in individual harm or equipment damage, that statement should not only be set conspicuously apart from the body of the text but be boxed in as well:

> **WARNING:** DO NOT IMMERSE
> THIS APPLIANCE IN WATER

Caution boxes are almost always rectangular, the width generally twice the height—although this can vary greatly.

Background or Field Shading

Shading is an effective way to highlight a visual or a block of text that is already boxed apart from the mainstream text. The only precaution to keep in mind here is that the shading should not overpower. Only the lightest grades of toning (whether in color or monochromatic) should be used.

Icons

With all the "clip-art" software around, it is tempting to use icons for everything; but, as with other design elements, these can be easily overused, marking the writer as an amateur. Icons almost invariably add a spirit of levity to the document. One exception is the use of icons in visual aids (see Figure 11.9B), where they serve as effective space-saving alternatives to written-out identifiers. Another is the use of icons in computer systems, as in Macintosh systems, where one has icons within icons; for example, each file is represented by a stack of pages; a group of files is collected under a folder icon; folders and files all are part of the desktop icon.

USING A DESKTOP PUBLISHING PROGRAM

If you have access to a computer with a hard drive and at least four megabytes of RAM, you can do what is known as **desktop publishing,** a document-layout program that has revolutionized the way magazine, newsletter, and brochure pages are designed. If you are not familiar with such a program, it is not at all difficult to learn; however, you need to give yourself plenty of time to become familiar with its numerous features.

You will be amazed—and gratified—by the document design feats you can conduct with these programs. They enable you to accomplish the following, as a few examples:

- View and edit dozens of pages at a time, reproduced in "thumbnail" size on the screen.
- Create visuals of all kinds, along with text, virtually all at the same time.
- Design borders of all kinds.
- Enlarge or miniaturize your visuals to conform to available space.
- Experiment with different column sizes and shapes; existing text will automatically reformat itself.

- Choose from dozens of type fonts, styles, and sizes; you can also customize your fonts.
- Rotate text and graphics elements in fractions of degrees.
- Instantly replace one type font, style, or size throughout the document.
- Incorporate or adapt ready-made formats in the program.
- Create striking design effects in minutes that would take a talented artist hours to complete.

Clearly, learning desktop publishing can be a valuable investment, especially if you anticipate producing large documents such as user manuals, newsletters, and annual reports; or working on the production staff of a magazine or book publisher.

DESIGNING A DOCUMENT: A CHECKLIST

☐ Have I sketched out a rough design of the layout of each page?

☐ Have I selected appropriate type fonts, sizes, and stylistic variations within a particular font?

☐ Did I format my text in a manner that facilitates readability and referability?

☐ Are my visuals integrated with the text in both a functional and eye-appealing manner?

☐ Have I made use of textual markers (bullets, response boxes, numbers, etc.) where useful?

■ CHAPTER SUMMARY

Becoming familiar with the variety of document design elements that exist is one thing; choosing and integrating them effectively is another. Like writing itself, this comes with steady practice and paying close attention to well-designed documents. The working out of a given document's design should be governed by simplicity, balance, and functionality. More specifically, the writer-designer should strive for ease of reading and reference, establishing a clear recognition of categories or hierarchical relationships, as well as establishing a coherent link between discussion and pictorial representation. The writer-designer also should be able to select appropriate type fonts, styles, and sizes; work out a textual layout pattern for each page of the document; and integrate text with visuals in a way that heightens comprehension of both elements and contributes to eye appeal.

■ FOR DISCUSSION

1. What contributes to "eye appeal" in a page layout? Why is eye appeal an important element in document design?

2. Study the advertisement below from an engineering magazine (Figure 10.9) and discuss the effectiveness of its typography, layout of textual elements, and text-graphics integration.

3. Visit the bound periodicals section of your library and examine pre-1930 copies of scientific and technical magazines such as *Popular Mechanics, Scientific American,* and *Electrical World.* What similarities and differences do you see between document design strategies then and now?

4. Which of the text-visual layout options in Figure 10.10 is the most effective? Why?

Why Mass Flow?

Today, hundreds of industries use gas mass flow meters to optimize their processes. Why? Because on the process line and the bottom line, mass flow makes sense.

Why mass flow? Because no technology better shows the real picture: you work in pounds of air or gas to conserve feedstocks and ingredients, and to optimize process chemistry.

you must buy additional temperature and pressure transmitters to convert the volumetric flow signal, you lose more than just money; you lose accuracy and you lose reliability.

Today you need high performance, cost effective instrumentation – and thermal mass flow meters have been proven to meet this criteria in a wide range of process applications.

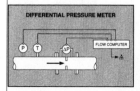

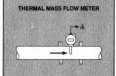

Call and let us send you one or more of Sierra's Industry Application Guides. Covering many processes, they show how others have put the advantages of thermal mass flow technology to work!

Mass Flow is the quantity of direct interest whenever the *molecules* of the gas are the control variable in your process – such as combustion, chemical reactions, ventilation, product drying and a variety of other process applications.

Conventional flow meter manufacturers know that you need mass flow, because it is independent of temperature and pressure. They do their best. But when

Sierra thermal mass flow meters directly monitor gas mass flow, and offer you these benefits:

■ *Accurate, Direct Reading of Mass Flow. No temperature or pressure compensation required.*

■ *Lowest Cost of Ownership. No need for additional devices, easy installation, and neglible operating costs.*

■ *Widest Turndown. With turndowns 10 times that of ΔP meters, thermal mass flow meters offer unmatched flexibility and exceptional low flow sensitivity.*

■ *Fast Response. 1 second.*

CALL TOLL FREE TODAY
800-866-0200
EXTENSION 147

SIERRA
INSTRUMENTS, INC.

5 Harris Court • Monterey, California 93940 • (408) 373-0200 • Fax (408) 373-4402

FIGURE 10.9 Text-graphics design in a magazine ad.

When in doubt, ask a police officer

When in doubt, ask a police officer

Lorem ipsum dolor sit amet, consectetur adipscing elit, sed
diam nonnumy eiusmod tempor incidunt ut labore

When in doubt, ask a police officer

et dolore

FIGURE 10.10 Trial layouts.

■ FOR YOUR NOTEBOOK

1. As you work on a technical document, keep a running log of the way you plan the finished layout. Do you sketch out a preliminary design of each page? Do you work by trial and error? What influences you to alter original design plans?

2. Brainstorm with others in class on possible design strategies for your documents in progress. Jot down everyone's suggestions. Later, experiment with all of them and make note of what works and what does not.

3. Keep your eyes open for effective and poor examples of document design. Make copies, organize them into categories, and tape them into your notebook for future study.

■ DOCUMENT DESIGN PROJECTS

1. Working with another student, redesign the front page of an existing newsletter or pamphlet. The idea here is not to improve upon the original, but to work out an alternative variation which, ideally, will be just as effective.

2. Prepare two alternative page designs for the first page of an imaginary newsletter, say, for the campus Astronomy Club or Engineers' Club. Examples of newsletter design may be found in Chapter 21.

3. Work out an effective page layout for the following text; indicate placement of visuals with boxed spaces:

Parts of a tree branch:

Terminal Bud: The fat bud at a branch tip will always grow first and fastest if you leave it. Cut it, and several buds will grow behind it. Leaf Bud: Flattish triangle on the side of a branch. To make one grow, cut just above it. Choose buds pointing outward from the trunk so the growing branch will have space and light. Flower Bud: Plump compared to leaf buds and first to swell in spring. On stone fruits they grow alone or beside leaf buds. On apples and pears they grow with a few leaves. Spurs: Twiglets on apples, pears, plums, and apricots. They grow on older branches, produce fat flower buds, then fruit. Don't remove them. Bud Scar: A ring on a branch that marks the point where the terminal bud began growing after the dormant season. The line marks the origin of this year's growth.

—From *All About Growing Fruits & Berries.* San Francisco:
ORTHO Books, 1982: 37.

11

Creating Tables and Figures

You have seen that technical documents require a total coordination of the writing, design, and illustration so that practical information may be conveyed quickly, clearly, and efficiently. In the preceding chapter, you saw how document design—formatting of text; selection of type fonts and sizes; layout of text and visuals on a page—contributes to efficient reading and referencing of documents. In this chapter you will focus on the designing and use of visual aids, such as tables, graphs, charts, photographs, and diagrams.

VISUAL AIDS IN SCIENTIFIC AND TECHNICAL WRITING: AN OVERVIEW

Anyone who has studied Euclidean geometry probably realizes that the use of technical illustrations to help readers visualize abstract relationships goes back many centuries—in the case of Euclid's *Elements,* to 300 B.C. (Figure 11.1). And one of the earliest known uses of technical illustration goes back many more centuries: Egyptian architects around 2000 B.C. prepared illustrated technical documents to help them erect temples, fortresses, irrigation systems, and burial chambers such as the famous pyramids (see Figure 1.2 in Chapter 1). During the Renaissance, all facets of culture—artistic, scientific, religious, philosophical, technological, mercantile—were not only practiced but celebrated; it was not uncommon to find master artists like Leonardo da Vinci and Albrecht Dürer creating technical drawings (Figure 11.2).

The universality of technical illustration is not hard to explain. In order to be fully *con*ceived (understood), data often need to be clearly and accurately *per*ceived (visualized). It is a psychological fact that when eye and mind work interactively, greater understanding is reinforced than when one operates separately from the other. Whenever it is necessary to report large quantities of data or relatively complex relationships among related sets of items, writers should look for the possibility of including an appropriate visual aid to capture pictorially what is being described or analyzed verbally. At the same time, writers should determine what type of visual aid would work best. Should the information be tabulated (informal or formal table)? graphed (line graph or bar graph)? structurally or hierarchically organized (organizational or tree chart)? broken down as elements in a process (flow chart)? or as portions of a whole (pie chart)? Should an object be depicted realistically (drawing or photograph)? according to its assembled parts (exploded diagram)? or according to its abstract function (schematic)? Most of the time, it is easy to make a choice; other times, several kinds of visual aids could do the job, making it tough to decide which type would work best.

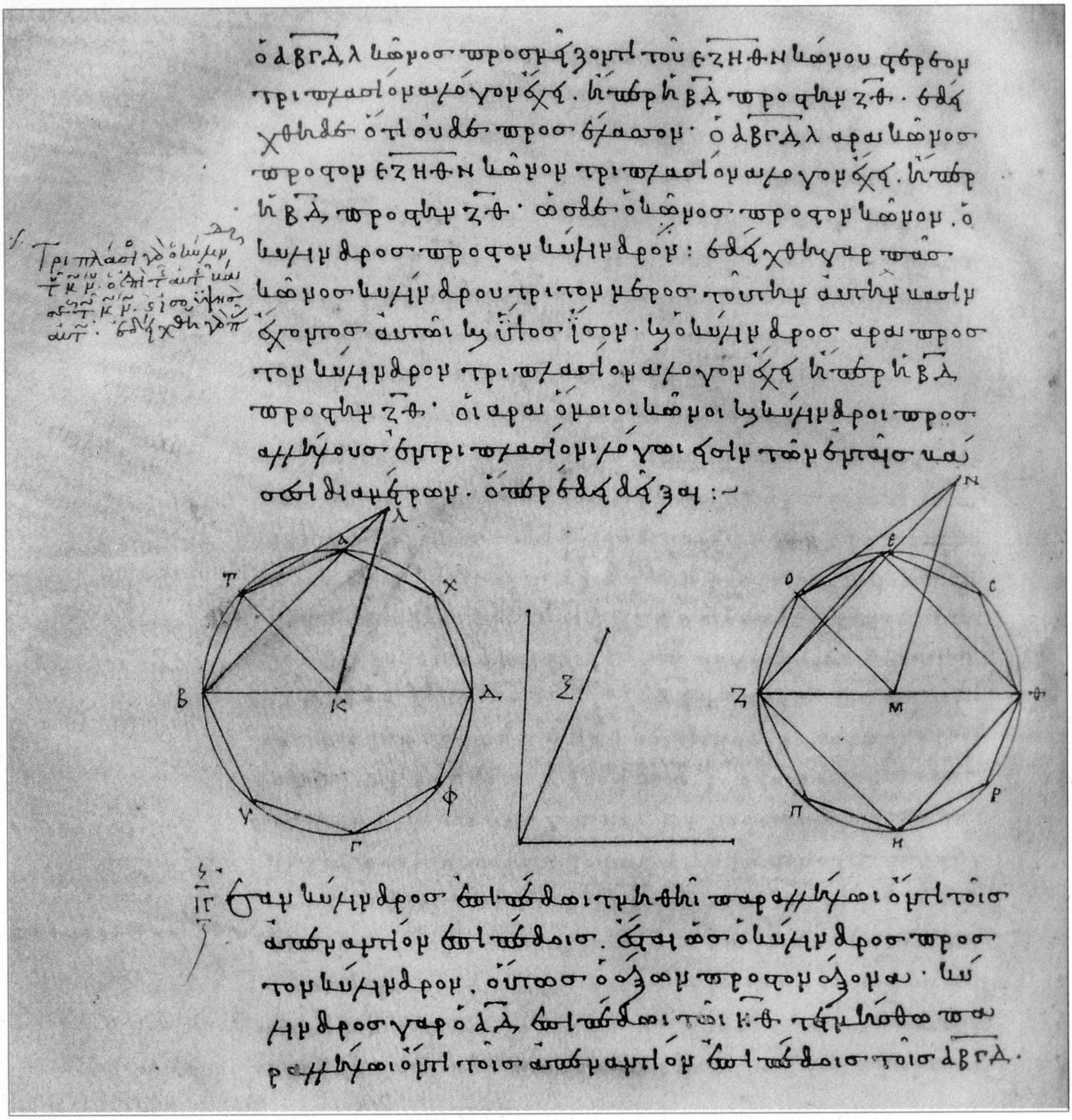

FIGURE 11.1   Facsimile of a page of the Bodleian Ms. of the *Elements;* copied A.D. 888.

CREATING TABLES

Tables are data cross-referencing devices. One set of data (such as units of time) is placed in columns; another set of data (such as sales quantities of different products) is arranged in rows that correspond to the units of time. To find the information for a given time unit, one simply locates that time unit. A table's use-

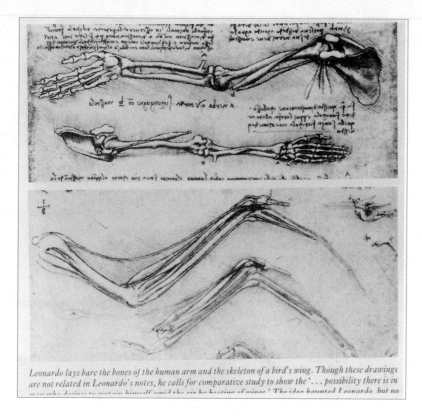

Leonardo lays bare the bones of the human arm and the skeleton of a bird's wing. Though these drawings are not related in Leonardo's notes, he calls for comparative study to show the '. . . possibility there is in man who desires to sustain himself amid the air by beating of wings.' The idea haunted Leonardo, but no

FIGURE 11.2   Examples of Leonardo da Vinci's many technical illustrations.

fulness lies in the *rapidity* with which the data are perceived and configured. To demonstrate this, read the following passage about the nutritional contents of a one-ounce bowl of shredded wheat:

> A 1 oz. bowl of shredded wheat, without milk, has 90 calories, and contains 3 grams of protein, 23 grams of carbohydrates, less than one gram of fat, 0 milligrams of cholesterol, 0 milligrams of sodium, and 120 milligrams of potassium.

> A 1 oz. bowl of shredded wheat, with one-half cup of skim milk, has 130 calories, and contains 7 grams of protein, 29 grams of carbohydrates, less than 1 gram of fat, 0 milligrams of cholesterol, 60 milligrams of sodium, and 320 milligrams of potassium.

Now study the same information in the table on top of p.222. In this table, the nutritional units are cross-referenced with units of the product's quantity: one ounce of the cereal by itself; and one ounce of the cereal plus one-half cup of skim milk. Instead of describing the data, a table *depicts* the data (which is why

Shredded Wheat Nutrition Information

	1 oz	with ½ cup skim milk
Calories	90	130
Protein	3 g	7 g
Carbohydrate	23 g	29 g
Fat	*	*
Cholesterol	0 mg	0 mg
Sodium	0 mg	60 mg
Potassium	120 mg	320 mg

*Contains less than 1 gram.

tables are considered visual aids), using lines and columns to create a cross-referencing system. One could say that the description is embedded in the arrangement itself.

Elaborate Tables

The preceding table showing the number of grams or milligrams of listed substances per ounce of shredded wheat constitutes a relatively simple table, so it would not be necessary to incorporate the data into a formal table grid; for more complex systems of data, however, such a layout should be used, as in Figure

	Low calcium diet (n = 6)		High calcium diet (n = 6)	
	Low sodium (supplement 22 mmol sodium/day)	High sodium (supplement 178 mmol sodium/day)	Low sodium (supplement 22 mmol sodium/day)	High sodium (supplement 178 mmol sodium/day)
Energy intake (MJ)*	13.7 ± 22.6	13.5 ± 2.8	13.0 ± 3.1	12.9 ± 2.9
Creatinine (mmol)	14.97 ± 1.52	15.08 ± 1.71	15.06 ± 2.75	14.78 ± 2.34
Sodium (mmol)†	8.19 ± 0.91	18.51 ± 1.98	8.53 ± 1.76	18.39 ± 3.17
Potassium (mmol)†¶	6.82 ± 1.10	7.47 ± 0.96	7.92 ± 1.09	7.95 ± 1.14
Calcium (mmol)§	0.250 ± 0.064	0.309 ± 0.138	0.281 ± 0.163	0.334 ± 0.161
Phosphorus (mmol)‖	2.31 ± 0.27	2.44 ± 1.56	2.44 ± 0.44	2.52 ± 0.28
Nitrogen (mmol)	6.41 ± 0.75	6.72 ± 0.80	6.95 ± 1.13	7.25 ± 1.29
Hydroxyproline (mg)	2.35 ± 0.26	2.44 ± 0.34	2.08 ± 0.32	2.08 ± 0.41
Blood pressure (mm Hg)				
Diastolic	67 ± 8	64 ± 9	62 ± 5	63 ± 5
Systolic**	116 ± 9	116 ± 7	110 ± 5	109 ± 8

[Amer. J. Clin. Nutr. 41, 1985, 56, © Am. J. Clin. Nutr., with permission]

FIGURE 11.3 Sample formal table grid with complex subdivisions and subheadings.

Type of test conducted	Test results					
	A	B	C	D	E	F
Type 1	xx	xx[1]	xx	xx	xx	xx
Type 2		yy	yy	yy	yy	yy
Type 3	zz	zz	zz[2]	zz	zz	zz

Stub — Row headings — Column headings — Column subheadings — Field

1 Footnote
2 Footnote
Source of Table: _____

FIGURE 11.4 Schematic grid for a formal table.

11.3. Notice the two section headings above the column headings and how the individual cells correspond to each subdivision marked by a column head.

Guidelines for Preparing a Table

1. Using a straightedge, draw the table grid, following the standard design shown in Figure 11.4.

2. Be sure that the column headings and the row headings are properly labeled, so that the reader immediately understands the nature of the data being tabulated. Study (A) and (B) versions of the table in Figure 11.5 to get an idea how improperly presented data can be difficult to perceive and, thus, to understand.

3. Align the data carefully; misaligned tabulations will be confusing (especially if decimals are used), and they will look amateurish as well.

4. Number and title each table. The title should be as brief and as informative as possible.

5. Introduce the table in the body of your document, preferably as close to the table as possible. Do not assume that a table (or any other visual aid) can serve as its own introduction.

6. Individual bits of data sometimes need to be annotated. Such notes must have corresponding superscript numbers or symbols within the table and

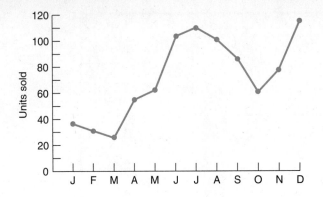

FIGURE 11.6 Line graph showing monthly camcorder sales for 1994.

as keys to the footnotes, which go directly underneath the table rather than appear with notes that are part of the text.

7. If you are using data from a published table, you must acknowledge that fact in a source note. The source line appears directly beneath the footnotes, if any.

CREATING FIGURES

Any visual aid that is not a table is designated as a "figure." These are customarily divided into graphs, charts, photographs, and diagrams.

Graphs

Graphs are used to depict trends—that is, relationships between variable events and the time during which these events occur. To depict this relationship between two variables, one needs to plot the occurrences on a coordinate grid,

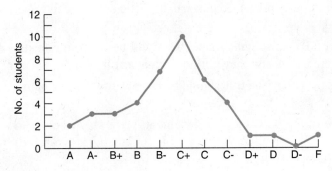

FIGURE 11.7 Line graph showing grade distribution for a class of 40.

which consists of a vertical axis to represent the frequency or quantity of events, and a horizontal axis to represent the preestablished variables, such as time. If you are interested in how many camcorders your retail electronics store has sold during 1994, you divide your horizontal axis into twelve equal sections, each representing a month, and you divide your vertical axis into appropriate quantity units (say of 10), beginning with zero in the lower left corner. You then **plot** each occurrence—mark the spot on the graph that represents the quantity of camcorders sold for January, for February, and so on.

Line Graphs

A **line graph** reproduces to a precise degree the shape of a distribution of variables, such as the number of camcorders sold over a given period of time (Figure 11.6) or a criterion distribution (e.g., the "grade curve," Figure 11.7).

Commonly two, three, or more sets of variables are plotted on a single graph. If a company wishes to compare its sale of computer systems with its sale of peripheral computer hardware (printers, hard drives, etc.), it would ask for a **multiple-line graph,** such as those in Figures 11.8A, B. To give vivid emphasis to the differences in trends among the trends depicted, while at the same time depicting the portion that each trend contributes to the whole, one might construct an **area graph** as in Figure 11.9A. Some area graphs plot the relative parameters of data subsets (Figure 11.9B).

Sometimes when large quantities of data need to be plotted in a line graph, it isn't possible to represent an accurate trend. In this case a **scatter graph** is created, wherein the individual bits of data are marked at their proper coordinates, leaving a "trend" subject to the viewer's own perception (Figure 11.10). A trend line, representing the *average* trend of the plotted data, can be superimposed upon the data field, however.

Bar Graphs

Like their line-graph counterparts, bar graphs depict comparative trends and distributions, and they can compare one set of distributed components with another. However, bar graphs place more emphasis on the individual, isolated units: a bar for January camcorder sales, another bar for February sales, and so on.

The most common types of bar graphs include the **simple** (or **single**) bar graph, which depicts a trend or distribution of one unit of data—for example, the monthly fluctuations in interest rate on commercial mortgages (Figure

Quantity × 10 lbs	Dry Chemicals produced			
	1990	1991	1992	1993
100	magnesium		magnesium	potassium sulfate
200	potassium sulfate	magnesium / potassium sulfate	potassium sulfate	magnesium
300		calcium carbonate	calcium carbonate	
400	calcium carbonate			calcium carbonate

(A)

Chemical	Quantity produced × 10 lbs			
	1990	1991	1992	1993
Potassium sulfate	200	200	200	100
Calcium Carbonate	400	300	300	400
Magnesium	100	200	100	200

(B)

FIGURE 11.5 (A) Improperly prepared table followed by (B) corrected table. The stub should contain the variable items for which information is sought.

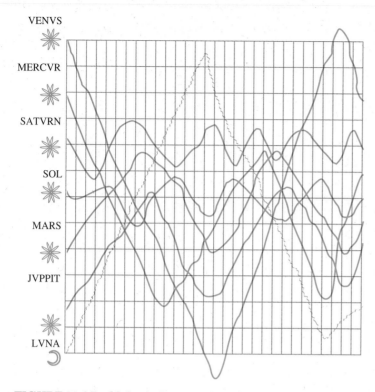

FIGURE 11.8A Multiple-line graph from the eleventh century (earliest known).

11.11), or the most frequently performed surgical procedures in a given year (Figure 11.12); the **multiple** bar graph, for comparing two or more units of data (Figure 11.13); the **one hundred percent** (or **divided**) bar graph, for breaking down a grand total into percentage units (Figure 11.14); and the **deviation** bar graph, for depicting positive (above-zero) and negative (below-zero) units (Figure 11.15).

Using one of today's versatile computer-graphics programs, one can design rather stunning, multicolored, three-dimensional graphs and charts. Such visuals can add excitement to a technical presentation that is top-heavy with facts and figures. Audiences will leave such a presentation with the images (that embody the facts and figures) firmly in mind.

Guidelines for Preparing a Graph

Although there are many types of graphs, they all share some common structural principles, reflected in the following guidelines:

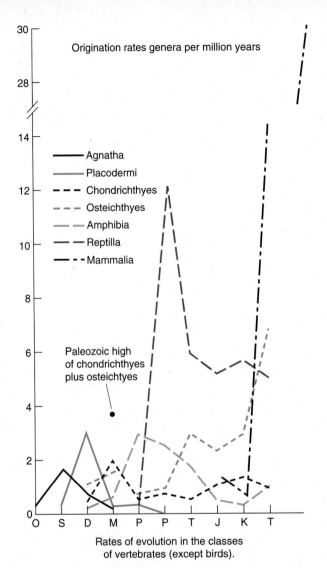

FIGURE 11.8B
Modern multiple-line graph. Note the interrupted scale at the top.

Origination rates genera per million years

Agnatha
Placodermi
Chondrichthyes
Osteichthyes
Amphibia
Reptilla
Mammalia

Paleozoic high
of chondrichthyes
plus osteichtyes

Rates of evolution in the classes
of vertebrates (except birds).

1. Use graph paper to prepare your graphs.

2. Establish fixed or predetermined variables, such as time periods or units that are being subject to analysis, on the horizontal axis; plot collected data, such as quantities or frequencies of occurrence, on the vertical axis.

3. Avoid distortion of representation by keeping the intervals between quantities on the vertical axis properly spaced so that large changes do not seem smaller than they are, or that small changes do not seem larger than they are (see Figure 11.16).

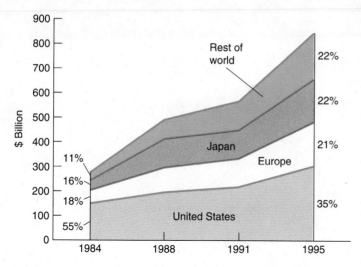

FIGURE 11.9A Area graph emphasizing comparative differences in circuit engineering trends.

4. When preparing multiple-line graphs or multiple-bar graphs, use distinctly different and conspicuous colors (e.g., red, green, blue) to represent each line or each set of bars corresponding to one element. If your graph is monochromatic, use conspicuously different designs instead of, say, dashes of different lengths or bars of different degrees of shading.

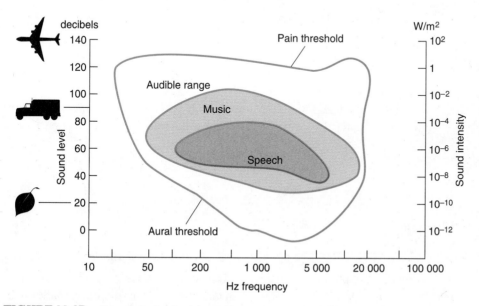

FIGURE 11.9B Area graph plotting auditory thresholds. Note the use of icons on the vertical axis.

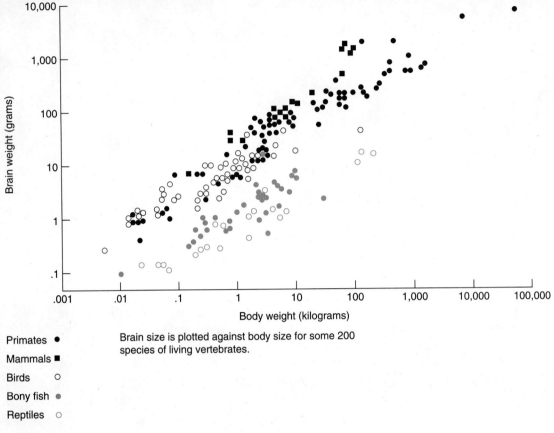

Primates ●

Mammals ■

Birds ○

Bony fish ●

Reptiles ○

Brain size is plotted against body size for some 200 species of living vertebrates.

FIGURE 11.10 Example of a scatter graph.

5. Indicate each data point on the graph with a solid mark. Connect each data point with carefully drawn straight lines that intersect the points perfectly.

6. Make sure each axis is labeled in a sufficiently informative manner as close to the outer axis line as possible and aligned with it.

Charts

Charts show distributions, and interconnections, either static (as in the case of an organizational, tree, or pie chart) or dynamic (as in the case of a flowchart). Unlike graphs, charts do not present data as sets of coordinate variables, but rather as proportional, hierarchical, spatial, or functional relationships.

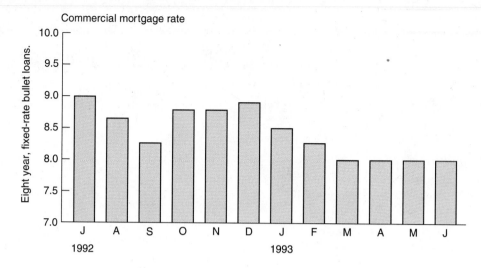

Commercial mortgage rate

FIGURE 11.11   Simple bar graph.

Pie Chart

A pie chart divides any whole (represented by a circle) into percentage segments (or slices, to maintain the pie metaphor*). They are commonly used to
show the relative breakdown of physical or economic entities (for example, the
chemical elements that make up the earth's crust, or the breakdown of expenditures in a household budget. Pie charts work best when not too many segments
are involved; otherwise, the segments become too hard to perceive clearly
(Figure 11.18A).

   Pie charts with more than three or four segments may also be a problem if
they are reproduced monochromatically; the different segment designs or degrees of shading may not be distinct enough for easy reference. One alternative
in such a case is to create an exploded or three-dimensional pie chart, as is
shown in Figure 11.18B.

Guidelines for Preparing a Pie Chart

1. Avoid gratuitous pie charting. Anything divided into thirds, for example,
   need not be represented by a chart. Use charts to depict uneven divisions.

---

*The pie chart metaphor is often an occasion for humor, probably because this visual aid is sometimes used
in stereotyped ways, as in "Your tax dollar," recently spoofed in a *New Yorker* cartoon (Figure 11.17).

The 10 most frequently performed surgeries

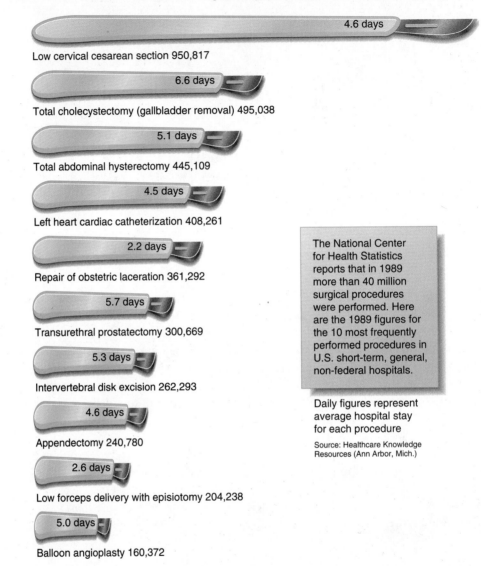

4.6 days

Low cervical cesarean section 950,817

6.6 days

Total cholecystectomy (gallbladder removal) 495,038

5.1 days

Total abdominal hysterectomy 445,109

4.5 days

Left heart cardiac catheterization 408,261

2.2 days

Repair of obstetric laceration 361,292

5.7 days

Transurethral prostatectomy 300,669

5.3 days

Intervertebral disk excision 262,293

4.6 days

Appendectomy 240,780

2.6 days

Low forceps delivery with episiotomy 204,238

5.0 days

Balloon angioplasty 160,372

The National Center for Health Statistics reports that in 1989 more than 40 million surgical procedures were performed. Here are the 1989 figures for the 10 most frequently performed procedures in U.S. short-term, general, non-federal hospitals.

Daily figures represent average hospital stay for each procedure

Source: Healthcare Knowledge Resources (Ann Arbor, Mich.)

FIGURE 11.12   Pictorial variation of a simple bar graph.

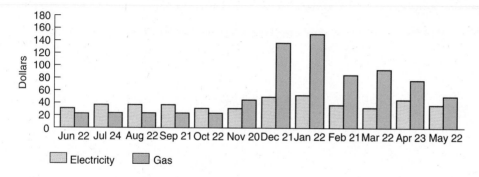

FIGURE 11.13   Multiple-bar graph showing comparative costs of gas and electricity for an individual household during 1992.

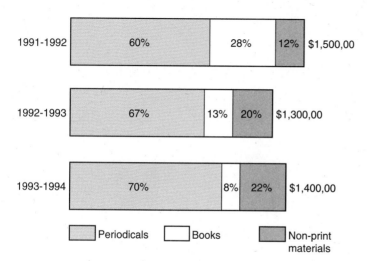

FIGURE 11.14   One hundred percent bar graphs depicting changes in library purchases.

On the other hand, charts with too many slices are difficult to represent and to perceive clearly.

2. Uneven portions (e.g., 3.13%) are difficult to depict clearly or accurately unless plotted on a computer (see Figure 11.19B). When designing the chart manually, either round off to the nearest half percent or use a different visual aid.

3. As with graphs, make the different segments as conspicuous as possible, using distinctly different colors or designs.

4. Group large segments with large segments; small with small.

5. Make certain the segments total 100%, or explain why they do not, as in Fig. 11.19B.

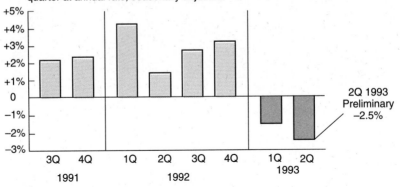

Productivity

Non-farm business productivity, percent change from previous quarter at annual rate, seasonally adjusted.

FIGURE 11.15 Deviation in bar graph.

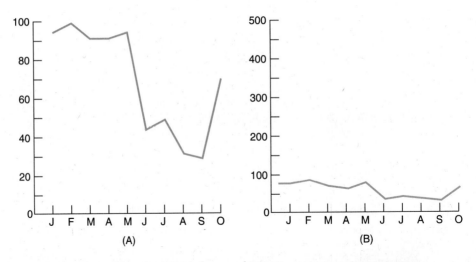

FIGURE 11.16 Distortion in graphs. Graph A shows a dealer's computer sales over ten months. Graph B shows the same results, except that the quantity of each unit has been increased fivefold, thus "flattening" the drop in sales.

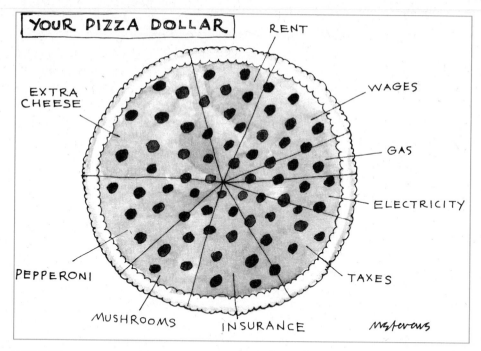

FIGURE 11.17 Cartoon satirizing gratuitous use of pie charts.

Drawing by M. Stevens; © 1992 The New Yorker Magazine, Inc.

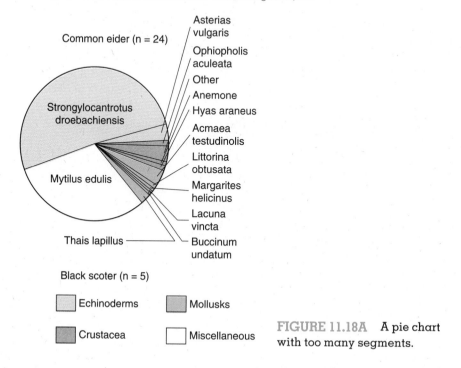

FIGURE 11.18A A pie chart with too many segments.

Organizational Chart

To represent the hierarchy of offices or departments in a given organization, be it a corporation, college, medical facility, or political entity, one would prepare an organizational chart (Figure 11.19). The purpose of such a scheme is to help readers perceive the hierarchical relationship of one part of the organization to all the other parts. Examining such a chart can serve as a guide for helping employees decide which office should be consulted.

Tree Chart

A tree chart is ideal for representing the complex organization of a physical system, such as the "evolutionary tree" of primates shown in Figure 11.20; or of an intellectual system (such as that of the "sentence tree" in Figure 11.21). Each radiating branch in the tree is clearly tied to a larger branch, which in turn is connected to an even larger branch, and so on. In a single glance, readers can perceive hierarchical relationships in a given system, no matter how numerous or complex the branchings are.

Flowchart

In order to depict progression or movement from one place to another, such as the flow of electric current from generator to household appliance—or from one developmental state to another, such as the transformation of hydraulic energy into mechanical energy in, say, a water meter—one would design a flowchart. Flowcharts may be symbolic, consisting of mere boxes and arrows, or they may be pictorial, like David Macaulay's imaginative water meter flowchart in Figure 11.22. As in organizational and tree charts, it is important to see how each element of the system is related to the system as a whole.

When designing a computer program, writers commonly work out a type of flowchart known as an **algorithm chart** (Figure 11.23). Algorithm charts can be exceedingly useful in working out the "logic circuit" of a software program's often complex sets of variables.

Photographs

Although a photograph is the most realistic reproduction of an object, it is usually not the ideal choice of visual aid. With a photograph it is difficult to be selective about the visual information you wish to present to your audience. For example, if you wanted to introduce the electrical system of an automobile, you

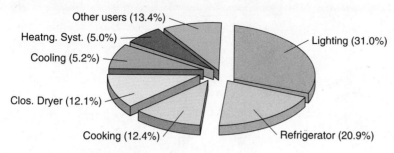

Annual Electricity Consumption

Other users (13.4%)

Heatng. Syst. (5.0%)

Cooling (5.2%)

Clos. Dryer (12.1%)

Cooking (12.4%)

Lighting (31.0%)

Refrigerator (20.9%)

Due to rounding, percentages may not add up to 100%

Source: Pacific Gas & Electric

FIGURE 11.18B Pie chart with easily perceivable segments. Note how "exploding" the segments makes them especially easy to see.

would most likely opt for a line drawing that *highlights* the system comprised of battery, distributor, spark plugs, starter, ignition, and headlights, rather than a photograph that shows everything under the hood, without any means of highlighting the system being discussed. On the other hand, a photograph rather than a diagram might work best for helping beginners locate a certain part under the hood.

Photographs are also necessary for presenting the accurate "real life" appearance of an object. Photographic reproduction of anatomical features—normal, abnormal, or diseased—is essential for medical texts. Geology and astronomy texts rely heavily on photographs for depicting terrestrial and extraterrestrial phenomena. In most sciences the study of minute or microscopic structures is essential; close-up photographs and photomicrographs, such as the one in Figure 11.24, are routinely used.

Guidelines for Creating Useful Photographs

1. If close-ups are needed, make sure you use the proper "macro" lenses and that the image is sufficiently magnified to reveal the details in question.

2. High-contrast photographs are necessary for machine or electrical parts; be sure such variables as lighting, film speed, aperture, f-stop, and shutter speed are used appropriately.

3. Remove all extraneous elements from the field, if possible; keep the image as simple as possible.

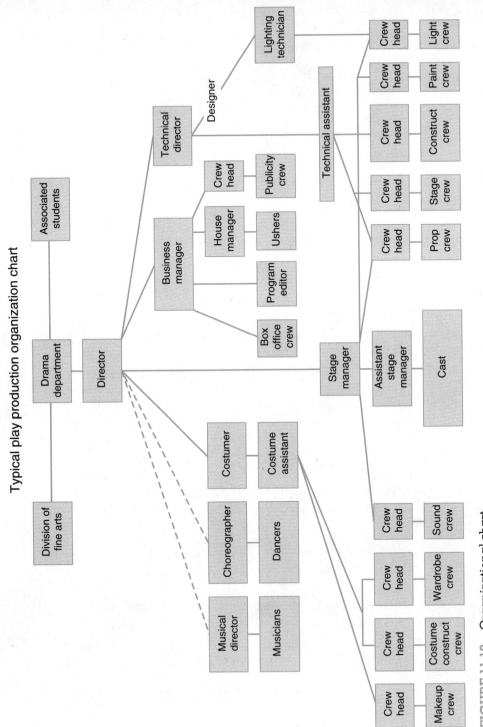

Typical play production organization chart

FIGURE 11.19 Organizational chart.

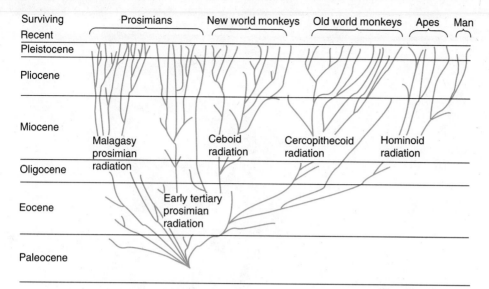

Surviving Recent

Prosimians | New world monkeys | Old world monkeys | Apes | Man

Pleistocene

Pliocene

Miocene

Malagasy prosimian radiation | Ceboid radiation | Cercopithecoid radiation | Hominoid radiation

Oligocene

Eocene

Early tertiary prosimian radiation

Paleocene

A representation of primate history and the origin of man as a series of adaptive radiations in space and time. (The selected lines of descent suggest a theoretical over-all pattern, but most lines are omitted and those shown are simplified and generalized.)

FIGURE 11.20   Tree chart depicting primate evolution.

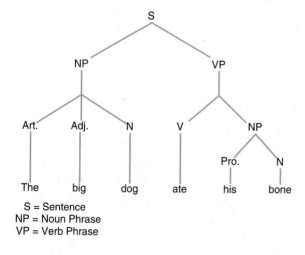

S = Sentence
NP = Noun Phrase
VP = Verb Phrase

FIGURE 11.21   Tree chart showing relationship among the grammatical elements of a sentence.

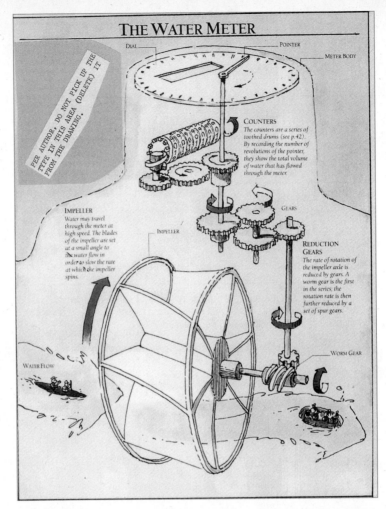

THE WATER METER

DIAL — POINTER — METER BODY

COUNTERS
The counters are a series of toothed drums (see p. 42). By recording the number of revolutions of the pointer, they show the total volume of water that has flowed through the meter.

IMPELLER
Water may travel through the meter at high speed. The blades of the impeller are set at a small angle to the water flow in order to slow the rate at which the impeller spins.

IMPELLER

GEARS

REDUCTION GEARS
The rate of rotation of the impeller axle is reduced by gears. A worm gear is the first in the series; the rotation rate is then further reduced by a set of spur gears.

WORM GEAR

WATER FLOW

PER AUTHOR: DO NOT PICK UP THE TYPE IN THIS AREA (DELETE) IT FROM THE DRAWING.

FIGURE 11.22   Flowchart using imaginative and realistic illustrations.

4. When planning a sequence of photographs, make sure the images form a coherent progression, each one corresponding exactly to its textual referent.

5. Superimpose any necessary markings such as arrows or circles on the photograph.

Diagrams

A diagram is a visual reconstruction, representational or abstract, of an object. Unlike a photograph, which usually reveals too much visual information about the object for the purpose, a diagram simplifies and highlights essential components. It is not unusual for a diagram to accompany a photograph of the same object; the photograph presents the holistic appearance of the object, while the diagram highlights important features that do not show up well on the photograph. It is often necessary to label the key features in a diagram, and, as with a photograph, to provide a concise explanatory caption underneath.

There are three categories of diagrams—representational, cutaway/exploded, and schematic ("block").

Representational Diagrams

A diagram that resembles the actual appearance of the object in question is a representational diagram, or drawing. Such a drawing will approximate the object's appearance enough to highlight key features, as in Figure 11.25. Maps, another type of representational diagram, approximate the actual topography or distributional features of a region just enough to emphasize the structures or features the writer-illustrator wishes to call attention to, such as earthquake occurrences along fault lines on a map of the San Francisco Bay Area (Figure 11.26).

Cutaway/Exploded Diagrams

These are in one sense varieties of representational diagrams, except that they depict what is not immediately available to the eye: the internal anatomy of an

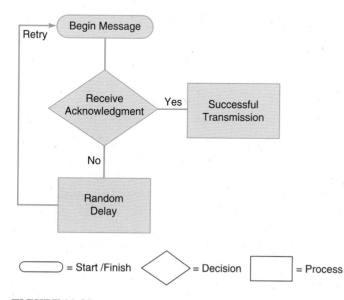

FIGURE 11.23   Simple algorithm chart depicting a communication link.

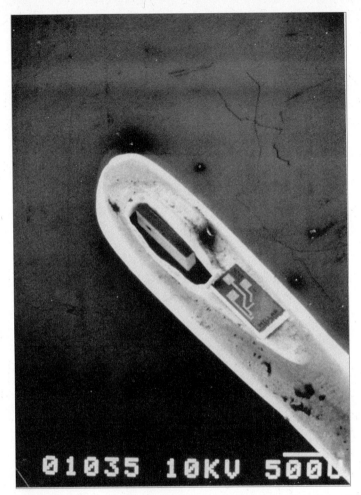

FIGURE 11.24 Photomicrograph of a pressure sensor smaller than a needle's eye.

object. Cutaways can vary from simple, often stylized representations (Figure 11.27) to complex, high-precision representations (Figure 11.28). Exploded diagrams, such as the one in Figure 11.29, exaggerate the spaces between parts so that the reader can better see and understand how the parts are assembled.

Creating representational and cutaway/exploded diagrams requires the use of precision drawing tools or, better yet, computer-aided drawing or drafting (CAD) programs, which range in level of difficulty from "startup" to engineering-graphics programs for professionals. The tiniest parts must be precisely rendered, their dimensions and positions relative to each other corresponding to their actual counterparts.

Schematic ("Block") Diagrams

In cases where the writer-illustrator wishes to highlight an object's function rather than its appearance, a schematic or block diagram is ideal. Here, no attempt is made to capture the external appearance of the structure—only the operational scheme is of concern. Schematics depict the function and layout of complex systems, such as the circuitry of an electrical device (Figure 11.30); they can also be used to demonstrate an interaction of elements, such as the interaction of air and airplane wing to create lift (Figure 11.31).

CREATING VISUALS: A CHECKLIST

- ☐ Is the type of visual I have selected the most appropriate for my purpose?
- ☐ Have I been careful to avoid distortion in my visual aid(s)?
- ☐ Have I made clear textual references to each visual aid?
- ☐ Does each visual aid I use actually contribute to my reader's understanding of the concept in question?
- ☐ Have I labeled all the necessary elements in my photographs or diagrams, and provided captions where necessary?
- ☐ Are my photographs or diagrams free of distracting elements?
- ☐ Did I remember to place any footnotes related to the figure or table directly underneath the visual aid itself?
- ☐ If the visual aid is not my own, did I remember to cite the source directly beneath the visual (beneath the footnotes, if any)?

■ CHAPTER SUMMARY

The most important principle for using visuals is that they should be regarded as integral parts of the document, rather than as supplements which can be pasted in or tagged on. Because technical writing is practical and demonstrative writing, visuals play an important role in translating abstract references into visible, tangible elements which reinforce comprehension and doability.

A great many types of visual aids exist. Tables are used to correlate one set of variables with another; for example, a table detailing the calories and nutrients (one set of variables) in various foods (another set of variables). Figures consist of graphs, for depicting the shapes of trends; charts, for depicting hierarchical structures or relationships among variables; photographs, for depicting

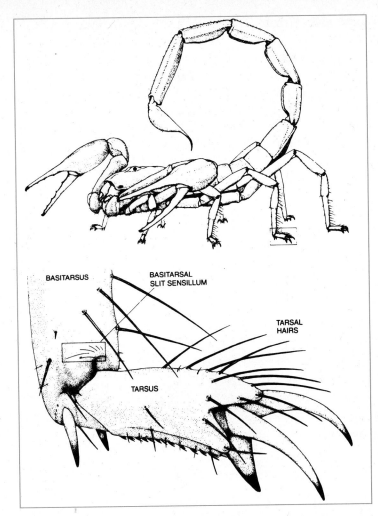

BASITARSUS

BASITARSAL
SLIT SENSILLUM

TARSAL
HAIRS

TARSUS

Illustrations by
Tom Prentiss,
from "Prey
Detection by the
Sand Scorpion,"
by Phillip H.
Brownell, copy-
right ©
December 1984
by Scientific
American, Inc.
All rights re-
served.

FIGURE 11.25 Representational drawings of a scorpion.
The aim here is accurate but selective representation.

realistic detail used in many sciences; and diagrams, for depicting the outer appearance, inner mechanism, or function of objects.

■ FOR DISCUSSION

1. Describe the difference between a table and a graph; a line graph and a bar graph; a graph and a chart; an organizational chart and a tree chart; a schematic diagram and an exploded diagram.
2. How is it possible to distort information in a bar graph? In a line graph? In a pie chart?

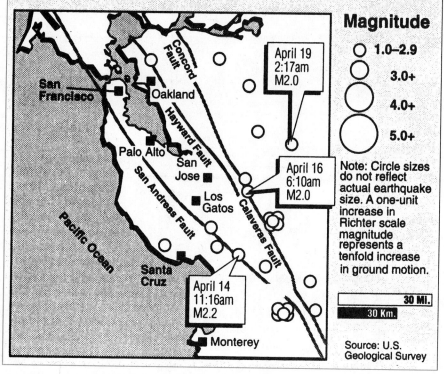

FOR THE SEVEN-DAY PERIOD ENDING MIDNIGHT, APRIL 20, 1994

Quake Watch: the week in review

It was a busier week in the Bay Area, with 25 earthquakes recorded. Three were as large as 2.0. For the daily USGS earthquake update, call (415) 329-4025 after 10 a.m.

April 19 2:17am M2.0

April 16 6:10am M2.0

April 14 11:16am M2.2

Magnitude

1.0–2.9
3.0+
4.0+
5.0+

Note: Circle sizes do not reflect actual earthquake size. A one-unit increase in Richter scale magnitude represents a tenfold increase in ground motion.

30 MI.
30 Km.

Source: U.S. Geological Survey

San Francisco
Oakland
Palo Alto
San Jose
Los Gatos
Santa Cruz
Monterey
Concord Fault
Hayward Fault
San Andreas Fault
Calaveras Fault
Pacific Ocean

FIGURE 11.26   Map depicting earthquake occurrences and magnitudes during a given week.

3. How do visual aids enhance comprehension of technical concepts? Refer to specific examples. Also give examples of ways that visuals may be used gratuitously.
4. Discuss ways in which a document with visual aids can be more persuasive than a document without them.

■ FOR YOUR NOTEBOOK

1. Begin collecting a variety of well-designed tables, graphs, charts, diagrams that you come across in periodicals, government documents, and

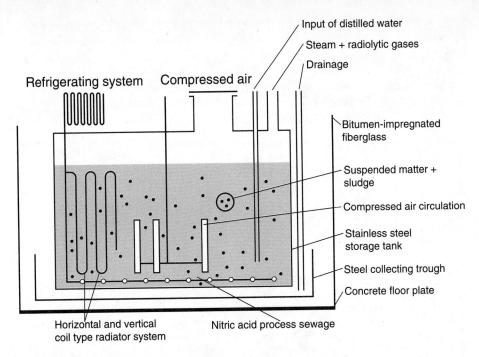

Refrigerating system   Compressed air

Input of distilled water

Steam + radiolytic gases

Drainage

Bitumen-impregnated fiberglass

Suspended matter + sludge

Compressed air circulation

Stainless steel storage tank

Steel collecting trough

Concrete floor plate

Horizontal and vertical coil type radiator system

Nitric acid process sewage

FIGURE 11.27   Simple cutaway diagram.

books. Copy these visuals into your notebook, and repeatedly examine them for their technique.

2. Use your notebook to experiment with two or more different kinds of visual aids for depicting a particular relationship. For example, construct a line graph, a bar graph, and a table to illustrate the change in your heart rate before, immediately after, and ten minutes after a 20-minute workout.

3. Practice drawing objects that require precise detail; for example, the circuitry of an electronic device; an exploded diagram of a telephone receiver.

◼ TECHNICAL ILLUSTRATION PROJECTS

1. Design an appropriate visual aid that shows the amount of rainfall in September, October, and November for any two cities in the United States. Use an almanac to locate the necessary rainfall statistics.

2. Create a flowchart to illustrate one of the following:
   a. The propagation of a radio signal from the source of transmission to the receiver, as follows: An oscillator produces a carrier wave which

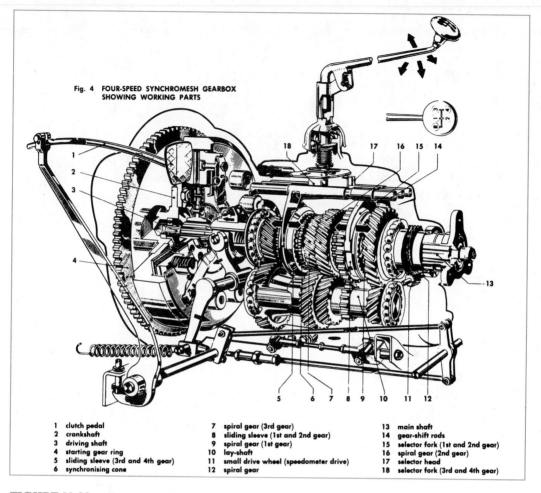

Fig. 4 FOUR-SPEED SYNCHROMESH GEARBOX
SHOWING WORKING PARTS

1	clutch pedal	7	spiral gear (3rd gear)	13	main shaft
2	crankshaft	8	sliding sleeve (1st and 2nd gear)	14	gear-shift rods
3	driving shaft	9	spiral gear (1st gear)	15	selector fork (1st and 2nd gear)
4	starting gear ring	10	lay-shaft	16	spiral gear (2nd gear)
5	sliding sleeve (3rd and 4th gear)	11	small drive wheel (speedometer drive)	17	selector head
6	synchronising cone	12	spiral gear	18	selector fork (3rd and 4th gear)

FIGURE 11.28   Complex cutaway diagram of a four-speed transmission.

"carries" the sound signal voices and music from a broadcast to an amplifier, then to a transmitter, and finally to individual receivers which separate the sound signal from the carrier signal. The sound signal is amplified and then broadcast by the loudspeakers.

b. Your own process of writing a paper, from initial idea to final draft.

c. The interaction, in a computer system, of input, processor, output, computer program, and memory.

d. The process of electricity generation and transmission: **Generation.** Coal is delivered to the plant where it is crushed and burned; steam created by the heat from the burning coal as it passes over water pipes drives turbines that generate 11,00 volts of electricity.

Lawn Edger and Trimmer Parts List

Order parts from McLane by specifying Model and Part Number

Ref. No.	Part No.	Description
1	2001	Handle
2	1011	*Pin, cotter, 1/8 × 1/2 (2 used)
3	1008	*Nut Lock hex 7/16" 14 (4 used)
4	2004	Spring
5	1007	Washer, plastic
6	2006	Bolt, hex 7/16" 14 × 2"
7	2007	Rod, Clutch and depth
8	1009	Grip handle, used (3)
9	2009	Lever, clutch and depth
10	1013	Lever, remote throttle control
11	2011	Clip, throttle cable
12	1014	Cross brace for handle
13	1015	*Nut 1/4-28 hex (10 used)
14	1018	*Bolt 1/4-28 hex × 7/8 (6 used)
15	2015	Frame, right hand
16	2016-6	Wheel, 6" (4 used)
16	2016-7	Wheel, 7" (4 used)
16	2016-8	Wheel, 8" (4 used)
17	2017	Engine
18	1044	*Bolt, 5/16"-24 × 1-1/2" (4 used)
19	1074	*Bolt, hex 5/16"-24 × 78" (5 used)
20	2020	*Nut, hex 7/16-14 (3 used)
21	2021-6, 7	Axle Front
21	2021-8	Axle Front
22	2022-6	Spacer Wheel
22	2022-7	Spacer Wheel
22	2022-8	Spacer Wheel
23	1075	*Nut Hex 5/16-24 (9)
24	2024-6, 7	Frame Crossmember (2 used)
24	2024-8	Frame Crossmember (2 used)
25	2025-6, 7	Axle, rear
25	2025-8	Axle, rear
26	2026	Bracket, engine pulley guard
27	2027	Guard, engine pulley, back belt
28	2028	*Screw, #10 × 1/2 sheet metal (2 used)
29	2029-6, 7	Frame, left hand
29	2029-8	Frame, left hand
30	1063	Pulley, Engine
31	1065	Set-screw 5/16"-24 × 5/16"
32	1064	Key, 3/16 × 3/16 × 5/8
33	2033	Pulley and shaft, cutterhead
34	2034	*Screw, 1/4-20 × 1/2 hex
35	2035	Guard, cutterhead front belt
36	2036	Bearing, cutterhead (2)
37	2037	Mandrel, head
40	2040	Clamp, blade guard
41	2041	*Bolt, 1/4-28 × 1-1/2 hex (2 used)
42	2042	Guard, main
44	2044	Washer, friction (2 used)
45	2045	*Nut, 1/2-20
46	2046	Deflector, dirt and debris
47	2047	U-Clamp
49	2049	Bolt, pivot hardened and ground
50	1120	*Flat Washer 7/16
51	2051	Spring, head recoil
52	2052	Cup
53	2053	Body, cutterhead
54	2054	Lever, angle shifting
55	2055	Pin, 5/16 × 2
56	2056	Spring
57	2057	Brace, cutterhead
58	2058	Drive Belt
59	2059	Cutting Blade
60	2060	Blade Wrench (accessory not illustrated)
61	2061	1" Pin
62	2062	U Clamp Spacer
63	2063	Grease Fitting

REPLACEMENT BEARING

Part No.	Bearing No.	No. Used	Description
2009	1117	1	Lever, Clutch & Depth
2016-6	1117	8	Wheels
2016-7	1117	8	Wheels
2016-8	1117	8	Wheels
2053	1117	1	Body Cutter

BRIGGS & STRATTON

Part No.	Description
2017-2R	100-2R-6-7 Model 60102, Type 0226-01
2017-3RB	100-3RB-7 Model 80332, Type 0474-01
2017-3RP	800-3RP-8 Model 80202, Type 0301-2
2017-3RB	800-3RB-8 Model 80332, Type 0474-01

KOHLER

2017-4K	4K-7 Model K 91
2017-4K	4FK-7

HONDA

2017-3.5	100-3.5 R-7 3.5 HP Model G150

Lawn Edger and Trimmer Parts Illustration

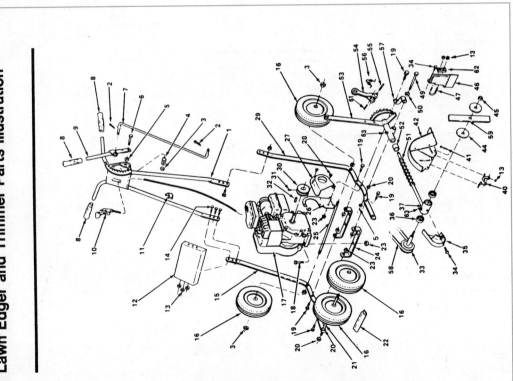

FIGURE 11.29   Exploded diagram with parts list.

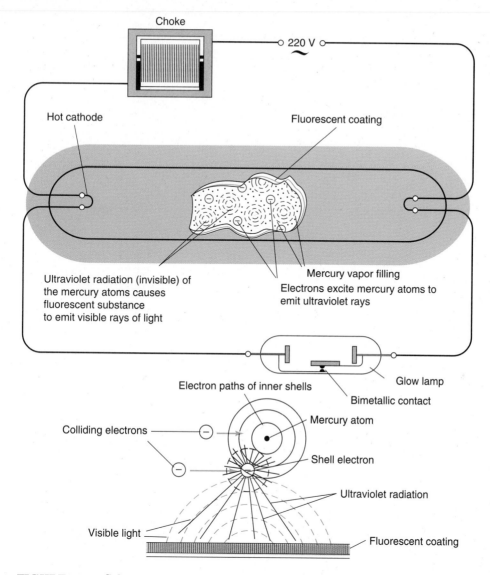

Choke

220 V

Hot cathode

Fluorescent coating

Ultraviolet radiation (invisible) of
the mercury atoms causes
fluorescent substance
to emit visible rays of light

Mercury vapor filling
Electrons excite mercury atoms to
emit ultraviolet rays

Glow lamp

Bimetallic contact

Electron paths of inner shells

Mercury atom

Colliding electrons

Shell electron

Ultraviolet radiation

Visible light

Fluorescent coating

FIGURE 11.30 Schematics used to depict the structure and function of a fluorescent lamp.

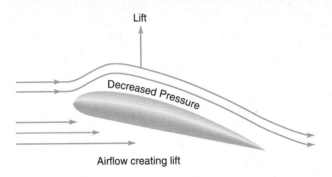

Lift

Decreased Pressure

Airflow creating lift

FIGURE 11.31   Simple schematic diagram depicting airflow over a wing.

Transmission. Transformers increase the voltage to 66,000 volts so that it can be transmitted. At the transmission destination, reverse transformers reduce the voltage to 22,000 volts. It is then transmitted to residential distribution areas where small transformers (placed atop telephone poles) further reduce voltage to 115–230 volts for individual residential customers.

3. Construct a line graph that depicts the changes in global population as follows:

Year (A.D.)	Population (in millions)
1	250
1000	275
1500	446
1600	486
1750	694
1800	919
1850	1091
1900	1571
1950	2495
1970	3632
1975	3967
2000 (est.)	6253

—Adapted from Dyno Lowenstein, *Graphics*. New York: Franklin Watts, 1976: 1.

4. Translate the information in the following table into a suitable graph:

Daily Caloric Consumption: Six Highest (A) and Six Lowest (B) Countries, 1988.

Rank	Country	Average Per Capita Daily Caloric Intake
	(A)	
1	Libya	3,812
2	Ireland	3,699
3	East Germany	3,689
4	Italy	3,688
5	Greece	3,668
6	United States	3,641
	(B)	
1	Chad	1,762
2	Maldives	1,765
3	Ghana	1,769
4	Uganda	1,784
5	Ethiopia	1,793
6	Bangladesh	1,837

Source: New Book of World Rankings, 1991. Reprinted in Donald G. Kaufman and Cecilia M. Franz, *Biosphere 2000: Protecting Our Global Environment.* New York: HarperCollins, 1993: 176.

5. Prepare an appropriate visual aid to represent each of the following sets of data:
   a. Municipal solid waste is comprised of paper (42%), organic waste (23%), glass (9%), metal (9%), plastic (7%), and miscellaneous items (10%).
   b. World oil, coal, and natural gas (ng) used between 1970 and 1990 included (in million-bbl equivalents): 48/oil, 32/coal, 15/ng in 1970; 55/oil, 29/coal, 16/ng in 1975; 66/oil, 34/coal, 17/ng in 1980; 54/oil, 32/coal, 18/ng in 1985; and 64/oil, 33/coal, 21/ng in 1990.
   c. Your annual college-related expenses, excluding clothing and entertainment.

6. Create a short procedural document in which each step is accompanied by a photograph or precision-detailed diagram or a combination of the two.

II

Writing in Work Environments

12

Corresponding with Memoranda and Letters

"The two workhorses of on-the-job writing, the memo and the letter, are the most frequent forms of written communication."

—Lorie Roth, "Education Makes a Difference: Results of a Survey of Writing on the Job." *Technical Communication Quarterly* 2 (Spr. 1993): 178.

Is it any wonder? Even small organizations often are scenes of frenzied activity, of many specialists interacting with other specialists to meet common goals: to provide services or products that reflect the successful integration of such diverse expertise. Letters and memos are effective ways of transmitting detailed information quickly—and in the case of electronic mail (e-mail), or facsimile mail (faxes), instantaneously—whether across the office or across the world.

KINDS OF CORRESPONDENCE-RELATED ACTIVITIES

Professional people, including technical professionals, often devote a substantial amount of time each day to the following activities, which involve the writing of letters or memos:

- Requesting information, or responding to questions about, a product or service.
- Calling and setting up agendas for meetings.
- Taking minutes for a meeting.
- Informing colleagues, associates, clients about new product developments.
- Proposing, or requesting proposals for, new company-related activities.
- Promoting a product or service or soliciting funds for a nonprofit organization.

Because it is not unusual to write several memos or letters in a single day, learning to write such documents with facility and skill is part of a professional person's basic training.

GENERAL CHARACTERISTICS OF MEMOS AND LETTERS

Correspondence refers to relatively short documents (usually one or two pages) directed to a single person or to a distinct group of persons (such as the members of a committee or research team). Their purposes may range from a two-sentence acknowledgement of receipt for goods to a multipage letter reporting work in progress or assessing the results of a test. Most memos and letters, however, do not exceed two pages.

But regardless of their purposes, memos and letters in the workplace follow established formats. It is important to know these formats because document appearance sends an immediate signal to the recipient: that the writer (and the organization he or she represents) is clearly an experienced professional.

The following distinction between a memorandum and a letter should be helpful: A memo tends to be impersonal, serving merely to report information (usually to several persons at once) in a fairly telegraphic manner. Memos typically convey reminders of meetings, deadlines, policy or procedure changes. Letters, on the other hand, can be formal or informal; but even when formal, they are not impersonal. Indeed, the personal note is struck with the salutation *Dear* so-and-so. No such salutation appears on a memo. No matter how formal the occasion for a letter may be, it communicates in a personable way: one individual addressing another. This is not to say that a letter cannot include reportage; it certainly can, but the reportage is framed within a personal context. Notice the difference between a memo reminding recipients about an upcoming meeting (Figure 12.1) and a letter conveying that same information from a personal context (Figure 12.2)

In the memo, the writer's principal aim is to convey a message as objectively as possible. Of course, a sense of urgency is implied, but no personal view is disclosed. To do so would not be in keeping with the purpose of a memorandum, not even when the writer is well acquainted with the recipients. The memo is also cordial ("Please note" instead of "Faculty are reminded that"). Objectivity does not mean that you have to sound like a robot.

Now compare the memo with the letter on the same topic, written to a member of the department who is on sabbatical leave. Note that the author has adopted a more personal tone in the letter and has shifted emphasis: he has taken the recipient's particular situation and needs into account.

Of course, letters often have to be written to strangers, but this does not warrant an impersonal tone. On the contrary, a personal tone becomes a key virtue when the writer is not acquainted with the recipient: it plays an important role in developing a healthy, trusting relationship between personnel and clientele.

GUIDELINES FOR WRITING A MEMORANDUM

To write a successful memo, you first need to have a clear idea of the exact message you need to send; you must then articulate that message as concisely as possible in a single line, known as the subject line (signaled by "RE," the

SANTA CLARA UNIVERSITY

Dept. of English Memorandum

May 1, 1993

TO: English Department Faculty

FROM: Fred White, Faculty Senate Council Representative

RE: University Library Systems Task-Force Planning Session

Please note that this important meeting takes place on Tues.,
May 11, at 4:00 in the Adobe Lodge. The meeting will give us
an opportunity to share our reactions to ideas about transforming
our university library from a traditional repository of books,
periodicals, documents and other materials, into an "electronic
information gateway." The more feedback the library task force
receives from concerned faculty and staff, the more informed
their final recommendations will be.

A wine-and-cheese reception will follow the meeting.

FIGURE 12.1   Departmental memorandum.

abbreviation for "regarding"), which appears underneath the DATE/TO/FROM headings:

November 12, 1993
TO: All drafting personnel
FROM: J. Oaks, Supervisor
RE:

Let's say you want to alert your technical staff to the latest upgrade of a CAD program. Your subject line would simply read:

RE: VISITECH 7.2 upgrade

Anything more specific would be unnecessary. However, anything less specific (e.g., "VISITECH" or "Upgrade") could cause momentary indecision about whether the memo warrants attention at all. Keep in mind that in the workplace time pressure and heavy workloads often mitigate against memos being read at all, *unless* the subject line grabs attention at once.

To begin the text of the memo, you simply write out a fuller version of the subject line—something like this:

This is to inform you that the latest VISITECH upgrade (7.2) has arrived . . .

The only thing left is to provide any elaborations of the message, such as what your staff must do to receive their upgrade:

To obtain your upgrade diskette, simply turn in your present diskette and sign the release form in my office.

Any concluding remarks? Perhaps you want to emphasize the importance of taking care of this business quickly:

Please get your upgrade a.s.a.p.; the new version includes modulating capabilities that are necessary for the next project we'll be working on.

Memos can be written quickly and efficiently when you think of this four-step process:

1. Formulate the key message as a kind of topic sentence, and present its subject on the subject line.
2. Begin the text of the memo with an elaboration of the subject line.
3. Add any *necessary* elaborations and qualifications to the message.
4. Proofread it carefully, then sign your name (or initials) beside your printed name alongside the FROM line.

SANTA CLARA UNIVERSITY

Department of English
Santa Clara, CA 95053

May 1, 1993

Professor Jonathan Smith
P.O. Box 0001
Santa Clara, CA 95050

Dear Jon:

   I hear that you've been making dramatic progress on your book during your released time from teaching this quarter. I know how valuable several uninterrupted weeks away from academic duties can be for such intense work. Certainly, I would not consider interrupting you now, except that the upcoming meeting of the Library Systems Task Force (see memo enclosed) is one I'm sure you would not want to miss, given your strong views about preserving our library's traditional identity while at the same time upgrading it to include state-of-the-art information systems. I hope you can attend, because the library task force needs the kind of feedback you can provide.

   By the way, a wine-and-cheese reception will immediately follow the meeting.

Best wishes,

Fred White

encl.

FIGURE 12.2   Business letter.

GUIDELINES FOR WRITING A FORMAL LETTER

Formal letters follow a standard format, although several minor variations exist. The following list covers the basic elements, proceeding from top to bottom of the page (Figure 12.3; see also Figure 12.2):

1. *Letterhead.* Formal letters should appear on quality stationery with the institution letterhead at the top. The individual need only type his or her address at the top (centered or flush right) if it differs from or is not included in the letterhead.

2. *Date.* The month should not be abbreviated or enumerated. Either the European style of dating (day month year) or the American style (month, day, year) is acceptable. Note that with the European style, commas are not used at all.

3. *Inside Address.* Place the recipient's name and address 2 spaces underneath the date. This will be identical to the name and address placed on the mailing envelope. NOTE: If at all possible, address the letter to an individual, not "To whom it may concern" or "Dear Sir or Madam." A depersonalized letter, remember, is a contradiction in terms. All you need to do is telephone the company or check The National Business Directory to find out the name of the appropriate recipient.

4. *Salutation.* In formal business letters, one always uses the salutation Dear + title (Mr., Ms., Dr., Reverend, Professor, President, etc.) + the recipient's surname. Only if the writer is well acquainted with the recipient is it proper to use the recipient's first name.

5. *Text of the Letter.*

   Opening paragraph: Unlike memos, letters do not need to get down to business immediately; they often begin with a greeting of some sort, roughly equivalent to small talk when beginning a conversation with someone. Such openings usually include an expression of thanks for a previous letter and a response (or reference) to the contents of that letter.

   Main paragraph(s): Here is where the key information is stated and elaborated. These paragraphs are equivalent to the "body" or "discussion" section of a critical paper or report.

   Concluding paragraph: Conclusions briefly summarize the points made in the main paragraph(s), and they end with an expression of appreciation for the opportunity to do business with the recipient and a hope that the professional relationship will continue.

6. *Complimentary Close.* Usually place this about 2 spaces below the last sentence of the concluding paragraph. Use one of the several conventional wordings, such

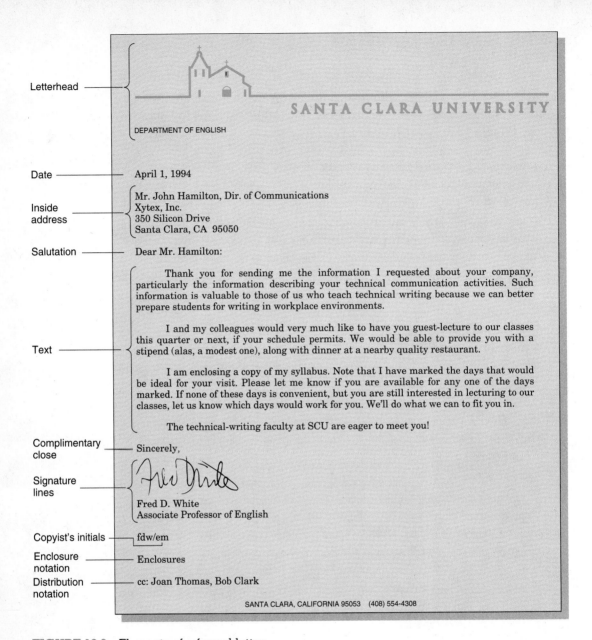

Letterhead

SANTA CLARA UNIVERSITY

DEPARTMENT OF ENGLISH

Date

April 1, 1994

Inside address

Mr. John Hamilton, Dir. of Communications
Xytex, Inc.
350 Silicon Drive
Santa Clara, CA 95050

Salutation

Dear Mr. Hamilton:

Text

Thank you for sending me the information I requested about your company, particularly the information describing your technical communication activities. Such information is valuable to those of us who teach technical writing because we can better prepare students for writing in workplace environments.

I and my colleagues would very much like to have you guest-lecture to our classes this quarter or next, if your schedule permits. We would be able to provide you with a stipend (alas, a modest one), along with dinner at a nearby quality restaurant.

I am enclosing a copy of my syllabus. Note that I have marked the days that would be ideal for your visit. Please let me know if you are available for any one of the days marked. If none of these days is convenient, but you are still interested in lecturing to our classes, let us know which days would work for you. We'll do what we can to fit you in.

The technical-writing faculty at SCU are eager to meet you!

Complimentary close

Sincerely,

Signature lines

Fred D. White
Associate Professor of English

Copyist's initials

fdw/em

Enclosure notation

Enclosures

Distribution notation

cc: Joan Thomas, Bob Clark

SANTA CLARA, CALIFORNIA 95053 (408) 554-4308

FIGURE 12.3 Elements of a formal letter.

as Sincerely, Best wishes, Yours truly, Cordially, or Respectfully. Unconventional or overly personal closes (Ever yours, Fondly, With fondest wishes, etc.) are inappropriate for formal letters.

7. *Signature Lines.* Type your name 4 spaces below the complimentary close and sign (blue or black ink) directly above the typed name.

8. *Enclosure Notation* [enclosure, encl., encls]. Use only when one or more items are included with the letter.

9. *Copyist's Initials.* The person typing or word processing your letter (if not yourself) will place his or her lowercase initials to the right of your uppercase initials, divided by a slash.

10. *Distribution Notation* [cc: + name(s)]. Use when you are sending a copy of the letter to the individual(s) named.

11. *Postscript* [P.S.]. Postscripts should be used sparingly, because they tend to give a formal letter a touch of informality that may not be desired. If any substantive information has been inadvertently omitted from the text of the letter, it is far better to rewrite the letter than to present it as a P.S. The most appropriate use for a postscript is to comment on a matter that would seem to fit comfortably with the rest of the letter.

FORMAL LETTER-WRITING DICTION

For some reason, many people assume that one must sound as formalistic as possible in a formal letter; the result is verbosity and an impression of affectation or pretense. The following list includes some common examples of affected prose, along with their simpler equivalents.

Instead Of . . .	Write This:
It has come to my attention that	I have learned that
Please be advised that you are now on our mailing list	You are now on our mailing list
I would like to thank you for your letter of June 16	Thank you for your letter of June 16
Enclosed please find . . .	I have enclosed . . .
As per our telephone conversation	As I mentioned during our telephone conversation
We are in receipt of	We have received

Please accept my gratitude in advance for attending to my aforesaid request at your earliest possible convenience.	I will be grateful if you can act on my request as soon as possible.

COMMON TYPES OF FORMAL LETTERS

Most formal letters fall into one of the following categories:

- application (solicited; unsolicited)
- inquiry
- response to inquiry
- complaint
- response to complaint
- publicity and solicitation
- transmittal

Letter of Application (With Resume)

In today's highly competitive job market, first impressions matter. When your application is one among dozens—sometimes even hundreds—a poorly written letter of application could promptly place you in the "reject" pile, no matter how impressive your resume.

A letter of application is more than a mere "courtesy cover letter" to accompany your resume; it reveals something about your personality, your earnestness, and of course your command of language and clarity of thought. The ideal tone is one that is neither overeager nor cut-and-dried, but enthusiastic yet restrained, self-confident without being overconfident or lacking in confidence. That is, effective letters of application should convey the writer's interest in and enthusiasm for the position but place emphasis upon qualifications. Notice how this tone is managed in the following letters of application. The first letter (Figure 12.4) is an example of a *solicited* (i.e., requested) letter of application; that is, the writer, Ann Noguchi, is responding to a published job notice or to a personal invitation to submit an application. The second letter (Figure 12.5) is *unsolicited:* Ann has no way of knowing whether or not a job vacancy exists at the company to which she is writing.

You will also want to study the first draft of Ann's second letter, which—Ann quickly realized after she reread it—contained many of the errors beginning job seekers make.

1200 Lincoln St. #202
Santa Clara, CA 95050
June 4, 1993

Mr. Ryan Haskill, Personnel Manager
Siltek Corporation
1000 Valley Boulevard
San Jose, CA 95120

Dear Mr. Haskill:

   I enjoyed speaking with you over the phone last Tuesday, and appreciate
your invitation to submit my application for the position of technical editor
for your company publications.

   As you have requested, I am enclosing samples of my work along with
my resume. These features were written while I was an editorial assistant
and staff writer for *Silicon Valley News*. During my two years with this
magazine I learned a great deal about editing, and about writing for readers
with a particular interest in our South Bay industrial community.

   The writing and editing duties I would assume, as you briefly described
them to me over the phone, sound like the duties I am especially qualified to
handle. They are certainly the duties I would find professionally satisfying.

   I look forward to hearing from you and to learn more about Siltek. Do
not hesitate to contact me if you require additional information.

Sincerely,

Ann Noguchi

Ann Noguchi

enclosures

FIGURE 12.4   Letter of application (solicited).

Ann begins on a cordial note; at the same time that she is thanking Mr. Haskill for the opportunity to submit an application, she is also of course *reminding* him that he had invited her to do so. It is always a good idea not to assume that your initial contact will be remembered. Next she highlights her most relevant work experience. Even though her enclosed resume (see Figure 12.7) will include it, mentioning it in the application letter places important emphasis on what will surely be a key consideration. Finally, Ann expresses enthusiasm for the position.

Ann's unsolicited letter to Moreno Electronics, unlike her solicited letter to Siltek, gets down to business at once—appropriately so, because the recipient, Ms. Moreno, would be wondering, Who is this person and why is she writing to me? As in her solicited letter, Ann reflects confidence and enthusiasm for the kind of work she is seeking.

As with any kind of writing, the words and the structure often do not come out right the first time. Study the first draft of Ann's letter to Amelia Moreno and see if you can spot the problems before reading the commentary that follows (see Figure 12.6).

One can see quickly why Ann rejected this initial draft. First of all, the tone is excessively diffident, and that can project a lack of self-confidence. (Of course, sounding overconfident can project a different kind of negative impression, one of arrogance!) Second, Ann is calling attention to her concerns about her experience being inadequate (*Silicon Valley News* as an obscure publication), rather than to the fact that she *has* experience writing for a metropolitan newspaper.

Writing a Resume

A resume is easy to write badly. Resumes by inexperienced writers include either too much information, inappropriate information, or not enough information.

Remember that a resume is simply a succinct overview of your background, not an autobiography. Ideally, you will include—on a single page—the following information, in the following order, carefully formatted:

- Name, address, telephone.
- Education (all colleges attended, most recent first). Summarize all degrees received, major(s) and minor, key courses taken, major collegiate activities
- Work experience (current or most recent first). If you've worked many jobs, select only those most relevant to the position for which you're applying. Provide brief summaries of each job.

1200 Lincoln St. #202
Santa Clara, CA 95050
June 8, 1993

Ms. Amelia Moreno, President
Moreno Electronics, Inc.
Tempe, AZ 85282

Dear Ms. Moreno:

As a UC Santa Cruz graduate (B.S., 1991) having double-majored in Journalism and Technical Communication, and as one who has worked for two years as a full-time editorial assistant and staff writer for *Silicon Valley News* (1991–93), I am eager to advance my career. Your firm has always struck me as an ideal place to work, ever since I read about it in *Fortune* Magazine. Northern California may be the "high-tech" capital of the U.S., and offer me more opportunity, but Arizona is where I was born (Yuma), and where I would like to live.

If you currently have a vacancy for a writer-editor, or are anticipating such a vacancy in the near future, please consider me for the job (my resume is enclosed). Editorial work and technical communication are my professional goals, and I am eager to continue this satisfying work. Please contact me if you require additional information.

Sincerely,

Ann Noguchi

Ann Noguchi

enclosure

FIGURE 12.5   Letter of application (unsolicited).

Dear Ms. Moreno:

I hope you don't mind my writing you this unsolicited letter, but I am eagerly looking for work in the area of in-house publication editing and writing. I hold a B.S. in Journalism and Technical Communication from UC Santa Cruz and have worked for two years as an editorial assistant and staff writer for *Silicon Valley News*. Although this is an obscure publication and it was my first job out of college, I feel that I am now ready to work at a major company such as your own.

My resume is enclosed. Please let me know if you have an opening.

Sincerely,

Ann Noguchi

Ann Noguchi
enclosure

FIGURE 12.6   Rejected first draft of a letter of application.

- Personal background information (recreational activities, languages fluent in, whether willing to relocate, community service). NOTE: It is illegal for an employer to inquire about your age, religion, nationality, ethnicity, or marital status; you may, if you wish, disclose this information voluntarily.
- References (two or three names of persons outside your family who are willing to recommend you for the job). Be sure to ask them if they are indeed willing to do so before listing them. NOTE: You have the option of omitting the names of references. Simply write "References furnished upon request."

Let us now take a look at Ann Noguchi's resume (Figure 12.7), which she encloses with every letter of application she sends, whether solicited or unsolicited.

Letter of Inquiry

Obtaining information from a given source may involve no more than a quick telephone call, but there are occasions for which telephoning would be inappropriate. When in doubt, write. Also, enclosing a stamped, self-addressed envelope (S.A.S.E.) is a courtesy that will not go unappreciated, and it will likely yield a speedier response. (When querying companies outside the United States, you will have to enclose international reply coupons—available at any post office—with the return envelope, since U.S. postage cannot be used outside the country.)

In a letter of inquiry, explain your request as clearly and in as much detail as possible; provide a rationale for the request; and conclude by thanking the recipient for taking the time to respond to your request. See Figure 12.8.

Response to a Letter of Inquiry

As a professional, you will probably have as many occasions to respond to inquiries as to write them. The format here is rather routine: first you thank the sender for the inquiry; then you respond to the inquiry as directly as possible. If you are unable to assist the sender, suggest someone who can, if possible. Conclude by inviting the writer to contact your company again if the need should arise. See Figure 12.9.

Letter of Complaint

Why *write* a complaint? Wouldn't it be more effective to confront the manager in person and voice your problem? Not necessarily. In a letter, one can more

RESUME OF ANN OKURA NOGUCHI
1200 Lincoln St. #202
Santa Clara, CA 95050
(408) 988-8901

EDUCATION
1989–1993

University of California, Santa Cruz
Santa Cruz, CA
B.S. in Journalism and in Technical Communication
(Double Major), Cum Laude.
GPA: 3.5
Coursework in technical journalism, technical
communication systems, graphic design, newspaper
editing.
Managing Editor of *The Santa Cruzan* (campus
newspaper), 1991–1993.
Member of the Journalism Society, 1990–1993;
Treasurer, 1992–1993.

WORK EXPERIENCE
(OFF CAMPUS)
1992—

Silicon Valley News, San Jose, CA
Journalism intern.
Write news features on new computer and software
products on the market; assist with page design.

1989–1992

Copi-Pro Services, Inc., Santa Cruz, CA
Copy documents, send faxes, for business clientele
(part-time work during school year; full-time during
summer).

FOREIGN LANGUAGE

Fluent in French.

PERSONAL

Raised in St. Paul, MN. Also lived in Fargo, ND and
Chicago, IL. California resident since 1987.
Participated in UCSC's Summer Study Abroad Program,
having studied French History at the University of Paris
(Summer, 1990).
Recreational interests include sailing
and filmmaking.

REFERENCES

Available upon request.

FIGURE 12.7   Resume.

123 Adamson Place
New York, NY 10000

Sept. 30, 1993
Dr. Clarissa Harlowe, Director
Technical Communication Program
Santa Clara University
Santa Clara, CA 95053

Dear Dr. Harlowe:

I am doing research for a term paper on possible ways of making
the metric system better known to the general public in the U.S.
My professor, Dr. Whetstone, recommended that I contact you for
information on this topic.

Here are the questions I would like your opinion on:
1. Should the metric system be taught separately, or in connection
   with our existing system of measurements?
2. What kinds of teaching tools would be most effective in teaching
   the metrical system?
3. Is the metric system being taught at SCU, outside of science classes?
   In what contexts?

I will appreciate any information you can give me on these matters. Of
course, I will properly cite you, as well as acknowledge your assistance, in
my paper.

Sincerely,

Kenneth Stearns

FIGURE 12.8   Letter of inquiry.

American Heart Association
National Center
7320 Greenville Avenue
Dallas, Texas 75231

Oct. 15, 1993
Ms. Alice Jensen
P.O. Box 543
Santa Clara University
Santa Clara, CA 95053

Dear Ms. Jensen:

Thank you for requesting information about the relationship of salt
to heart disease. We are pleased to send you our booklet, *Salt, Sodium
and Blood Pressure: Piecing Together the Puzzle*, recently published
by the A.H.A., originally developed by the Chicago Heart Association,
one of our affiliates.

I call your attention to pp. 12–16 of the booklet, which compares
high-sodium and low-sodium diets.

I hope this information will help you with your research project, and
thank you for contacting the A.H.A. Please let us know if we can be of
further assistance.

Sincerely,

J. Smith

J. Smith, Assistant Director
A.H.A. Information Center

FIGURE 12.9   Response to a letter of inquiry.

fully describe the situation, more carefully persuade the reader why an injustice or oversight has occurred. In an in-person or telephone confrontation, one runs the risk of becoming emotional, or of forgetting to mention important points.

Good letters of complaint tend to exhibit the following characteristics. They

- introduce the nature of the complaint clearly, objectively, concisely;
- provide a clear and convincing justification for the complaint;
- are neither angry nor excessively deferential; rather, are cordial, firm;
- do not threaten, either with legal action or with a harsh tone, such as, "I will never do business with you again!"

Study the letter of complaint in Figure 12.10.

Response to a Letter of Complaint

If you find yourself having to respond to complaint letters, there are three important principles to keep in mind: (1) you must always be civil and businesslike, no matter how uncivil the complaint letter; (2) not all complaints are justified or compensable; and when they are not, offer some alternative; (3) when a complaint is justifiable and compensable, mere compensation by itself is not enough. Let's consider each principle in turn.

(1) As one who represents your company with every word you write, you cannot ignore or respond in a reactionary manner to a customer, regardless of how ill-mannered the customer might be. Instead, your tone must be one of understanding, patience, and willingness to make some kind of reparation. How would you respond to the following letter?

> To whom it may concern:
>
> I have had it with you people! Of the dozen audio tapes I've purchased from your store in the last four months, two had defective sound tracks, one got stuck in my tape-player, and one was completely blank! Each time I requested my money back, the clerk rudely told me the tapes were nonrefundable. Maybe you'd like it if I started telling everyone I see to do their business elsewhere.

One must not, of course, read this as a personal attack; the writer has no idea to whom he or she is writing.

(2) Instead of ignoring the letter or writing something equally nasty in return, you need to respond in a sympathetic manner. However, since you have

P.O. Box 1349
New York, NY 10007

November 1, 1993
Customer Service Dept.
Hermes Shoe Company
100 Mountain Ave.
Fairbanks, AK

Dear Sir or Madam:

I purchased a pair of your TRAK-STAR running shoes (size 9), which I ordered from a catalog last month and received three days ago. After trying them once, I was extremely disappointed. They pinched my feet terribly and caused blisters. Also, I discovered a wedge of plastic protruding inside the left shoe.

I have always been happy with Hermes Shoes in the past, but this time I am not. Please send me a full refund ($69.95). I am returning the shoes to you under separate cover. Could you also reimburse me for the postage?

Thank you very much.

Sincerely,

Darla Dempsey

Darla Dempsey

FIGURE 12.10   Letter of complaint.

no immediate way to verify the validity of the claims made in the complaint letter, you may decide to inform the writer how he or she can obtain a refund.

(3) You want to counter the writer's hostility with tangible proof of goodwill; you need to promote the writer's continued business; and yet, you cannot immediately send a large refund. How does the following response satisfy the three requirements?

Dear _____:

Thank you for your letter of [date]. We at the Record Barn are sorry to hear of your run of bad luck with defective audiotape purchases. Let me assure you that any time you are dissatisfied with a purchase from us for any reason, you may receive a refund as soon as you show us your receipt of purchase. I have already warned my staff that rudeness to any customer for whatever reason will not be tolerated.

Please accept the enclosed $20 certificate, good toward any purchase in our store. Again, we regret the unfortunate purchases you have made, and sincerely hope you will continue to do business with us.

Sincerely,

A little bit of tolerance and civility can go a long way.

Letter of Solicitation or Promotion

As you may have noticed, solicitations from nonprofit organizations and product or service promotions from businesses are almost always in letter format (see Figures 12.11 and 12.12). This makes sense because letters convey a friendly, personal dimension to the communication. The reader feels as if the author is addressing him or her individually, even when a certain amount of technical information is being explained.

When composing a letter of promotion or solicitation, keep the following principles in mind:

- Persuade with facts and figures rather than with "rhetoric."
- Keep the tone cordial, but do not pretend that you are acquainted with the recipient.
- Answer questions the recipient would want to have answered; e.g., what percentage of a donation would be tax deductible? Does the organization report on the way the funds are disbursed, and if so could the recipient obtain a copy of the report?

Memorial Sloan-Kettering Cancer Center
1275 York Avenue, New York, New York 10021

July 13, 1992

Mr Fred D White
6746 Cielito Way
San Jose, CA 95119-1738

Dear Mr White:

You may never need us...but right now we need you.

Here at Memorial Sloan-Kettering we are making significant advances in the fight against cancer. Our scientists have:

* Armed monoclonal antibodies with cancer-killing drugs to seek out and destroy cancer cells.

* Pioneered new methods of breast reconstruction that minimize disfigurement.

* Developed a new chemotherapy protocol called M-VAC that represents a significant advance in the treatment of bladder cancer.

* Launched a major new program to find more effective ways of treating ovarian cancer.

* Devised a new treatment for testicular cancer that improves the survival rate while preserving potency.

* Investigated the links between nutrition and cancer, especially the role of diet in preventing colorectal cancer.

* Developed a drug called gallium nitrate that counters cancer-related hypercalcemia, a potentially life-threatening excess of calcium in the blood that affects 10 to 20 percent of cancer patients.

That's why I am urgently seeking your help.

(over, please)

Memorial Sloan-Kettering Cancer Center
1275 York Avenue, New York, New York 10021

July 13, 1992

Mr Fred D White
6746 Cielito Way
San Jose, CA 95119-1738

Dear Mr White:

You may never need us...but right now we need you.

Here at Memorial Sloan-Kettering we are making significant advances in the fight against cancer. Our scientists have:

* Armed monoclonal antibodies with cancer-killing drugs to seek out and destroy cancer cells.

* Pioneered new methods of breast reconstruction that minimize disfigurement.

* Developed a new chemotherapy protocol called M-VAC that represents a significant advance in the treatment of bladder cancer.

* Launched a major new program to find more effective ways of treating ovarian cancer.

* Devised a new treatment for testicular cancer that improves the survival rate while preserving potency.

* Investigated the links between nutrition and cancer, especially the role of diet in preventing colorectal cancer.

* Developed a drug called gallium nitrate that counters cancer-related hypercalcemia, a potentially life-threatening excess of calcium in the blood that affects 10 to 20 percent of cancer patients.

That's why I am urgently seeking your help.

(over, please)

FIGURE 12.11   Letter of solicitation.

**SANTA TERESA
DENTAL CENTER**

Roger A. Foulk, D.D.S.

6950 Santa Teresa Blvd., Suite D • San Jose, CA 95119 • 225-4158

June 5, 1992

Dear Patients,

     Recent television reports on Prime Time Live and Street Stories about AIDS and the dental office have inundated our office with questions from patients, especially about our infection control procedures. While I disagree with the sensationalism and half-truths of these programs, this is an important topic organized dentistry has been side-stepping for too long.

     In our office we treat everyone as if they are HIV (Human Immunodeficiency Virus) carriers. Some may take offense to this, but it is the only way to be sure all patients and staff are protected. The following are among the many steps we take to insure your protection.

- All areas of the treatment rooms, from lights to chairs, buttons to work surfaces are double wiped with an approved disinfectant after each patient; or covered with plastic wrap which is discarded after each patient.

- All plastic parts and instruments are thrown away after each use, **never** disinfected and **never** used again.

- All metal instruments are cleaned in a high-powered ultrasonic bath. We then use a high pressure and heat autoclave to sterilize, which kills 100 % of all disease producing organisms. The instruments are kept in the sterilized autoclave bags so that no hands touch them until ready to use.

- Hand pieces are autoclaved after each patient. The cups for cleaning and polishing (prophy cups) are plastic, and are disposed of after each use. Our ultrasonic scalers for cleanings are heat sterilized and again kept in the sterile bag until use.

- We have all been vaccinated against the Hepatitis B virus (HBV), which is a condition of employment. I recently tested negative for HIV. It is important to note that the above precautions are much more important to prevention of transmission than a negative result in a HIV test.

     We thank you all for your trust. Long term patients know we have been on the leading edge of dentistry which includes infection control. If you or any of your friends would like to see our procedures, please call us. We have all taken continuing education courses in order to keep abreast of the latest recommendations.

     Excellent, comfortable dentistry can be performed in a virtually risk free manner. We are committed to improving the safety and quality of our service each and every day.

Very Truly Yours,

Roger A. Foulk, D.D.S.
Steven A. Rose, D.D.S.
Donald M. Foulk, D.D.S.
and Staff

FIGURE 12.12   Letter of promotion.

Letter of Transmittal

Formal proposals and reports should always be accompanied by a letter of transmittal (known in more informal contexts as a cover letter). Such letters are more than a mere courtesy or formality; they remind the recipient of the fact that the proposal or report had been ordered; they describe its contents in a sentence or two; and they acknowledge those persons who have assisted in the preparation of the report.

    For discussion of letters of transmittal for particular kinds of reports, see Chapter 13 and Appendix C.

USING ELECTRONIC MAIL
(E-MAIL) AND FAXES

Because it is necessary to communicate with colleagues separated by large distances—and because conventional mail communication, by computer-technology standards, is impractically slow, even considering overnight delivery services—nothing developed to date can exceed the transmission speed of an e-mail message.

An electronic mail message is composed on the computer, using an e-mail program, and then transmitted to the recipient's computer (wherever it is, anywhere in the world) via modem. To receive and send e-mail messages, you need an account, an e-mail address (both yours and your recipient's), and a secret password so no one but yourself can read your mail. Technically, there are no limits to the length of an e-mail message, but the shorter your message, the better. Also, the system should be used discreetly; it is easy to become flooded with e-mail messages if you give your address, let's say, to a national organization with a penchant for networking with members.

Once you have set up your e-mail system, you will know when incoming mail arrives because your computer will flash a message or your telephone message light will blink.

Electronic mail is always in memorandum format because that is how the system is set up. All you need to do is concentrate on the message. Also, if you have trouble proofreading directly on the screen, you will want to take extra time to examine the text of the document before dispatching it.

A particularly versatile feature of e-mail is that messages may be sent to several persons at once (for example, to members of a committee), using a group name or a "cc" prompt. Such a feature saves time, duplicating costs, and paper.

Finally, electronic mail is sometimes informal, quite like a telephone conversation. It is done rapidly, spontaneously; and although it is possible to save e-mail messages (ones you send as well as ones you receive), they are usually discarded (sometimes a bit too hastily; always think twice before discarding a message!) Print it out if you're not sure. If more formality is required for a message, then the logical choice would be to use a fax machine.

A fax machine is essentially a photocopier that digitizes literally every bit of information on a sheet of paper, transmits it over the telephone lines, and reproduces an exact copy on your recipient's fax machine. A fax message, therefore, can be prepared with the same care you would give to a conventionally distributed technical document. To send a fax on your fax machine, you need to know the recipient's fax number, which is simply a telephone number (including area code).

DRAFTING MEMOS AND LETTERS: A CHECKLIST

☐ (Memo) Have I stated the purpose clearly and concisely, both in the subject line and in the opening paragraph?

☐ (Memo) Is my tone objective (without sounding disembodied)?

☐ (Letter) Have I avoided "letterese" (stilted, cliched phrases)?

☐ (Letter) Have I made the purpose of the letter clear after a cordial introduction?

☐ (Letter of Application) Does my letter reflect my enthusiasm for the target job?

☐ (Letter of Application) If the letter is unsolicited, have I included sufficient background information? Is my rationale for writing convincing? If the letter is solicited, have I reminded the recipient, in the opening paragraph, of earlier contact?

☐ (Letter of Complaint) Did I explain the problem objectively? Have I maintained a civil tone throughout?

☐ (Response to Letter of Complaint) Have I responded courteously to the recipient? If unable to satisfy the recipient's request, did I explain why and offer some alternative?

☐ (Letter of Promotion or Solicitation) Did I emphasize the factual elements in the promotion? Have I addressed the recipient directly?

■ CHAPTER SUMMARY

Correspondence—letters and memoranda—is the most common form of writing in the workplace. Memoranda are always formal, addressed to several persons at a time, and always include TO / FROM / SUBJECT lines at the top of the page. Memos are used for announcing meetings or new products, requesting or presenting information to employees or colleagues. Letters, on the other hand, are addressed to one person (or two or more persons who are closely associated) and are more personal in tone. Common types include letters of inquiry, letters of complaint or criticism, promotional letters, and job-application letters. The newest modes of correspondence, electronic mail (e-mail) and fax transmission, are the natural outcomes of our computer-based culture. Messages, documents, can now travel from one computer screen or fax machine to another anywhere in the world.

FONT BANK PRIVATE OFFER

250 FONTS JUST $49.95

100% COMPATIBLE WITH ALL YOUR MACINTOSH APPLICATIONS

TigerDirect 800 Douglas Entrance, Executive Tower, 7th Floor, Coral Gables, FL 33134 Call toll-free 800 666-2562 or fax 305 529-2990

Dear Mac Enthusiast:

In 1453, Johann Gutenberg and his associates changed civilization forever by publishing the Mainz Bible, the world's first printed book. It was a milestone event in the history of man; the ability to print books in large numbers. And it hinged on the concept of movable type.

Today, most of us have more sophisticated tools on our desktops than Gutenberg could have ever imagined. The power of **publishing** within our computers--with type, graphics, powerful software and high speed and resolution laser printers! Gutenberg would have spent months designing and printing a book---today, you can do it in a few minutes!

Here's an amazing offer on **250** display typefaces that Gutenberg would have loved! The entire premium-quality FontBank collection (compatible with all Mac software) is yours for just pennies per typeface! TigerDirect and the makers of FontBank combine to invite you to participate in this PRIVATE, INVITATION-ONLY OFFER!

<u>INTRODUCING FONTBANK: THE ULTIMATE TYPE LIBRARY: JUST $49.95!</u>

Now, you can own a complete designer library--250 foundry-quality typefaces--**for just $49.95!** And this is **not** a group of second-rate fonts. Each face is designed and drawn to "display quality"--then carefully digitized, ready for use in your Mac applications. Completely scalable, FontBank faces print with stunning, razor-sharpness at any size.

The finest display type library in the world--at just pennies per typeface!

<u>USE THESE TYPEFACES WITH ANY MAC PROGRAM</u>

The FontBank $49.95 offer is the perfect addition to your system--with your word processor, page layout program, spreadsheet, fax software and loads of other programs. Installation is streamlined; just a few minutes and you're ready to print!

You get screen fonts as well as printer fonts--PLUS a free bonus that I'll tell you more about in a moment...

next page please...

■ FOR DISCUSSION

1. What is the customary purpose of a memorandum? A formal letter? In what ways, if any, do the two types of correspondence overlap? In what ways are they most different?

2. Critique the unsolicited job-application letter on pages 281–282, pinpointing its strengths and weaknesses.

A TYPEFACE FOR ANY--AND EVERY--OCCASSION.

The first thing you'll notice about this remarkable type collection is its diversity. There are typefaces here for just about any mood or feel. We've even prepared a giant type chart and enclosed it in this mailer so you can see for yourself the enormous range of possibilities. Take a look and see for yourself just how extensive this library is...

250 HIGH-QUALITY TYPEFACES--PLUS A FREE GIFT!

When you order FontBank, Tiger includes a FREE copy of **BorderBank**, one of the handiest programs I've ever owned--what a difference a creative border can make to your page! This value-packed collection includes **100 ready-to-use borders** that you can use at any size. These borders were designed by desktop publishing guru **T. Craig Smith** and are compatible with all leading page layout programs.

BorderBank is a $149 value (and easily worth the price of this bundle alone)--but it's yours <u>with our compliments</u> when you participate in the FontBank PRIVATE OFFER. This offer is valid to addressee only, and is not transferrable.

THE BEST VALUE IN TYPE!

You simply cannot find a more value-packed bundle on quality, scalable type anywhere! The FontBank PRIVATE OFFER is the perfect way to start your library--or expand your existing portfolio of fonts for an unheard-of low price.

How long have you been running the same tired typefaces? Isn't it time to express yourself with words--and professional looking pages?

Since you've been selected to participate in this offer, I've reserved a copy of FontBank and BorderBank for you. Just **call us today at** (800) 666-2562 and take advantage of this exclusive offer...

Regards,

Carl Fiorentino

Carl Fiorentino   President, TigerSoftware

P.S.: Why not call **(800) 666-2562** or fax **(305) 529-2990** and order yours today? FontBank is the best value in type and ready to use with any Mac software in the popular PostScript format, the industry standard. TrueType version is also available, please specify when ordering.

[Unsolicited job-application letter]

125 Orchard Street
San Jose, CA 95129
(408) 233-0202
August 9, 1993

To whom it may concern:

My name is Elliott Stearns and I have just graduated from Gramercy College with a B.A. in Journalism. I heard that you are interested in hiring someone for an entry-level position in writing user manuals for your computer products. This sounds like the kind of job I would enjoy doing. While at Gramercy College I was an assistant sports editor and occasionally wrote articles. I also got high grades in all my English and Journalism courses. I am familiar with MS-DOS, Windows, PageMaker, and Lotus. Also, I once worked part-time at a software company editing manuscript copy for user manuals.

I am available for immediate hiring, and can come in for an interview at any time.

Sincerely yours,

Elliott Stearns

3. Study the promotional letter on pages 280–281. Why do you suppose the promoters have chosen a letter format over, say, a conventional advertisement, to publicize their services? To what degree does the letter follow the conventions of the formal business letter? To what degree does it differ from these conventions?

4. Critique the following letter of complaint:

October 7, 1993

Conrad Jones, Associate Chair
NCAA Committee

Dear Mr. Jones:

This past year, the number of games that NCAA teams were allowed to play decreased from 34 to 28 games. I wanted to express to you how I felt about this decision from an athlete's viewpoint.

The whole purpose of the competition, it seems to me, is to play as many games as possible. If you get rid of a certain number of games, that means that practice time is increased. When this happens, students are still not spending the time cut from games on schoolwork.

If you want to shorten the seasons so students will have more time to study, then don't you think it would be better to shorten the practice time, rather than the number of games allowed? Please take this under consideration.

Sincerely yours,

Anita Kreski
SCU Junior

■ FOR YOUR NOTEBOOK

1. Collect old memos or business letters you or your friends and relatives have received. What kinds of "letterese" expressions—clichés and stilted or verbose phrases—do you find?

2. Compare the opening and concluding paragraphs of several business letters. Take note of their strengths and weaknesses.

3. Practice writing job-application letters to different kinds of employers. Do you emphasize parts of your work experience or education differently with different recipients? Why?

■ LETTER AND MEMORANDUM WRITING PROJECTS

1. You are the C.E.O. of a software company that has just successfully produced and marketed a new computer game. Write a letter to your design department head, urging him or her to create two more games for two different kinds of users.

2. Drawing from your own experience, compose a letter of complaint. Some possibilities: poor service in a restaurant, barbershop, or on an airplane; purchasing a defective product; inadequate repair work on your car or bicycle.

    Exchange letters with someone in class. Assuming the role of complaint adjuster, respond as realistically as possible to the complaint you've received.

3. Write an accurate job resume, then write a letter of application, directed to a real or imaginary company you would enjoy working for. Next, review your resume in light of the application letter. What changes in emphasis, in either the resume or the letter or both, do you feel you need to make, and why? Now exchange resume and letter with those of a classmate. Each of you role-play the intended recipient. Give reasons for any changes you may suggest.

4. Write a 2-page promotional letter on behalf of a nonprofit organization of your choice.

5. Evidence has surfaced that some highly classified information in your company has leaked to the public, including competitors. Write a memorandum to all of the company's employees, announcing new security measures. Company identity, your role in it, and the nature of the classified information are up to you.

CHAPTER

13

Proposing a Project

The Purpose of Proposal Writing

Proposal Writing in Action

Types of Proposals and Their Audiences

Creating Proposals in an Organization

Planning the Proposal

Getting the Proposal Research Underway

Drafting the Proposal

Proposing a Project: A Checklist

A s a college student, you have written at least one proposal of sorts—and a successful one at that: your application for admission. As with other formal proposals, admission applications are competitive documents: you attempt, in this case, to assure the reader (the admissions officer) that you can meet the academic standards of the college in question. Given the fact that many (often the majority) of applicants cannot be accepted, you need to do more than express enthusiasm or flaunt your high school A's. In professional life, similarly, you will need to do more than what is expected—to convince the proposal solicitors, using sufficient evidence, that you can complete a proposed project more efficiently and economically than your competitors. Perfecting your skill at this kind of writing will prove to be rewarding in more ways than one!

THE PURPOSE OF PROPOSAL WRITING

Proposals are written for the purpose of acquiring administrative and financial support of the project being proposed. They can range from informal requests for travel expenses to formal point-by-point descriptions (with accompanying explanations, assessments, and quality-control monitoring systems) of a billion-dollar aerospace production project.[1]

Imagine that you are an environmental chemist, part of a team investigating the possibility of reclaiming a lake that has been badly polluted by manufacturers. Because you cannot undertake such an expensive and time-consuming project without outside funding, you and your staff are obliged to write up a *proposal* in which you . . .

- *State the problem:* The pollutants in Elk lake, which empties into a stream that feeds a major aquifer, are contaminating the drinking water of the Elk River community.

- *Explain the need (rationale) for solving the problem:* Unless work is done at once to eliminate the Elk Lake pollutants, serious health hazards, can break out in the Elk River community, and possibly in other communities downstream.

---

[1]For example, as this is being written (August 1993), Motorola Inc. has just awarded Lockheed Missiles & Space Co. in Sunnyvale, California, a $700 million contract to build 125 communications satellites for their Iridium Telecommunications System over the next ten years. The technology involved will revolutionize cellular telephone communications. Without a team of enterprising proposal writers at Lockheed, this contract would not have been landed.

- *State the means for solving the problem:* To determine the least expensive means of ridding Elk Lake of its pollutants, three water-decontamination methods will be analyzed and compared.
- *Describe the benefits resulting from the solution:* The former purity of the water supply will be virtually 100% restored.
- *Determine costs:* Estimate all costs for each decontamination method; select the most effective method at the lowest cost.
- *Provide a timetable for completion of the work:* It can only be an estimate, of course, but it should be logically determined, rather than simply guessed at haphazardly.

Proposals are, in effect, a recipe for the work to be done, and are sufficiently detailed to serve as a gauge for evaluating the final outcome. That is why contracts are awarded—or rejected—on the basis of proposals alone.

Not only is proposal writing a way of life in business, science, industry, and education; it shapes their very evolution. In any scientific discipline, for example, research proceeds only when proposals are accepted; research is stillborn when they are rejected. As rhetorician Greg Myers explains, "The writing of proposals, which takes up such a large proportion of the active researcher's time, is part of the consensus building essential to the development of scientific knowledge."[2] In other words, proposal writing deepens writers' understanding of the subject matter, better preparing them to overcome potential obstacles that interfere with attaining consensus with other scientists or engineers.

PROPOSAL WRITING IN ACTION

A proposal is a kind of script, one to which the proposer (analogous to the actor) as well as the sponsor (director) is able to refer whenever necessary. When complex concepts are written down, the likelihood of something being overlooked is greatly reduced. As with procedural documents (see Chapter 17), proposals are consulted frequently once the proposed project gets underway. That is why most proposals have a set format: it expedites the selection process.

Before any project can be funded, the sponsor has got to be convinced of its importance and potential for success. Proposals narrow the guesswork, the risk factors; they give proposers the opportunity to demonstrate on paper not only what needs to be done but how it will be done. If the proposal is in com-

---

[2]Greg Meyers, *Writing Biology: Texts in the Social Construction of Scientific Knowledge.* Madison: University of Wisconsin Press, 1990: 42.

petition with other proposals (and most are), then it will include reasons why its approach to the problem is the most sensible and cost-effective.

Even if the proposal in question is "internal" (targeted to an office within the proposer's own company), the proposer should still think in terms of competing—not with other proposals, but with other potential uses for the company's funds.

TYPES OF PROPOSALS AND THEIR AUDIENCES

As suggested above, proposals may be external or internal. External proposals are addressed to organizations that are independent from (outside of) the proposer's organization. Internal proposals, by contrast, are addressed to individuals, groups, or departments within a given organization. This is not always a clear-cut distinction, however. A large organization may have many subsidiaries or franchises, each involved in very different kinds of work. These subsidiaries may or may not be considered "external" to each other, depending upon the administrative infrastructure.

Also, proposals may differ in degree of formality. A professor of biochemistry, for example, writes a brief proposal in the form of a memorandum to his or her department chair, requesting additional funds to conduct a relatively inexpensive experiment to observe the effects of cell growth when subjected to ultraviolet radiation. Or, that same professor may wish to design a much more elaborate and expensive experiment, one that would be conducted in space during a space shuttle flight, thus requiring a more formal and elaborate external proposal for NASA.

CREATING PROPOSALS IN AN ORGANIZATION

Proposals for designing and manufacturing a product or rendering a service may be solicited or unsolicited. An established company whose products or services are valued—a company that possesses a strong reputation for delivering goods as promised—will likely be invited to submit proposals for additional projects desired by the potential client. The document drawn up for such an invitation is called a Request for Proposal (RFP). For example, NASA, pleased with the quality of the heat sensors they had requested an electronics company, Sensidyne, to design and build for them five years ago, now submits an RFP to

Sensidyne for an upgraded heat sensor—one that must be three times as sensitive as the previous model and be ready in two months. Sensidyne administrators immediately sit down with the design engineers and ask: Can we do it? If the answer is yes, a great many internal questions must be answered before Sensidyne commits to writing up a formal proposal for NASA:

- How much research and development must go into the project?
- Is the time frame realistic, considering the time needed for adequate testing?
- Can we price the new sensor competitively?
- Will we be able to guarantee the product unconditionally?

If the answer is yes to these questions, writers and project supervisors are brought in, and everyone begins brainstorming about what to include in the proposal—particularly, what to emphasize about Sensidyne's particular approach to the project that would most likely beat out the competition.

Participating in a Proposal Brainstorming Session

If you were part of a group involved with proposal brainstorming, what would you be listening for? What questions would you most likely want to ask? Here are some possibilities:

1. Do we have enough hard data about the proposed project to sound sufficiently knowledgeable and authoritative? If the answer is no . . .
2. Should we finance a feasibility study to acquire the hard data we need?
3. What should be the key selling features of our approach to the project?
4. How can we cut expenses to a bare minimum without compromising the integrity of the proposal?

PLANNING THE PROPOSAL

Because formal proposals are expensive to prepare and are fundamental to the reputation and future endeavors of the company, they must be carefully planned and organized. Although every company will develop its own proposal strategies, we can recognize the following general elements common to all of them:

1. *What* the problem is that the proposed project will attempt to solve.
2. The problem in context: *why* the problem needs to be solved.
3. A step-by-step overview of *how* the work of the project will be conducted, including descriptions of required equipment, materials, and personnel (including resumes of qualifications).
    a. [for external proposals] Special features of your particular approach to the work (How is your approach better than those of potential competitors?).
4. An estimate of *how much* the project will cost.
5. A list of *who* will be working on the project, with brief descriptions of each employee's qualifications.
6. A schedule showing *when* each stage of the project will be completed.

Let us now consider a case in point in which all of these elements come into play:

> Alan Medina, a recent graduate (B.S. in Psychology and Public Health) intent on pursuing a career in geriatric health care, is spending the summer working as a nurse's aide at Grandview Residence, a health-care facility for the aged. One of Alan's greatest concerns is the lack of what he considers to be sufficient activities for the 60 residents. The "activities room" consists of a TV, a few card tables, and these items:
>
> two decks of cards
>
> one checkers set
>
> one chess set
>
> one croquet set
>
> two shelves of books (mostly tattered, small-print paperbacks)
>
> The residents are bored and listless. They watch TV in the cramped room for an hour or so, until they doze off; only a few are able to play cards, checkers, or chess at a time; the croquet set is good only during daytime. Few of them bother to read because the selection of books is poor and the print too small to read with comfort.
>
> Alan asks the board of directors to consider a proposal for investigating the best means of improving the recreational opportunities at Grandview. They reluctantly agree but warn Alan that his proposal would be rejected unless he could convince them that the problem he perceives is warranted and urgent; that is, he would need to justify how the residents (and eventually the reputation of Grandview Residence itself) would benefit from the improvements.

GETTING THE PROPOSAL
RESEARCH UNDERWAY

Before beginning work on the proposal document itself, Alan decides to mingle with the residents and get a clearer sense of the recreational activities they would enjoy. As he converses with them, he takes copious notes. Alan realizes at once what an important step this is, because much of what they tell him he has not anticipated; for example, they voiced their discomfort with existing chairs, with wobbly tables, with too small a television screen, with the recreation room's poor acoustics, and with improper ventilation. Several of the people are animal lovers and would greatly enjoy the opportunity to have a cat or dog, at least for a short time. Other residents used to be active theatergoers, and they would love to attend a play or musical performance.

Next Alan assimilates these notes into a rough outline, tailoring the five basic elements of a proposal to the situation at hand:

1. *What is the problem?* The residents are bored and unhappy as a result of too-limited activities. Even though they are "old," they have a lot of interests and energies that are not being tapped.

2. *Why does the problem need to be solved now?* When an individual's physical or mental resources are not drawn upon, those resources tend to dissipate. That is what I see happening to the residents. Many have grown despondent and weaker because of a lack of activities. If the current situation is allowed to continue much longer, the residents' health will be seriously compromised.

3. *What is the purpose of my proposal?* To investigate the possibilities for improving the recreational facilities; and to recommend what I consider to be the best options.

4. *How would I carry out my investigation?* Contact those who are responsible for purchasing more recreational equipment and games; for bringing in performers (singers, theater and dance companies, magicians, comedians, etc.) to entertain the residents on a regular basis. Arrange to have humane-society volunteers bring dogs and cats to the residents. Inquire into the possibilities of launching a fund-raising campaign to acquire the necessary funds.

5. *How much time will I need?* How long do I need to allow in order to interview residents, contact resource persons, obtain cost estimates, organize resources and community support networking, prepare a final report on the project? How much time do I need to add to guard against unforeseen delays?

DRAFTING THE PROPOSAL

Once Alan feels confident that he has a clear (if tentative) approach to his proposed investigation, he now rolls up his sleeves and prepares a full-fledged draft of the proposal. He writes quickly, not pausing to worry about whether he is including everything. Two or three times he reviews his notes to remind himself of key concerns. After he finishes the draft, he reviews his original source material carefully for thoroughness, allowing himself at least a day to think over what else he could add, or delete as impractical or irrelevant. This proves an important step because he suddenly realizes that he needs to return to observe more carefully the reading habits of the residents. Finally, he completes the draft, rereads it and makes marginal notes, then shows it to two of his colleagues, who make several additional suggestions. He then revises the draft, incorporating several of the new suggestions.

Alan lets four of his associates (the two who read the first draft plus two others) read the revised draft and offer additional criticism. Once again, he heeds the suggestions he thinks are useful, ignores the ones that are not, then does a line-by-line scrutiny of the manuscript: Are the sentences clear and concise? Are the terms and computations accurate? Are the grammar, spelling, and mechanics correct? When he's satisfied that they are, he prepares a final printout and proofreads it. Alan's last two tasks are to write an abstract and a letter of transmittal.

The proposal that Alan submitted appears on pages 292–295.

PROPOSING A PROJECT: A CHECKLIST

- ☐ Have I researched the proposed project to allow for a factually sound proposal?
- ☐ Have I stated the problem and purpose clearly?
- ☐ Have I solicited feedback on my proposal plans from others?
- ☐ Are all relevant facts and figures included and documented?
- ☐ Have I worked out the budget carefully, striving to keep it as low as possible? Have I justified every proposed expenditure?
- ☐ Does the proposal rationale anticipate objections from opponents?
- ☐ Is my writing clear, concise, coherent?

■ CHAPTER SUMMARY

Proposals are reports of work intended to be conducted, submitted to requesters for the purpose of receiving sponsorship, which includes financial support nec-

A PROPOSAL TO IMPROVE RECREATIONAL FACILITIES
AT THE GRANDVIEW RESIDENCE FOR RETIRED CITIZENS,
GRANDVIEW, MICHIGAN

Prepared for the Board of Directors,
Grandview Residence
By
Alan Medina, Nurse's Aide

January 10, 1994

[Letter of Transmittal]

GRANDVIEW RESIDENCE

1055 Overlook Drive
Grandview, Michigan 48977

January 10, 1994

Gerald Larson, M.D., President
Grandview Residence

Dear Dr. Larson:

I am pleased to submit my proposal for conducting a study to improve the recreational opportunities at Grandview Residence.

As you know, the purpose of my proposed investigation is to find a way to enhance the quality of life for our residents. Lack of physical and mental stimulation can seriously undermine the general health and well-being of these senior citizens. By finding a cost-efficient means of enriching their lives during their residency here, our reputation as leaders in residential care for the aged will be assured.

I am grateful to my supervisor, Mary Ortiz, R.N., for her valuable assistance, and to you, Dr. Larson, for your support and encouragement.

Sincerely,

Alan Medina

Alan Medina
Nurse's Aide

ABSTRACT

Recreational opportunities for the senior citizens at Grandview Residence are inadequate. The majority are forced to spend much of their time being inactive: watching TV or napping excessively. Because there are clear connections between lack of activity and deterioration of mental and physical health among the aged, a feasibility study urgently needs to be conducted to determine the best strategy for improving recreational opportunities. Such a study can be conducted within a month and at a relatively low cost.

Purpose of this proposal

The senior citizens of Grandview Residence lack adequate recreational facilities. The purpose of this proposal is to request funding to conduct a formal feasibility study that would determine which recreational improvements would work best for our residents.

Nature of the problem

For a facility that is responsible for the welfare of sixty residents, most of whom are emotionally stable and intellectually alert, Grandview possesses seriously inadequate recreational opportunities. It is common knowledge that lack of mental stimulation often results in physical dysfunction. According to Mary Ortiz, R.N. (my supervisor), "Prolonged inactivity can aggravate existing cardiovascular and respiratory problems and lead to debilitating muscular atrophy." Lack of the kinds of stimulation that recreational activities afford can also lead to depression and diminishment of cognitive abilities.

This proposal will first describe the existing recreational opportunities at Grandview, then describe the kind of feasibility study that would determine how best to improve these opportunities. Finally, the cost of such an investigation will be discussed.

Existing Recreational Opportunities at Grandview

For our sixty residents, we have one television, two

decks of playing cards, one checkers set, one chess set, one croquet set, and about thirty paperback books (mostly novels) in tattered condition. Clearly, these materials are inadequate, even if we had larger quantities of them on hand.

What a formal feasibility study of the problem can accomplish

The goal of the study will be to take as much guess-work as possible out of determining which recreation-al activities our residents would truly find useful and stimulating. A formal study will involve inter-viewing the residents, discovering what recreational activities are available, visiting other facilities for the aged to observe their recreational facili-ties, and finally, acquiring some recreational equip-ment for a trial run.

Cost analysis

A feasibility study of the best ways to improve the recreational facilities at Grandview would incur the following costs:

- Assistant to help conduct survey to determine residents' preferred recreational activities.....................$ 50.
- Catalogues of recreational equipment and supplies...................................75.
- Local travel to other health-care facilities to discover how others have dealt with this problem.................50.
- Three assistants to supervise trial activities...................................150.
- Purchase of trial equipment.................200.
- Preparation of the feasibility report........50.

TOTAL...............$575.

Timeframe

Research.............3 weeks
Trial activities......1 week
Writing the report....2 weeks
TOTAL......6 weeks

essary for completing the project. They range from brief requests for work-related expenses to highly elaborate descriptions of industrial projects that require years to complete. A good proposal clearly states the problem to be investigated, the need or rationale for solving the problem, methods to be used for solving the problem, the benefits resulting from the proposed project, the costs involved, and a timetable for completing the work.

■ FOR DISCUSSION

1. Critique the draft of Alan Medina's proposal.
2. Study Carl Sagan's "The Way to Mars" Proposal, (Fig. 13.1) and evaluate its persuasiveness. Has Sagan sufficiently anticipated his opposition?
3. Critique the student proposal on pages 299–306 to examine the decline in sea otters:

■ FOR YOUR NOTEBOOK

1. Jot down ideas for campus-improvement projects over the next several days. Look for articles in the campus newspaper, take note of any campus problems you encounter directly, talk to faculty and staff. Here are some areas you might look into:
    food service
    textbook buyback policies
    class registration procedures
    availability of evening classes
    special-interest groups/clubs
    campus safety
    library noise policies
    computer or printer availability or reliability
    campus activities (quality/frequency of evening lectures, concerts, dances, films, excursions, etc.)
    parking
    building accessibilty for disabled students, faculty and staff
2. Write out a detailed proposal outline for some of the campus-improvement projects you have noted and would like to see carried out.

THE MARS DECLARATION

Mars is the world next door, the nearest planet on which human explorers could safely land. Although it is sometimes as warm as a New England October, Mars is a chilly place, so cold that some of its thin carbon dioxide atmosphere freezes out at the winter pole. There are pink skies, fields of boulders, sand dunes, vast extinct volcanos that dwarf anything on Earth, a great canyon that would cross most of the United States, sandstorms that sometimes reach half the speed of sound, strange bright and dark markings on the surface, hundreds of ancient river valleys, mountains shaped like pyramids and many other mysteries.

Mars is a storehouse of scientific information—important in its own right but also for the light it may cast on the origins of life and on safeguarding the environment of the Earth. If Mars once had abundant liquid water, what happened to it? How did a once Earthlike world become so parched, frigid and comparatively airless? Is there something important on Mars that we need to know about our own fragile world?

The prospect of human exploration of Mars is ecumenical—remarkable for the diversity of supporting opinion it embraces. It is being advocated on many grounds:

• As a potential scientific bonanza—for example, on climatic change, on the search for present or past life, on the understanding of enigmatic Martian landforms, and on the application of new knowledge to understanding our own planet

• As a means, through robotic precursor and support missions to Mars, of reviving a stagnant U.S. planetary program

• As providing a coherent focus and sense of purpose to a dispirited NASA for many future research and development activities on an appropriate timescale and with affordable costs

• As giving a crisp and unambiguous purpose to the U.S. space station—needed for in-orbit assembly of the interplanetary transfer vehicle or vehicles, and for study of long-duration life support for space travelers

• As the next great human adventure, able to excite and inspire people of all ages the world over

• As an aperture to enhanced national prestige and technological development

• As a realistic and possibly unique opportunity for the U.S. and the [former] USSR to work together in the spotlight of world public opinion, and with other nations, on behalf of the human species

• As a model and stimulant for mutually advantageous U.S./Russian cooperation here on Earth

• As a means for economic reconversion of the aerospace industry if and when massive reductions in strategic weapons—long promised by the United States and the [former] Soviet Union—are implemented

• As a worthy application of the traditional military virtues of organization and valor to great expeditions of discovery

• As a step toward the long-term objective of establishing humanity as a multi-planet species

• Or simply as the obvious response to a deeply felt perception of the future calling

Advances in technology now make feasible a systematic process of exploration and discovery on the planet Mars—beginning with robot roving vehicles and sample return missions and culminating in the first footfall of human beings on another planet. The cost would be no greater than that of a single major strategic weapons system, and if shared among two or more nations, the cost to each nation would be still less. No major additional technological advances seem to be required, and the step from today to the first landing of humans on Mars appears to be technologically easier than the step from President John F. Kennedy's announcement of the *Apollo* program on May 25, 1961, to the first landing of humans on the Moon on July 20, 1969.

We represent a wide diversity of backgrounds in the fields of sciences, technology, religion, the arts, politics and government. Few of us adhere to every one of the arguments listed above, but we share a common vision of Mars as a historic, constructive objective for the technological ambitions of the human species over the next few decades.

We endorse the goal of human exploration of Mars and urge that initial steps toward its implementation be taken throughout the world.

THE WAY TO MARS

By Carl Sagan

FIGURE 13.1   Proposal for pursuing the exploration of Mars.

■ PROPOSAL WRITING PROJECTS

1. Working with one or two other students, imagine yourself part of an athletic-equipment manufacturer's proposal writing team. You have received an RFP from a major-league baseball (or soccer or football) corporate office inviting you to propose quality uniforms (and protective gear) for all team members. Prepare a competitive proposal for them.

2. Write a proposal for a feasibility study or scientific investigation you wish to conduct for your final report. Include an abstract and letter of transmittal.

3. Write a proposal for a new class you would like to see added to the department of your major. Be sure to identify the problem(s) that not having the course is causing, and how the new course would solve those problems—not just for you but for other students in the major. You may also wish to consult the department chair on matters of cost.

4. Develop the outline you have written in For Your Notebook, topic 2, into a complete proposal.

5. Prepare a formal proposal, complete with abstract and letter of transmittal, based on Carl Sagan's "The Way to Mars."

WHERE HAVE ALL THE SEA OTTERS GONE?

A Proposal to Examine the Decline in Sea Otter Populations off the West Coast of the United States

Prepared for
The Monterey Bay Aquarium Sea Otter Rescue
and Care Program Staff and Members
Monterey, California

by
Tristen Moors
Marine Biology College Intern
Santa Clara University
Santa Clara, California

May 20, 1991

Project Summary

A recent sea otter census taken along the west coast of the United States indicates that the population has dropped slightly over the past five years. The sea otter population is currently not endangered but is listed on the critical list, so any decrease in population size is a cause for concern. Because the Monterey Bay Aquarium has been involved with the rescue and care of sick, injured, and orphaned otters for the last six years, this proposal recommends that research be conducted to determine the reason for the otter population decrease. It is hypothesized that many otters are dying as a direct result of harmful pollutants in the water. The study would attempt to prove this hypothesis by conducting thorough autopsies on dead otters and analyzing toxic chemical concentrations contained in their systems. By recording the general area where the dead otters are found and examining which chemicals may be behind their death, it will be possible to determine the source of the pollution. The Aquarium would then have a basis upon which to proceed for proposed cleanup measures in order to ensure the safety of both their own otters and the entire otter population.

Project Description

In 1985, the Monterey Bay Aquarium established a sea otter rescue and care program designed to rehabilitate unhealthy or injured sea otters to a point where they could be released back into the wild. Each visiting otter was examined by veterinary doctors throughout their stay at the Aquarium and after their release. Most documentation indicates that a majority of the otters brought to the Aquarium were young pups who had either been abandoned or separated from their mothers during periods of high surf. A few cases, however, showed evidence of illness which could have been blamed on a number of factors. As the number of otters brought to the Aquarium increases, the variety of illnesses and injuries increase as well. The otters that make it to the Aquarium, however, are only a small sample of the entire west coast operation.

I propose that your institution commission me to carry out a study to evaluate the feasibility of pinpointing the case for the overall otter depletion. I will investigate the equipment and methods needed for the study (although I believe they will be minimal) and report the feasibility of possible cleanup projects based upon the criteria of cost, overall timeframe, public interest, and otter benefits if pollution is indeed found to be the cause of the recent otter deaths. If another source is found to be the reason for the otter population decline, suggestions to combat the problem will be offered and feasibility reported in hopes of preventing further otter deaths in the future.

Rationale and Significance

On March 24, 1989, an oil tanker named <u>Valdez</u> spilled more than 10,000 gallons of oil into Alaskan waters. Because the spill was so large, cleanup crews and environmentalists had difficulty in containing the spilled oil. As a result, all types of marine wildlife were affected. Oil covered many birds' feathers, making it impossible for them to fly or dive for food; fish died by the thousands from their exposure to the toxins contained in the oil; and hundreds of otters froze because they were unable to clean their thick fur. Smaller subsequent oil spills have occurred since the <u>Exxon Valdez</u> accident, and many more animals and organisms have been affected. Because the otter population is relatively small, however, it is likely that they have suffered more than any other marine animal. Exxon continues to clean the Alaskan waters in hopes of ridding the entire area of oil, but the process is slow and sea otters continue to die.

The oil spill in Alaska explains the drop in otter count in the northern part of the west coast, but it does not give researchers any clues as to the reason for the decrease in otters along the central and southern west coast. Otters play a vital role in the ecosystem by controlling the sea urchin population. The sea otter must eat up to one-fourth of its own body weight in a single day, and it relies on

the sea urchin for its main source of food. Otters'
large appetites help to eliminate large numbers of
sea urchins, which are capable of reproducing incredibly fast. Without the predatory habits of the sea
otter, the sea urchin has no direct competitor and
the sea urchin population expands rapidly. Increased
numbers of sea urchins cause detrimental harm to the
coastal environment because the animals eat away at
the kelp which both protect the coastline and offer
small animals shelter from large predators. In the
1800s, sea otters were hunted almost to the point of
extinction. Their thick fur coats attracted fur trappers from all over the world and left the ocean void
of large otter populations. During that time, the sea
urchin population began to overgraze their algal food
sources and prevent kelp establishment. Without the
kelp, many fishes and other marine invertebrates had
no safe place to hide from larger fish and sharks.
With the loss of an important habitat, the entire
coastal ecosystem began to shift to accommodate large
marine animals but deplete the small invertebrates.
Although the larger animals flourished at first,
their numbers soon began to drop as well, due to the
lack of food available for them to eat. The removal
of one animal had caused an entire region to undergo
a dramatic change which was both harmful and life
threatening for many species.

   Once the otter problem became apparent, laws were
passed to protect the animals from fur trappers and
other hunters. As a result, the otter population has
grown significantly, but it is still not safe from
the threat of extinction. Humans are once again forcing the otters to utilize their survival skills
because of the pollution, toxic chemicals, and plastics dumped into their environment. The problems
affecting the otters have already been identified in
a general sense but not sufficiently studied in terms
of detailed information. Until the problems are examined fully, it will be difficult to remedy the situation and protect the otters from extinction. It has
been proven that the otter population is capable of
bouncing back after suffering severe loss if we allow
them the freedom and ability to do so. It is obvious

that preventative measures can be taken to provide otters with adequate living conditions, but it is up to all of us to take the time to study the otter problem before it is too late.

Plan of Work

The proposed study breaks down into the following steps:

1. Over the next year and hopefully continuing on to the rest of the decade, record and document every dead otter by presumed location of death, approximate age, sex, weight, and any other distinguishing characteristics such as:

   a. plastic six-pack can holder wrapped around mouth
   b. tar on fur
   c. bleeding limbs, possibly the result of an encounter with a boat
   d. drift or gill netting, covering the body
   e. fishing hooks or lines
   f. any other observable reasons for death

   Also, collect a small sample of water where the otter was found. Sometimes, otters may float for miles before washing up on the beach or being found by boaters, but it will be important to analyze the water so a correlation between otter death and water pollution can be made.

2. Perform autopsies, in the lab, on all otters and examine various chemical concentrations to determine probable and/or accessory causes of death. Perform water analysis on all collected samples.

3. Report all findings to the Monterey Bay Aquarium, where the information will be collected and analyzed. This will allow information from all marine wildlife agencies, such as Marine Mammal Fund, to be filtered to one location where the data can be analyzed on a wide scale.

4. Collate and analyze data, drawing conclusions as to probable cause(s) for the sea otter population. Suggestions will also be provided recommending ways in which the problem(s) can be solved. Prior reports on otter deaths have provided suffi-

cient information on individual cases but have not
addressed the entire population. The deaths may be a
result of regional disruption, such as caused by the
Alaskan oil spill; but looking at the overall effects
of the entire coastal population is important if we
are to understand how to protect the otters of all
areas. Conclusions drawn from regional studies can be
examined and analyzed so that "unaffected" areas
become aware of the problems they may face at some
point in the future.

Since the collection of dead otters and performance of autopsies are already being conducted, the
Aquarium will not need any additional equipment and
personnel to carry out these tasks. A simple telephone call to all marine wildlife agencies is all
that will be necessary to collect the findings. Data
can then be entered in a computer database program
at the Aquarium, where the results can be analyzed.
Water sample results could also be documented alongside each otter profile, making it easier to see if
the water environment contributed to the otter's
death. Again, the Aquarium would only be responsible
for gathering and collating the information, and correspondence will be the key to the operation. With
the labs' help and analyses, many conclusions could
be drawn as to why the otter population is decreasing on both a regional and coastal basis.

Because the Department of Fish and Game conducts
an otter census every year, the Aquarium would not
be responsible for counting the living otters. The
Aquarium, along with myself, would only be responsible for interpreting and evaluating received data
from dead otters. The Aquarium could potentially perform all of the above tasks on their own, but
because all of these duties are already being ful-
filled, it is not necessary for the Aquarium to take
them over. It is necessary, however, to use these
outside sources as a way to conduct research essential to the project. The project will be carried out
by several different agencies so it will not take
long to perform all of the necessary tasks. The real
problem involves waiting for the otters to die and
be collected. Overall otter count will be submitted

on a yearly basis, but regional observations about water contamination and its effects on the otters will be submitted every two months. The study should continue for more than a year to observe changes over a long period of time. Watching the situation closely will also help keep any discovered problems under control.

Facilities and Equipment

The expenses and facilities needed to conduct this research will be minimal for the Aquarium since most of the work will be done by labs outside of the Aquarium institution. Because most of the labs are governmentally funded, there may be no cost for the autopsies and water sample analyses. If, however, there will be a charge for the tests, it should be relatively inexpensive. I will simply need a computer (equipped with both database and word processing abilities) and telephone to gather all of the necessary information.

Sources

I plan to use many organizations in order to gather the information I need to draw some conclusions concerning the sea otter population decline. The U.S. Department of Fish and Game will be held responsible for not only yearly otter counts but also collection of otters and water samples. Many times, they are not the first to find the dead otters but are called to the scene to take care of the problem. They often take the otters to marine research labs such as Scripps in San Diego or Moss Landing near Monterey. Both of these institutions will be contacted on a regular basis to report their findings. Although these facilities will offer most of the information needed for this project, other nonprofit organizations, such as the Marine Mammal Rehabilitation Center in Marin, will be contacted so that they too are aware of the research being conducted. As secondary information sources, they will also contribute to the pool of data that will be necessary for the project's final prognosis. The Monterey Bay Aquarium's research specialists and library will serve as a final source of information and will play a large part in the analysis of the data itself.

Conclusion

The sea otters are in grave danger of extinction. They face many obstacles which are not yet fully understood but will be revealed in more detail with this study. Sea otters play an important part not only in the ocean ecosystem but also the land ecosystem. This study will provide necessary information which will help to save both marine and land environments from severe damage. Without this study, however, the sea otter's future, as well as our own, is uncertain.

Personal Qualifications

I am a junior biology major at Santa Clara University, where I carry a 3.8 grade point average (4.0 scale). Although my course work at Santa Clara is not marine based, I have a strong animal-biology background. I am familiar with sea otters and other marine mammals inhabiting Monterey Bay. In the summer of 1990, I worked for the Monterey Bay Aquarium as a tidepool diver and college intern, where I learned about the sea otters' population decline and the Aquarium's efforts to curtail the dropping otter numbers. As a college intern, I was often involved in the care and food collection trips for the resident otters. I also had a chance to interact with many wild otters which live in the kelp beds off the Aquarium. All of these experiences helped me to become better acquainted with the sea otters and the dangers they face both in the wild and in captivity.

14

Reporting Progress or Activity

J ust as a large project must first be proposed and accepted before it can be approved and undertaken, so must its progress be reported. Typically, a job will be planned in stages, and as each stage is completed, the project manager will prepare, usually on a regular (e.g., weekly, monthly, or quarterly) basis, a short report describing the progress that has been made. Some workplaces also require activity reports. Like progress reports, these describe work accomplished for a given period; but unlike progress reports, they include *everything* noteworthy (which could indeed include work on projects for which separate reports were filed) that was undertaken on the job over a given period.

This chapter describes both kinds of reports and presents guidelines for preparing them.

THE PURPOSE OF REPORTING PROGRESS

Reporting progress constitutes a kind of contractual relationship with the employer: the writer is explaining how he or she is allocating time for the assigned or contracted task in progress. This is important information for the employer, not simply because it reveals whether the allocated time is being used efficiently (thus helping to evaluate the employee's performance and to discourage unethical practices), but because it can be used as a basis for allocating similar projects in the future. And it is important for the writer to explain how the time is being spent because work on some projects may be more time-consuming than they seem. Administrators often have several projects going at once because they will interconnect at some higher level. A good example of this would be projects relating to a space-shuttle mission: separate endeavors such as deploying a satellite, conducting a biological experiment, testing the effects of certain exercise regimens in a zero-g environment—different from one another on one level, interconnected on another. Unless such projects are carefully coordinated and monitored, substantial additional expenses will certainly accrue. In an age of budget *cutbacks,* unforeseen expenditures could threaten the very existence of the project.

Clearly, it is very important to keep careful and accurate records of all activities related to a job in progress. Likelier than not, it will be relevant to the needs of those who will read your progress report.

You may recall from Chapter 2 the example of reporting progress on the testing of plant foods. The principal reader is the project supervisor for a commercial nursery, who needs to have detailed knowledge of projected versus actual expenditures for conducting the tests and for materials and equipment purchased. Any discrepancies must be accounted for so that the supervisor can prepare her own report to the owner of the nursery.

One segment of the experimenter's progress report might look like this:

Progress on the Testing of New Plant Food Products on Boston Ferns:

Time allotted: 7 days.

Time actually used: 9 days.

Reason for discrepancy: test ferns had been placed in the wrong size pots and had to be repotted. This required purchasing 25 new pots, costing $18.75.

Plant-food samples allotted: 10 lbs. each of Miracle Grow, Magic Formula, Sutter's Plant Food, and Fern Diet.

Plant food actually used: 15 lbs. of each brand.

Reason for discrepancy: Original allotments were insufficient for feeding all specimens properly.

Additional plant food and potting expenses incurred:

```
Miracle Grow................$12.50
Magic Formula...............11.00
Sutter's Plant Food........ 11.50
Fern Diet.........................14.75
25 New Pots....................18.75
TOTAL..........................$68.50
```

ELEMENTS OF A PROGRESS REPORT

Like any report, a progress report should have a distinct introduction, discussion, and conclusion: an introductory section to describe the purpose, background information, and scope of the report's contents; a discussion section to present detailed data regarding the progress made on the project, and any complications that have arisen; and an optional concluding section that summarizes the key points raised in the discussion and gives an overview of the progress made during the period in question.

More specifically, a progress report should include the following information:

Summary

• Summary of progress made since last report.

• Summary of expenditures.

Discussion

• List of activities completed.

- Time alloted vs. time actually spent on each task.
- Explanation of any time discrepancies.
- Explanation of any unforeseen problems.
- Revised estimate of time needed to complete the project.
- Discussion of estimated costs vs. actual costs incurred.

Conclusion (Optional)

- Summary of key points presented in the **Discussion** section.
- Overview of progress achieved (all that has been completed to date plus what remains to be done).

The project manager typically presents his or her progress report in the form of a memorandum, and structures it as an outline. Unlike more complex reports, such as proposals and research reports, formal progress reports are most successful when they are most telegraphic, yielding "cut-and-dried" data as clearly as possible. Progress reports truly are "just the facts" documents, not so much read as accessed.

Ingredients for a formal progress report include:

1. *The date.* This is the day on which the report is submitted, rather than written.
2. *To/From/Subject* lines. If assigned identification numbers, use them alongside or directly underneath the names. The subject should be stated as concisely as possible, reflect the identity of the project, and be no longer than two lines. If the project has a number, it may be placed here. Also indicate the number of the report (e.g., "Second of Four Quarterly Progress Reports").
3. *Summary of progress made.* Readers of progress reports want the "bottom line" information first; this includes a statement of the timeline status and the budget status.
4. *Discussion of progress.* Subdivide progress made on each of the recognized facets of the project; report the progress for each facet. Discussion should include:
   a. what has been accomplished
   b. what was scheduled for completion but remains incomplete, and why
   c. what is left to complete
   d. expenditures (unless regarded as a separate facet of the project)
5. *Attachments.* Any necessary tabulation of items and expenses may either be grouped in an appendix or inserted where referred to in the body of the report.

Using these guidelines as a gauge, study the following progress report:

March 21, 1980
AA-10,373

TO: C. D. Jarman
Sta. # 5579

FROM: E. L. Lewis
Sta. # 5188

Ext. # 7975

SUBJECT: Four Corners Units 4 and 5 Particulate Removal Project
Monthly Progress Report—February, 1980

This is the February, 1980, Monthly Progress Report on project activities:

1. Pilot plant is still headed for 4/17/80 in-service date.

2. Finalizing plan and schedule to get 12/82 service date on full-scale project.

3. No change in cost estimates.

4. Accounting recorded charges of $452,000 in February.

A. ENGINEERING

1. Significant Milestone Accomplishments Achieved During the Month:

   a. Completed study on project fire protection needs.

   b. Met with Joy-Western, Wheelabrator-Frye, Research-Cottrell, American Air Filter, and Buell Envirotech to discuss filter-house proposals.

   c. Provided recommendation and cost estimate to include restroom facilities in the new control rooms.

   d. Provided caisson survey notes to UE&C.

   e. Released bid package for construction yard.

   f. Reviewed the following UE&C specifications, studies, and documents:

      (1) Axial vs. Centrifugal Fan Study.

      (2) PDM Control Description.

      (3) Survey Services Specification.

      (4) Duct Fabrication Specification.

      (5) PRP and SO_2 Fan Study.

(continued)

(6) Demolition and Site Preparation Specification.

2. Areas of Work Involving Major Blocks of Personnel Hours or Major Decisions:

   a. Flue gas acid dew point temperature evaluation.

   b. Evaluation of filterhouse proposals.

   c. Site survey for permanent brass caps.

   d. Review of demolition specification.

   e. Study of project fire protection needs.

   f. Release of bid package for construction yard.

3. Summary of Activities That Are Behind Schedule: Relocation of construction facilities.

4. Summary of Key Activities Expected During the Coming Month:

   a. Complete comments on the following UE&C specifications and documents:

      (1) Site Preparation and Foundation Specification.

      (2) Concrete Specification.

      (3) Inspection and Testing Services.

      (4) Structural Steel Specification.

      (5) Structural Steel Erection Specification.

      (6) Drilled-in-Pier Specification.

B. CONSTRUCTION

1. Significant Milestone Accomplishments Achieved During the Month:

   a. Completed demolition of horizontal scrubber test module.

   b. Received bids for relocation of construction facilities.

   c. Reviewed specification for demolition and site preparation.

2. Areas of Work Involving Major Blocks of Manhours or Major Decisions:

   a. Installation of baghouse test module.

   b. Demolition of horizontal scrubber test module.

3. Summary of Activities That Are Behind Schedule:

(continued)

a. Installation of baghouse test module.

b. Relocation of construction facilities.

c. Bid package for Unit 5 precipitator control room HVAC.

4. Summary of Key Activities Expected During the Coming Month:

a. Continue erection of baghouse test module.

b. Start work on relocation of construction facilities.

c. Review bid package for Unit 5 precipitator control house HVAC.

C. COST CONTROL

1. Significant Milestone Accomplishments Achieved During the Month:
Reviewed and approved engineering economic factors for use by UE&C.

2. Areas of Work Involving Major Blocks of Manhours or Major Decisions:

a. Followed up on cost of construction and pre-operating power.

b. Completed and routed draft of Contract Change Order Procedure.

c. Started reauthorization of WA 19-2103 #5 Precipitator Upgrade.

d. Obtained preliminary insurance data from UE&C.

e. Analyzed January cost report.

f. Updated Cash Flow Report.

g. Submitted new unit numbers to Budget Services for employee redistribution.

h. Corrected accounting for Lodge-Cottrell costs incurred by UE&C.

i. Prepared three-year cash forecast for Construction Budget Administration.

j. Reviewed UE&C procedure manual.

3. Summary of Key Activities Expected During the Coming Month:
Update estimate to include cost of construction and pre-operating power.

(continued)

Table 1 Four Corners Units 4 and 5 Particulate Removal Project*
Comparative Cash Flow Report (thousands of dollars)

| 1980 | APS | | UE&C | |
	Current Month Est. Act. 0/(U)	Cumulative Est. Act. 0/(U)	Current Month Est. Act. 0/(U)	Cumulative Est. Act. 0/(U)
To 12/31/79	$ —	$ 323	$ —	$ 545
Jan.	194	517	148	693
Feb.	284	801	169	862
Mar.			186	1,049
Apr.				
May				
June				
July				
Aug.				
Sept.				
Oct.				
Nov.				
Dec.				

| 1980 | Total Project | |
	Current Month Est. Act. 0/(U)	Cumulative Est. Act. 0/(U)
To 12/31/79	$ —	$ 868
Jan.	342	1,210
Feb.	452	1,663
Mar.		
Apr.		
May		
June		
July		
Aug.		
Sept.		
Oct.		
Nov.		
Dec.		

*Includes P.O.s 105–T1 and 105–K4 (230KV Power Supply)

Table 2   Four Corners Units 4 and 5 SO_2 Removal Project Comparative Cash Flow Report (thousands of dollars)

1980	APS Current Month Est. Act. 0/(U)	APS Cumulative Est. Act. 0/(U)	UE&C Current Month Est. Act. 0/(U)	UE&C Cumulative Est. Act. 0/(U)
To 12/31/79	$ —	$ 37	$ —	$ —
Jan.	8	45	—	—
Feb.	68	—	—	—
Mar.				
Apr.				
May				
June				
July				
Aug.				
Sept.				
Oct.				
Nov.				
Dec.				

1980	Total Project Current Month Est. Act. 0/(U)	Total Project Cumulative Est. Act. 0/(U)
To 12/31/79	$ —	$ 37
Jan.	8	45
Feb.	68	114
Mar.		
Apr.		
May		
June		
July		
Aug.		
Sept.		
Oct.		
Nov.		
Dec.		

(continued)

D. INSURANCE

1. Significant Milestone Accomplishments Achieved During the Month:

   a. Strategy meeting with Project Manager, reviewing OCP approach, tentative implementation schedule and data needed from UE&C.

   b. Cost projections updated.

2. Summary of Key Activities Expected During the Coming Month: Upon receipt of PDM and other UE&C data, present underwriting data to Brokers.

E. PROCUREMENT

1. Significant Milestone Accomplishments Achieved During the Month:

   a. Reviewed specifications 20,100,313; 20,100,314; 20,100,315; 20,100,316; 20,100,317.

   b. Received and distributed copies of bids received on Inquiry No. 20,100,310 (500KVA transformer).

   c. Sent Inquiry #20,100,308 (Booster Fans) to prospective bidders.

   d. Made commercial evaluation of bids received on Inquiry No. 20,100,300 (Baghouse).

   e. Issued the following purchase orders for the Filterhouse Pilot Plant:

P.O. #	Vendor	Description
20,100,499	Elder Leasing	Mobile office
20,100,800	Morgan Drive Away	Move mobile office
20,100,801	Patent Scaffolding	Field supervision
20,100,802	Morgan Drive Away	Move mobile offices
20,100,803	Leeds & Northrup	Charts
20,100,804	Riley Co.	Relays & flasher
20,100,805	G.E. Supply	Motor starter heaters
20,100,806	Eastside Electric	Fuses

(continued)

F. FILTERHOUSE PILOT PROGRAM

1. Significant Milestone Accomplishments Achieved During the Month:
   a. Coordinated construction and engineering work by APS.
   b. Evaluated contractor bids for FPP testing.
   c. APS low temperature design work.
   d. Selection of bag material and bag manufacturer.
   e. Coordinated LCD and S/R engineering.
   f. Operating Manual—80% complete.
2. Areas of Work Involving Major Blocks of Manhours or Major Decisions:
   a. All of the above.
   b. Responding to field engineering problems.
   c. Reviewed UE&C technical and commercial evaluations.
3. Summary of Key Activities Expected During the Coming Month:
   a. Complete operator's manual.
   b. Sign contract with FPP test contractor.
   c. Determine requirement for low temperature FPP approach and method to achieve that.
   d. Track erection of FPP unit.
   e. Track I&C and electrical FPP checklist.

G. SCHEDULING— FILTERHOUSE PILOT PLANT

1. Significant Milestone Accomplishments Achieved During the Month:
   a. "Tie-ins" for both Units 4 and 5 completed.
   b. Module area support towers installed.
   c. Pre-Ass'y ductwork from module area to Unit 5 completed.
   d. Module area scaffolding engineering completed.
2. Summary of Key Activities That Are Behind Schedule:
   a. Installation of scaffolding from Unit 5 to Unit 4.
   b. Valve control wiring diagram (APS Engineering—Jan. 25 was late finish)

(continued)

c. Module area support details.

d. Module area scaffold installation.

3. Summary of Key Activities Expected During the Coming Month:

a. Begin duct insulation.

b. Ductwork installation—Module to Unit 5, Unit 5 to Unit 4.

c. Instrumentation installation.

H. PROJECT AGREEMENT

1. Significant Milestone Accomplishments Achieved During the Month:

a. Rescheduled two-day meeting from February to March 18 and 19. These two days should bring us to the final form of the Project Agreement with possibly only the final economic items remaining.

b. A report of the meeting will be delivered orally on March 20th to E. L. Lewis by R. W. Gimbernat and E. B. Dellaquila.

I. OTHER

Attached is a copy of UE&C's Monthly Progress Report for February, 1980.

E. L. Lewis, Project Manager
Emission Abatement Projects

ELL/iec
Attachment

cc: T. Woods   E. Chartier
     L. Mundth   E. Jochim
     A. Simko   W. Ekstrom
     B. Clark   Field Office
     F. Heacock   File: 000.1
     D/F: 3/21/80   204.3

Source: Arizona Public Service Co.; repr. in *Readings for Technical Writers*, ed. Debra Jourret & Julie Kling. Glenview, IL: Scott-Foresman, 1984.

Report Analysis

Probably the first thing you noticed about this report is that it is entirely in outline form. For a progress report, this is apt. Indeed, many companies give their project workers a progress-report template (often on disk) to fill out each time a report must be filed. The outline format enables the reader—usually the project supervisor or an auditor—to locate the key facts and figures.

The Particulate Removal Project requires monthly progress reports in several areas: Engineering, Construction, Cost Control, Insurance, Procurement, Pilot Program, Scheduling, and Project Agreement (i.e., how closely the original agreed-upon plans are being followed). Tables are used to present numerical data. Note that these are incomplete tables; when the final progress report is made, they will have been completed. The report has no introduction, conclusion, abstract, table of contents, or title page; these would be considered redundant in so short and skeletal a report.

ELEMENTS OF AN ACTIVITY REPORT

Having an accurate record of one's activities over a given period gives managers an important insight into institutional productivity and serves as a nonarbitrary basis for determining promotion and salary increase. Such reports are especially important in occupations such as teaching, where direct supervision is usually absent.

An activity report is organized into the different kinds of duties that constitute the employee's job repertoire; each duty is then broken down into specific activities. A college professor, for example, will report on his or her activities in the context of three major duties: teaching, scholarly work, and service. These duties generally are broken down as follows:

I. TEACHING
   A. Regularly Scheduled Classes
   B. Evening, Summer School, Extension Classes
   C. Special Classes

II. SCHOLARLY WORK /PERFORMANCE
   A. Print or Microfilm Publication
      1. Academic publishing
         a. Books, Monographs
         b. Articles
         c. Reports
      2. Nonacademic publishing

      a. Books

      b. Articles, Essays

      c. Fiction

      d. Poetry

      e. Dramatic works

      f. Musical compositions

      g. Public performance

   B. Nonprint/Microfilm Publication

     1. Unpublished papers presented at conferences

     2. Public lectures

III. SERVICE

   A. University

     1. Committee work

      a. University level

      b. College level

      c. Department level

     2. Special events

   B. Community

     1. Consultantships

     2. Community action programs

It is important to keep careful records of your activities because you may be the only one doing so. Even if the activity seems too trivial to mention, make a note of it, and then decide whether to include it when it is time to prepare the report. Also make a note justifying any activity that is not immediately self-evident. If you list attending a conference as an auditor rather than as a presenter, for example, you will need to justify the reason. If you are a machinist who attended a tool-designer's conference, your reason could be that the seminars you attended at the conference enhanced skills that are essential for your work. If you cannot justify an activity, do not include it in your report.

DRAFTING A PROGRESS OR ACTIVITY REPORT: A CHECKLIST

Progress Report

☐ Have I subdivided the project into distinct categories and reported progress on each?

☐ Does my opening summary describe the essential status of the project?

- [ ] Did I record the data accurately, using tables where necessary?
- [ ] Has all the work been described and evaluated in terms of the project's goals?
- [ ] Is my budget analysis accurate?
- [ ] Did I compare actual expenditures with projected ones?

Activity Report

- [ ] Did I divide report into the constituent duties that define my job?
- [ ] Have I justified activities not immediately self-evident? Omitted mention of activities that I cannot justify?
- [ ] Have I included all activities of relevance?

■ CHAPTER SUMMARY

Reporting progress on a given project, or reporting one's activities during a given period, by means of progress and activity reports plays an important role in helping managers plan and coordinate projects that are interconnected and in helping to determine promotions, salary increases, and the like. These reports also serve as records of how project funds are used. Were original estimates and allocations appropriate? Were there overlooked expenses? Were some expenses underestimated? Such records are important for determining the nature and amount of budget planning for future projects.

Progress reports are usually written as memoranda. The subject line identifies the project and the number of the particular progress report. Next comes a summary of the essential progress made during the period in question—most commonly for the preceding month or quarter. The discussion section subdivides the project into discrete areas, and the progress made in each area is reported therein. All progress reports must include a summary of costs.

Activity reports, like progress reports, summarize work completed during a given period; but unlike progress reports, there is not necessarily one pre-established project involved, toward which steady progress is expected to be made. Rather, activity reports summarize routine duties that may *include* work on short-term and long-term projects.

■ FOR DISCUSSION

1. Why are progress reports important?
2. According to technical communications specialist Cezar M. Ornatowski, "Technical documents . . . play an important role in the dramaturgy of

JONAITIS ENGINEERING COMPANY
1715 Mandel Road, Chicago Illinois 60646

August 1, 1978

Ms. Cecilia Roop, Chairperson
Commission for Recreation
14 Randolph Street
Chicago, IL 60601

Dear Ms. Roop:

    Subject: Progress of Feasibility Study for
               Chicago Sports Complex (July 1–July 31)

    This is a report of progress on the feasibility study
you requested. Soldier Field, the near West Side, and the
South Loop are being studied to determine which would be best
for a sports complex. Criteria for the study are traffic
accessibility and preconstruction costs, which include land
value, cost of razing, and cost of road modifications.

WORK COMPLETED

TRAFFIC ACCESSIBILITY

    Research on potential traffic patterns has been
completed for one of the three alternatives. The Soldier Field
site presents severe traffic problems. The only four-lane
access road in the area is Lake Shore Drive, and traffic tie-ups
for the large complex would be even worse than the ones that
occur when Soldier Field is used. No road modifications
would adequately solve the problem.
    Work on traffic patterns for the near West and South
Loop sites began in September 15 and is still in preliminary
stages.

PRECONSTRUCTION COSTS

    Except for the cost of traffic modifications at the near
West and South Loop, all the cost data have been gathered and
are presented in the following table:

COST–BREAKDOWN TABLE

	Near West Side	South Loop	Soldier Field
Land	$1,000,000	$1,500,000	$1,000,000
Removal of old structures	1,000,000	1,300,000	1,800,000
Improvement of traffic facilities	TO BE GATHERED	TO BE GATHERED	2,000,000

At this point, the Soldier Field site, which would cost a total of $4,800,000 and still not be adequate for reasons of traffic, has been tentatively determined unfeasible. Of the other two sites, the cost of land and razing at the near West is $800,000 less than at the South Loop, but traffic data has not been gathered.

WORK SCHEDULED (AUGUST 1–AUGUST 31)

TRAFFIC ACCESSIBILITY

Data for traffic patterns at the near West and South Loop are being gathered and processing will begin approximately October 15.

PRECONSTRUCTION COSTS

Figures for Soldier Field are complete, and cost data for the other sites are complete except for the cost of road modifications. Those data will be processed by November 1, when work on a draft of the final report will begin.

At present, this feasibility study is ahead of schedule, and a draft of the final report should be available prior to August 31.

Sincerely yours,

Robert W. Jonaitis

organizational life through which the organization and . . . the various departments, interests, or individuals, garner resources and advance their causes" ["Between Efficiency and Politics . . . , "*Technical Communication Quarterly* 1 (W '92): 100]. How does this apply to progress-report writing in particular?

3. What structural characteristics of the Particulate Removal Project progress report do you find problematic, if any?

4. Suggest an alternative format for the report. Justify your choice.

5. What is the difference between a progress report and an activity report?

6. Evaluate the progress report which appears on pages 322–323.

■ FOR YOUR NOTEBOOK

1. Think of yourself as an established professional in your chosen field, engaged in an important research project of your choice. First summarize the nature and purpose of the project, then outline a progress report that reflects the first quarter of your research. Your audience is a well-known grant-issuing agency in your field.

2. Take detailed notes on the progress you are making on your term project. Record research activities, interviews, moments of confusion, even casual conversations; take note of lab work, observations, people you've contacted for information, on or off campus; describe the planning and drafting stages of the project.

3. Keep an accurate record of all the school-related activities with which you are involved.

■ WRITING PROJECTS

1. Using the notes you have been taking while working on your term project (see For Your Notebook, topic 2), prepare a progress report covering the first quarter of your work.

2. Imagine that you are the project manager of a research team that had been awarded a contract for one of the proposals described in Chapter 13. Write a report detailing your progress for the first half of the project.

3. Drawing from your notebook entries (see For Your Notebook, topic 3), prepare an activity report for your preceding month's academic work.

15

Reporting Empirical Research

Advances in science and technology occur as rapidly as they do largely because information is disseminated so efficiently, not only among specialists working in the same or similar areas but often among larger communities of scientists, technicians, and administrators, as well as among the general public. Thanks to computer networking, information spreads everywhere at great speed, so that professional people anywhere in the world become participants in an ongoing electronic conversation of vast proportions.

Every week, innumerable documents from hundreds of specialized fields are published. Because a specialist needs to keep abreast of the findings of fellow specialists within his or her field, it is important that reports be coherently structured, that they follow established formats, be carefully documented, clearly and concisely written, and incorporate effective visual aids such as tables, graphs, charts, and drawings.

This chapter introduces you to the nature of empirical research, and to techniques for transforming the collected data into effective reports.

THE NATURE OF EMPIRICAL RESEARCH

Empirical research is the firsthand investigation of a phenomenon for the purpose of determining its nature and the effect it has on other phenomena. Empirical researchers obtain their data directly from the source instead of from the published or unpublished findings of other researchers. Here are just a few examples of empirical research activities:

- Investigating a possible correlation between eyestrain (or neuromuscular disorders) and the amount of time spent working in front of a computer screen.
- Conducting a poll to determine the nature and frequency of campus security incidents.
- Obtaining water samples from different points along an urban waterway and analyzing them to determine what contaminants, if any, are present and whether they pose a health threat.
- Observing students' problem-solving strategies and correlating what has been observed with learning difficulties.

As these examples illustrate, empirical research involves the researcher in "hands-on," information-yielding activities, either in the laboratory or in the "field"—physical sites (natural or artificial) where problems may be occurring, or in both places.

How does one conduct research? The question might seem obvious, but it can be easy for inexperienced researchers to overlook important elements. First of all, any research project presupposes that a hypothesis has been formulated. A hypothesis is a tentative explanation of a problem or phenomenon which has been brought into focus by a need to resolve the problem or understand the phenomenon. At your workplace, for example, if you wished to investigate a possible correlation between eyestrain and extent of computer use, you might already suspect that a correlation exists, perhaps because those employees who have logged in the most computer time at your place of employment have complained the most about eyestrain. To give another example: Imagine that you have been visiting a particular wilderness river every summer to fish and have consistently returned with a full creel of full-sized rainbow trout; but on your last visit, you caught but three scrawny and sickly specimens. At once you suspect that toxic effluent from the new lumber mill three miles upstream is the culprit. But before suspicion can become hypothesis, you need to gather some evidence in order to justify conducting a formal investigation.

Once a hypothesis is established, the next step is to determine a *method* for testing the hypothesis. It is not enough to say that you have found evidence of chemicals associated with the manufacture of paper, which might be the reason why trout populations are dwindling; it is not enough to see one instance of eyestrain directly related to excessive computer use. You must work out methods by which these hypotheses can be proven or disproven.

Working out a method of investigation brings one into the heart of scientific inquiry. For the eyestrain hypothesis, the investigator would conceive of a particular experiment, then recruit volunteers to comprise a test group and a control group; he or she would then perform the experiment by having the test group work *x* number of hours on, say, screens with glare protectors, and having the control group work the same number of hours on screens without glare protectors. For the stream-pollutant hypothesis, the investigator would take water samples at various sites along the stream, test for the existence of substances known to be toxic to fish, test both the sickly trout and the healthy trout to determine if such toxins are or are not present, and then determine if these toxins are indeed being discharged from the paper mill.

WHY CONDUCT EMPIRICAL RESEARCH?

Fundamental to any empirical research activity is *purpose:* there must be a compelling reason for conducting the research. Why is it important to study the correlation between eyestrain and computer use? If a correlation is shown to exist, the installment of glare protectors could result in greater employee efficiency, not to mention eye safety.[1] Why take a survey of students' opinions about campus residence-hall security? Perhaps because many residents have had personal belongings stolen or have been harassed by strangers who had somehow managed to get inside the building without a key. The residents' testimonies will constitute important evidence for the researcher, whose report could convince the administration to hire additional security officers and to reinforce security regulations. Why is it important to observe the time it takes for a rat to learn its way through a maze? In such experiments, sometimes referred to as "pure" science, the purpose is less pragmatic, often to play a hunch or even to grope in the dark: perhaps the experimenter senses that some correlation may exist between a particular stimulus and increased ability to learn the task. This may not be too compelling a need in itself, but the experiment might provide valuable insights into the motivational factors that trigger learning in human beings; then again, it might not. That is what empirical research is all about.

Purpose and need presuppose another important facet of empirical research: analyzing and interpreting the results. It is naive to think that data by themselves automatically communicate their meaning. The purpose for which the data have been gathered will have a strong influence on the way the data will be interpreted. Thus, a chemist who investigates a paper mill's production process for the purpose of determining how environmentally safe it is would examine the process quite differently from, say, an industrial chemist who investigates the same operation for the sole purpose of reducing manufacturing costs or of improving the quality of the product.

WHO READS EMPIRICAL RESEARCH REPORTS?

Knowledge has become so specialized that it needs to be made understandable to nonspecialists—those whose expertise lies outside the topic of the report—who need to understand the essence of the report for their own specialties. A hospital administrator, for example, may need to read a report on the effectiveness of a new technique of radiation therapy before purchasing the requisite equipment, which could run into millions of dollars. Another potential audience

---

[1]Studies conducted specifically to ascertain a causal link such as this in order to decide whether to change existing policy (e.g., to require all employees to use glare protectors on their computer monitors) are a type of empirical research known as feasibility research; see Chapter 16.

for such a report would be hospital benefactors, many of whom would have no medical background at all.

The first challenge in reaching multiple audiences is simply recognizing that not everyone will understand the technical language that may seem elementary to the writer. Besides specialized words or phrases (which can be defined in the document on a separate page known as a *glossary*), entire concepts and background information will need to be explained to nonspecialists. Often this requires collaboration with those outside the writer's field.

Writers also need to be sensitive to the differing aims of each audience type in reading the report. One research team explains some of the different aims involved in a particular case:

> The executive may want only the summary and conclusions of a report evaluating a new telecommunications system. This information might help the executive decide whether to adopt the system for the company. The resident engineer may need the technical sections of the report, including the primary statistical results, the functional explanations, and the procedures used to collect and analyze data. That information would help in understanding and implementing the telecommunications system.
>
> —V. Melissa Holland, Veda R. Charrow, and William W. Wright, "How Can Technical Writers Write Effectively for Several Audiences at Once?" *Solving Problems in Technical Writing,* ed. Lynn Beene and Peter White. New York: Oxford University Press, 1988: 33.

STRUCTURING THE RESEARCH REPORT

Empirical research reports typically contain the following sections (not counting the table of contents, abstract, or appendices):

> INTRODUCTION. Includes a statement of *purpose* (which includes an overview of the problem to be investigated), *background information* needed to understand the problem, *methods* used to investigate the problem, and the *scope* of the report (what is covered in the text of the report).
>
> COLLECTED DATA. Includes the *actual data* obtained from the investigation and detailed *explanations* of whatever the intended readers need to know to fully understand the data.

I. INTRODUCTION
    A. Statement of Purpose
    B. Overview of the Problem to be Investigated
    C. Background Information
        1. Origins of the problem
        2. Technical aspects of the problem
        3. Seriousness of the problem
    D. Methods of Investigation
        1. Sites visited
        2. Tests conducted; materials & equipment used
        3. Experts interviewed
        4. Surveys taken
        5. Secondary sources consulted
    E. Scope of the Report

II. COLLECTED DATA
    A. Presentation of Findings
    B. Analysis of Findings

III. CONCLUSION
    A. Summary of Collected Data
    B. Interpretation/Conclusion
    C. Recommendations

FIGURE 15.1 Elements of a typical outline for a research report.

CONCLUSIONS AND RECOMMENDATIONS. Includes a *summary* of the key aspects of the collected data, *interpretations* of the collected data and *recommendations* arrived at in light of those interpretations (see Figure 15.1).

PLANNING THE REPORT

As with any writing task, writing an empirical research report begins with identifying a problem and deciding on a strategy for investigating that problem.

Imagine that you're a public safety inspector who has been hearing rumors about fire hazards in the residence halls on your campus. Because you suspect that such rumors may have some basis in truth, you decide to investigate each of the five residence halls to see firsthand how well they adhere to fire-safety regulations. You will check out every piece of fire-safety equipment on every floor, then inspect several rooms at random on each floor for possible fire hazards. Your notes for one of the residence halls might look something like those in Figure 15.2.

Once you have all such data collected, a logical next step would be to review the fire codes in order to determine the extent of the violations observed

FIRE INSPECTION: WILSON HALL

Date of Inspection: January 10, 1994

I. EQUIPMENT

Type of Equipment	Status
Smoke Detectors (160)	All tested; 3 detectors defective (Rooms 112, 318, 330; batteries missing in 2 detectors (Rooms. 219, 523);
Fire Extinguishers: 12	All tested and in working order;
Fire Alarm	Tested and in working order

II. HAZARDS

Type of Hazard	Location
Wiring not up to code	Laundry room
Unsafe extension cords	Rooms 103, 115, 200, 227, 307, 408, 414, 425, 512, 516

FIGURE 15.2　Notes from a fire inspection

in the residence halls. It is also possible that some of the codes are not up-to-date (e.g., codes for proper outlets for computer equipment), which would mean expanding the report to include discussion of this problem.

You are just about ready to begin working on your report. First, of course, you want to work out as detailed an outline as possible. Remember that the very effort to work out a structure for the report will give you insights into what that structure should include. It is very common for writers to revise the outline many times—prior to, during, and following the drafting of the report.

Here is an example of what your first outline for the report on residence hall fire hazards might look like.

I. Introduction

Statement of Purpose   A. The purpose of this report is to determine the kinds of fire hazards, if any, that exist in Seacliff College's five residence halls.

Overview of the Problem   B. As this report will demonstrate, several fire hazards of a serious nature have been detected in three of the four residence halls.

| Background Information | C. Several complaints from resident students as well as from public safety officers had led the Director of Campus Safety to order a fire inspection of each residence hall. The Director also learned that a fire had actually broken out in the laundry room of Wilson Hall but was put out quickly by the students who were in the room at the time. The fire hazards detected are serious enough to warrant immediate attention. They involve (1) replacement of several electrical outlets; (2) replacement of five defective smoke detectors (one in Deacon Hall, two in Wilson Hall, one in Harriet Hall, and one in Fiske Hall); (3) rewiring of elevator control box in Wilson Hall. |

| Method of investigation | D. The writer of this report, following standard fire-inspection procedures, inspected every room (including utility rooms) and corridor in each of the four residence halls on campus. |

| Scope of this report | E. This report describes the fire hazard inspection conducted in each of Seacliff College's four residence halls, describes the condition of every piece of fire-safety equipment located in each residence hall, then identifies and locates every hazard encountered. The report concludes with specific recommendations. |

II. Collected Data
[Elaboration of the notes you took during the inspection]

III. Conclusion

| Evaluation of findings | A. The hazards described above are clear violations of campus and city fire-safety codes; moreover, they are among the most dangerous of fire hazards. |

B. This investigator recommends that the Director of Campus Safety order repairs or alterations to be made immediately to correct the hazards. Fire-safety regulations require compliance no later than 72 hours from the date of this report.

WRITING THE REPORT

As with any kind of writing, employing a set of heuristics will go a long way toward helping you produce a substantive draft. You have conducted your research and have taken pages of detailed notes. Your next step is to ask, How can I communicate my findings in the most coherent manner possible? The way to transform notes into coherent reportage is to ask the key *operative questions* that any report must answer:

1. What is the *purpose* of the investigation?
2. What *problem, need,* or *hypothesis* led to the investigation?
3. What *materials and equipment* are needed to conduct the investigation?
4. What *method* was used to yield the desired information?
5. What were the *results* of the experiment?
6. What is the *significance* of those findings?
7. On what *sources* did the researcher rely? (on which the reader may also need to rely)

These questions can then be used as a means of structuring your report.

Read the report on alternative mulching techniques for erosion control (pages 334–341). First read for comprehension, then read it again critically, paying attention to the way the author, Michael V. Harding, structures his report.

Analysis

"Erosion Control Effectiveness" presents the findings of an empirical research investigation. The introduction provides background information on the seriousness of soil erosion and resulting sedimentation and on the way that erosion leads to infertility of the subsoil; this leads to discussion of the available erosion control products and techniques and the lack of a uniform standard. At this point the author can introduce the purpose of the research, which was to quantify the performance of the materials to determine their value in erosion control, using rainfall simulation as a testing device. Harding uses short, concise paragraphs to describe each facet of the background, building logically toward his statement of purpose.

The collected data section is subdivided into Method, Results, and Discussion.

Method: Describes the rainfall simulation device, the mulch materials used for the simulation tests (Table 1, page 337), and the experiment design.

Erosion Control Effectiveness: Comparative Studies of Alternative Mulching Techniques

Michael V. Harding

ABSTRACT: *The wide range of revegetation and erosion control problems is paralleled by a diversity of products and techniques employed to address them. Current state specifications for erosion control products are as varied as the products themselves. Consequently, no uniform standard exists which can be used to select materials based on their erosion control effectiveness. Utilizing a rainfall simulator device, tests were conducted to determine the erosion control effectiveness of alternative materials and practices. Treatments were applied to a soil bed containing a Blount silt loam inclined to a 9% slope and subjected to rainfall intensities of 10.66 cm and 14.73 cm for one hour. Runoff water (kg) and sediment (g) were collected and weighed for each erosion control practice tested. Sediment concentration in the runoff water, water velocity reduction and soil loss from each practice was compared against values obtained from a bare soil control. The results appear to indicate a difference in the erosion control effectiveness of the alternative mulching techniques tested. Where water velocity was reduced by the mulch matrix, sediment concentrations in the runoff water and, ultimately, soil losses appeared to be correspondingly reduced. Overall, nylon monofilament materials did not appear to be as effective as natural organic materials, possibly due to their lack of moisture-holding capacity and general inability to trap soil particles as efficiently. Although laboratory evaluation of erosion control products may not be the best approach to determine performance and effectiveness in the field, the purpose of the rainfall simulation test was to demonstrate a method for evaluating alternative mulching techniques within the framework of erosion control parameters they were designed to address.*

KEY WORDS: *erosion control effectiveness, rainfall simulation, sediment concentration.*

INTRODUCTION

Land disturbing activities have the potential to develop severe erosion and sedimentation problems due to massive disturbance of the landscape and exposure of bare soil to weathering by wind and water. Soil

(continued)

erosion and the resultant sedimentation contributes to the overall cost of a project as slopes must be reworked to fill in gullies and deposited sediment must be removed from catch basins and traps for disposal or redistribution.

Short-term erosion and sedimentation problems can become long-term revegetation and maintenance problems. An eroding seedbed not only loses soil, but also nutrients and seed applied for revegetation. The resulting effects of erosion are not only aesthetically displeasing: sites can be left with infertile, droughty subsoil or with rills and gullies which limit accessibility to maintenance crews and delay permanent vegetation establishment and, hence, permanent erosion control.

The wide range of erosion control and revegetation problems is paralleled by a diversity of products and techniques employed to address them. If optimum growing conditions occurred after final seed preparation, many erosion control products and practices would be unnecessary. But the fact is that, in most parts of the country, bare soil is exposed to weathering for varying time periods ranging from weeks to months. Various mulching techniques and materials have been developed that provide temporary ground cover until mature, permanent vegetative ground cover is established.

Blown straw or hay mulch, mulch grown in-situ, hydroseeding, hydromulching, soil sealants, blankets of degradable organic materials, man-made fabrics, and matrices of nondegradable synthetic materials all have their specific applications.

Current state specifications and use of erosion control products and techniques are as varied as the products themselves. In most cases, final revegetation and erosion control is based on relative cost, on the assumption that the public interest is best served by providing the least costly alternative. Consequently, no uniform standard exists which selects materials based on their erosion control *effectiveness*. An exclusively economic orientation to erosion control is acceptable on large, flat areas where conventional seeding and mulching is effective in establishing permanent vegetation and thereby permanent erosion control. On steep slopes, critical areas, and under unusual circumstances, these lower-cost techniques typically are not effective.

In recent years, materials have been developed which, though costing more initially, are correspondingly more effective in controlling soil erosion and promoting permanent vegetation establishment. The purpose of this study was to quantify the performance of some of these materials under similar conditions in order to show justification for their use on the basis of erosion control effectiveness and to demonstrate rainfall simulation as a potential technique for the standardization and evaluation of erosion control materials.

(continued)

METHOD

The use of rainfall simulators has become an accepted practice in determining the erodibility of soils. Factors which influence soil erosion have been defined for undisturbed soils (Wischmeier and Smith 1978) and, in recent years, for reclaimed strip mine soils (Stein et al. 1983; Barfield et al. 1983). The amount of ground cover, either in the form of permanent vegetation or temporary ground litter/mulch directly affects the rate of soil erosion (Lee and Skogerboe 1984). Many studies have demonstrated the benefits of agricultural crop residues in the reduction of soil losses. Some work has been done to evaluate the effectiveness of manufactured erosion control materials. Ingold and Thomson (1986) quantified the relative performance of manufactured synthetic and natural fiber erosion control systems using rainfall simulation to simulate various storm events.

For the purpose of this study, a rainfall simulation device was used which produced storm intensities of 10.66 cm and 14.73 cm for one hour. Twelve manufactured erosion control materials were selected for testing based on their prior specification for critical area erosion control. In addition, conventional applications of 2,240 kg/ha and 4,480 kg/ha blown straw mulch were included in the evaluations. The straw was not crimped or tacked to the soil surface. Table 1 lists the materials tested.

A 1.2 m by 3.04 m soil bed containing a Blount silt loam was constructed beneath the rainfall simulator and inclined to a 9% slope. The soil bed was wetted before each test until field capacity was reached, at which time the material to be tested was placed on the bed according to the manufacturer's instructions. Each treatment was subjected to rainfall intensities of 10.66 cm or 14.73 cm for one hour. Runoff water (kg) and sediment (g) were collected and sediment was filtered from the runoff water, weighed, and compared against bare soil (control) losses for the respective rainfall intensity. Dye tests were conducted to determine the relative velocity of the runoff water from each test.

RESULTS

Water Velocity Reduction

All mulch treatments were compared to the bare soil (control) plots. To determine the decrease in water velocity afforded by each mulch material, the values obtained for each treatment were compared to

(continued)

Table 1 Materials List for Rainfall Simulation Tests

Mulch Material	Characteristics
1.	100% wheat straw/top net
2.	100% wheat straw/two nets
3.	70% wheat straw/30% coconut fiber
4.	70% wheat straw/30% coconut fiber
5.	100% coconut fiber
6.	Nylon monofilament/two nets
7.	Nylon monofilament/rigid/bonded
8.	Vinyl monofilament/flexible/bonded
9.	Curled wood fibers/top net
10.	Curled wood fibers/two nets
11.	Anti-wash netting (jute)
12.	Interwoven paper and thread
13.	Uncrimped wheat straw—2,242 kg/ha
14.	Uncrimped wheat straw—4,484 kg/ha

those from the bare soil to derive percent water velocity reduction, as reported in Figure 1.

Treatments which involved the use of organic materials in the form of straw, wood shavings, paper, and coconut or jute fiber exhibited a higher level of water velocity reduction ranging from 45%–78%. In contrast, synthetic nylon and vinyl monofilament materials reduced velocities by 24%–32% respectively; an exception to this trend was the netted nylon blanket, no. 6, which achieved a 74% reduction.

Sediment Concentration in Runoff Water

Concentration of sediment in the runoff water (reported as grams of soil per kilogram of runoff water) illustrates the effectiveness of the various treatments in reducing sediment delivery. Table 2 reports the materials tested under the 10.66 cm/hr and 14.73 cm/hr rainfall intensities.

Sediment concentration data parallel the relative order of the water velocity reduction results, but with a wider separation in the values. When comparing runoff values for each material to the values of the respective control, organic materials show a propensity to reduce sediment concentrations at a rate greater than the synthetic products. An exception is the netted nylon blanket, no. 6 (0.046 g/kg), for which sediment concentration was less than 1% of the control value, and the vinyl/monofilament, no. 8 (1.78 g/kg), which exhibited concentrations which were 7% of the control (22.09 g/kg). By contrast, no. 4 (0.38 g/kg), no. 5 (0.514 g/kg), and no. 1 (0.944 g/kg) reduced concentrations to less than 1 kg of soil per kg of runoff water.

(continued)

Soil Loss

As might be expected, a measurement of the actual soil loss occurring from each test plot mirrors the results of the sediment concentration tests (Table 3).

Reductions in soil loss were determined by subtracting the soil loss for each material tested from the soil loss for the control. The difference in soil loss was then divided by the control value and subtracted from 100 to determine percent soil-loss reduction. No. 6 (99.8%), no. 4 (99.5%), no. 2 (98.6%), no. 3 (98.7%), no. 5 (98.4%), and no. 1 (97.5%) exhibited reductions greater than 97.5%. All other materials exhibited a high degree of soil loss reduction (84%–93.5%) with the exception of no. 7 (53%).

DISCUSSION

The results of the tests appear to indicate a significant difference in the erosion control effectiveness of alternative mulching techniques. Materials which are fibrous in nature and interwoven so as to achieve

Table 2 Sediment Concentrations
(G Soil/kg Runoff Water)

Test 1 (10.66 cm rain)	g/kg
6. Nylon monofilament/two nets	0.046
4. 70% wheat straw/30% coconut fiber	0.113
2. 100% wheat straw/two nets	0.326
11. Anti-wash netting (jute)	1.50
8. Vinyl monofilament/flexible/bonded	1.78
9. Curled wood fiber/top net	1.91
13. 2,242 kg/ha straw	2.81
Bare soil control	22.09

Test 2 (14.73 cm rain)	g/kg
3. 70% wheat straw/30% coconut fiber	0.38
5. 100% coconut fiber	0.514
1. 100% wheat straw/top net	0.944
10. Curled wood fiber/two nets	1.99
12. Interwoven paper and thread	2.24
14. 4,484 kg/ha straw	3.14
7. Nylon monofilament/rigid/bonded	15.18
Bare soil control	30.06

(continued)

Table 3 Actual Soil Loss

	Test 1 (10.66 cm rain)			Test 2 (14.73 cm rain)		
Material	Soil loss (g)	% Reduction		Material	Soil loss (g)	% Reduction
6.	4.9	99.8		3.	57.3	98.7
4.	10.1	99.5		5.	70.9	98.4
2.	29.0	98.6		1.	110.4	97.5
11.	168.0	91.8		10.	278.3	93.5
9.	195.4	90.4		12.	303.0	93.0
8.	213.2	89.6		14.	458.9	89.3
13.	325.4	84.0		7.	2,033.6	53.0
Bare soil	2,032.3	—		Bare soil	4,268.5	—

a matrix high in surface area are more efficient in reducing rainfall impact, dislodgement and transportation of soil particles. This conclusion is illustrated by the fact that 1, 2, 3, 4, 5, and 6 are manufactured under the same process which produces an extended matrix and therefore rank as the top six materials for each test.

These materials showed a clear superiority over alternative

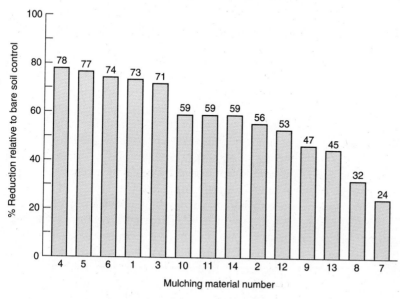

FIGURE 1   Percent water velocity reduction.

(continued)

mulching techniques in the reduction of water flow velocities. Lower water velocities carry less sediment and scouring of the soil surface decreases as water velocity decreases. It follows that if soil particles are not dislodged through rainfall impact and overland water flow, their concentration in runoff water will also be reduced. Sediment concentration data (Table 3) show mulch matrices also appear to trap soil particles carried by runoff water, further reducing sediment delivery, an important water quality consideration.

Finally, extended matrix materials such as those previously mentioned ultimately reduce soil losses; in this case up to 99.8%. But the key to their effectiveness is that they stay in place and maintain a high level of soil-fiber contact. The relatively poor performance of the conventional 2,240 kg/ha and 4,480 kg/ha straw mulching can be attributed to movement of the mulch downslope during the rainfall events, lack of contact with the soil surface, and dislodgement of particles by rainfall impact on exposed soil.

CONCLUSION

A variety of materials and techniques is available to control soil erosion. Manufactured materials vary in their physical composition, the manner in which they are applied and their cost. Low-cost alternatives are not always effective on critical areas, and some higher cost products do not guarantee a correspondingly high level of erosion protection. With increased public awareness of erosion, regulations enforcing its control and the development of new products designed to provide solutions, a uniform method for evaluating erosion control materials should be established.

American Standard Society for the Testing of Materials (ASTM) specifications which evaluate geotextiles as to their grab strength, flexibility, tensile strength or elongation do not apply to the effectiveness of a material to control erosion. Yet, state departments of transportation frequently use ASTM parameters to compare dissimilar erosion control products, such as nylon monofilament and straw.

Rainfall simulation is one technique which can be used to evaluate materials within the framework of conditions they were developed to address. Although laboratory evaluation of erosion control products may not be the best approach to determine performance and effectiveness in the field, rainfall simulation is an appropriate method for comparative evaluations of alternative mulching techniques under exact, controlled conditions. A general ranking can be inferred from such tests, permitting a design engineer or contractor to evaluate cost-

(continued)

effectiveness prior to installation.

Finally, uniform standards should be established which specify the use of erosion control materials based upon their effectiveness and appropriateness for a given set of conditions. It is incumbent upon members of the erosion control industry to work with organizations such as the Transportation Research Board and the International Erosion Control Association to develop guidelines for the establishment of uniform standards.

ACKNOWLEDGMENT

The author would like to acknowledge the work of Stephanie Schroeder, graduate student at Purdue University, for the collection and measurement of runoff and sediment for the test plots.

REFERENCES

Barfield, B. J., R. I. Barnhisel, M. C. Hirschi, I. D. Moore, J. L. Powell. 1983. *Erodibility and eroded size distribution of western Kentucky soil and reconstructed topsoil.* CRIS Project no. 907-15-2. Lexington, KY: College of Agriculture, University of Kentucky.

Ingold, T. S., and J. C. Thompson. 1986. Result of current research of synthetic and natural fiber erosion control systems. In *Erosion control: Protecting our future. Proceedings of Conference XVII of the International Erosion Control Association.* Steamboat Springs, CO: International Erosion Control Association.

Lee, C. R. and J. G. Skogerboe. 1984. *Restoration of pyritic soils and the quantification of erosion control.* Miscellaneous Paper EL-84-7, (October). Vicksburg, MS: U.S. Army Engineer Waterways Experiment Station, CE.

Stein, O. R., C. B. Roth, W. C. Modenhauer, and D. T. Hahn. 1983. Erodibility of selected Indiana reclaimed strip mined soils. In *Proceedings 1983 symposium on surface mine hydrology, sedimentology and reclamation.* Lexington, KY: College of Engineering, University of Kentucky.

Wischmeier, W. H. and D. D. Smith. 1978. Predicting rainfall erosion losses, a guide to conservation planning. U.S. Department of Agriculture, Science and Education Administration, Agricultural Handbook no. 537. Washington, DC: U.S. Department of Agriculture.

Michael V. Harding is President of Aspen Environmental Consultants, Inc. of Fenton (St. Louis), MO 63026.

John J. Berger, ed. *Environmental Restoration: Science and Strategies for Restoring the Earth*, Washington, DC: Island Press, 1990.

Results: Describes the outcome of the experiment. The organic materials showed a higher level of water-velocity reduction; the sediment concentrations parallel the runoff data; reductions in soil loss reflected the results of the sediment concentration tests.

Discussion: [Here the author evaluates the *significance* of the results.] A major difference exists in the erosion-control effectiveness of the different mulching techniques.

Finally, in the Conclusion, the author summarizes the collected data, making it clear that the purpose of the research has been attained: "A general ranking can be inferred from [the] tests"; the author recommends, based on the findings of his research, that "uniform standards should be established which specify the use of erosion control materials based upon their effectiveness and appropriateness for a given set of conditions."

WRITING THE ABSTRACT

An essential element of any investigative report is the *abstract,* a concise, objective overview of the contents, placed at the head of the document (usually on a separate page) so that it may be read quickly. Readers of technical reports (researchers, executives, project supervisors), in search of particular kinds of information, rely on abstracts to help them decide which reports to peruse in depth. For this reason, the abstract must cover all the major points of the report, including the key findings or recommendations.

Look back at the abstract that Michael Harding prepared for his report, "Erosion Control Effectiveness." This is a fine example of an *informative abstract,* in which all elements of the report are succinctly represented.[2] Note that the first three sentences are identical to key sentences in the Introduction (paragraphs 3 and 5). The next few sentences correspond to the key sentences in the Method section, and the next several sentences cover the key points presented under Results. The last sentence of the abstract is the key sentence from the Conclusion. Information that does not appear in the report is never included in the abstract.

Some reports, particularly those designed to help business executives make a decision of some sort, include an *executive summary* at the head of the report. Like abstracts, executive summaries give an overview of the contents of the report; but they also emphasize the rationale behind the decision that is advocated.

---

[2]A briefer type of abstract, known as a *descriptive abstract,* focuses more on the nature of the report itself than on what it contains. See the entry for "Abstract, Summary" in Appendix A.

When writing an abstract, keep the following principles in mind:

1. Extract the main idea from each section of the document; do not include specific data.

2. Make sure that the connections between each sentence are clear. This sometimes requires combining ideas that were originally stated separately.

3. After completing a draft of the abstract, double check the text of the report to make sure that all essential information is represented.

DRAFTING AN EMPIRICAL RESEARCH REPORT: A CHECKLIST

☐ Have I completed my research and collected all of the necessary data?

☐ Does my introduction include a clear statement of the problem, the purpose of the report, the background information, a description of the method of research, and the scope of the report?

☐ Have I analyzed my collected data in sufficient detail?

☐ Have I fully interpreted/evaluated the data?

☐ Are my conclusions clear? logically derived from the findings?

☐ If relevant, have I made appropriate recommendations?

☐ Did I include the necessary front matter? (See Appendix C for models.)
   title page
   letter of transmittal
   table of contents
   table of figures and tables
   informative abstract
   glossary (including explanation of symbols)

☐ Did I include the necessary end matter? (See Appendix C for models.)
   appendices
   acknowledgments
   end notes
   references (works cited)

☐ Are my tables and figures properly prepared, necessary, and well-integrated with the text?

☐ Have I formatted the report properly and attractively?

■ CHAPTER SUMMARY

Empirical research refers to the firsthand investigation of a phenomenon or activity in order to determine its nature and its effects upon other phenomena. Researchers first formulate hypotheses based on a perceived problem; they then work out methods (experiments) for determining the validity of their hypotheses. Detailed and accurate notetaking is essential. Finally the researchers write their reports: they describe the problem, explain the purpose of the study and provide necessary background information to help readers appreciate the importance and complexity of the problem, describe the methods they used to solve the problem, analyze and interpret the collected data, and finally state their conclusions and make recommendations. In writing the report, researchers must also keep their principal readers in mind—often a mixture of specialists and nonspecialists, technicians, administrators, and supervisors. It is important that the report be made accessible to all who will need to read it. An informative abstract, which covers the key points of the report—its purpose, methodology, results, and conclusion—is placed at the beginning of the report for quick reference.

■ FOR DISCUSSION

1. Critique the draft of a student empirical research report on pages 345–347. Use the preceding checklist on page 343 as a framework for your comments.

2. The professional report on pages 348–357 examines the nature of surface displacements in two of California's strongest earthquakes in this century—the 1906 San Francisco earthquake and the 1989 Loma Prieta earthquake. After studying the report, prepare to discuss the following:
    a. As is the nature of most short reports published in *Science,* subheads are not included. Assume that you are an editor wishing to reprint the article in your own science journal and wish to include subheads. Where would you insert them? How would you word them?
    b. Describe the intended audience for this report.
    c. Jot down any words or passages that you do not understand. Share these passages with others in class. Together, try to arrive at the specific basis for the difficulty (e.g., undefined technical language or concepts, unfamiliar mathematical or statistical references).

3. What difficulties, if any, did you have when reading "Erosion Control Effectiveness"? What might the author have done to prevent these difficulties? How might the report be made more accessible to an audience of laypersons?

English 179
Technical Writing
May 26, 1994

Field Inspection Report:
SCU's Training Room

Purpose

The purpose of this report is to evaluate the SCU training room, specifically focusing on ways to improve its operation.

Scope

The report first describes the training room and its equipment, then describes typical tasks of the personnel, with particular attention to rehabilitation procedures of a baseball player recovering from muscle damage, and finally evaluates these procedures and makes recommendations.

Location

The SCU training room is located in Leavy Center between the men's and women's locker rooms.

Description

The training room consists of: four taping tables; four stimulation and ultrasound tables; ultrasound, hydro-collator, and stimulation units; three whirlpools (2 hot, 1 cold); three stationary bicycles; a Cybex room (used for rehabilitation following leg or arm injuries); an ice machine; medical supplies; and two offices (one for each head trainer). The staff consists of: two head trainers, Ed Gonzales and Annette Stern; one chiropractor, Don Andrews D.C.; one physical therapist, Ronald Coleman; and eight student trainers (7 female, 1 male). The training room is open from 8 A.M. until the conclusion of the last practice or game.

Discussion

Observation of a student trainer indicated the various facets of his or her job. For example, a trainer may spend 5-15 minutes taping an athlete's ankle, knee, etc.; then assist a different athlete in stretching; catch up on paperwork that needs to be done; tape other athletes; do ultrasound on another athlete; and the cycle continues until all the athletes have been assisted. The trainer never seems to have a dull moment. Constantly, athletes are entering and leaving. Some come to report a recent injury, others need to be taped, others sit in the whirlpools, and still others report for their rehabilitation following a surgery or an injury.

After watching the general procedures of the training room for one and a half hours, I made an observation of the various steps involved in a rehabilitation process of a baseball player who was recovering from a quadricep muscle tear. Initially, for 15 minutes, heat packs from the hydrocollator were placed on his quad to loosen up the muscle. Next, ultrasound, deeper heating, was applied to the muscle for 5 minutes. Ultrasound stimulates the muscles 2-3 inches in depth. Following ultrasound, he did five minutes of stretching, rode the bike for 15-20 minutes to increase stamina and endurance, and then did an additional 5 minutes of stretching. Following stretching, the trainer assisted the athlete with the Cybex machine. The athlete worked on the Cybex machine for 20 minutes to increase his range of motion and strengthen his injured quad muscle. Next, the athlete applied ice to his quad and, finally, the trainer used the stimulation unit, which emits electrical pulses, on the athlete's quad muscle in order to work the muscle and stimulate it. Thus, the rehabilitation process requires approximately one and a half to two hours of the athlete's time and approximately 30-40 minutes of the trainer's assistance and supervision.

Evaluation/Recommendations

1. There seems to be a need for more student trainers, specifically, male trainers. For example, at times, there is apprehension from a male athlete who needs rehabilitation or therapy near the groin area or other sensitive areas. A female trainer gets uneasy and embarrassed to perform the necessary treatment. It would be a more comfortable situation for the male ath-

lete if there were a male trainer. With only one male trainer, it is hard for him to have to do all the work on male athletes without any assistance from other trainers.

Also, additional trainers would allow more personal attention to be given to each athlete, thus reinforcing the working relationship between athlete and trainer.

2. The training room would run more efficiently if the athletes set up appointments for treatment or taping. An athlete who needs a trainer to assist in rehabilitation has to wait until the trainer is done taping the other athletes. Vice versa, athletes who need taping have to wait until a trainer returns from guiding an athlete in rehabilitation. A possible appointment system could consist of informing the athletes that from 3:00–4:00 trainers would tape, and from 4:00–5:00 they would assist in rehabilitation and therapy. This would eliminate the confusion as to who has first priority, those who need taping or those who need therapy.

3. Another suggestion would be to update the Cybex equipment. There are advanced and computerized models which allow the athletes to improve their range of motion at a faster rate, thus shortening the duration of rehabilitation. The only problem with this recommendation is the need for the funding, and since the training room has a limited budget, it would be difficult to purchase the advanced equipment without a donation.

4. Finally, a significant part of athletics is the respect and support of other athletes. The training room seems to be a meeting place for athletes from various sports and it was important to see the camaraderie, support, and encouragement that male and female athletes gave to each other. An athlete's determination and sportsmanship can decline without the feeling that fellow athletes respect and support their efforts.

Surface Displacements in the 1906 San Francisco and 1989 Loma Prieta Earthquakes

Paul Segall and Mike Lisowski

The horizontal displacements accompanying the 1906 San Francisco earthquake and the 1989 Loma Prieta earthquake are computed from geodetic survey measurements. The 1906 earthquake displacement field is entirely consistent with right-lateral strike slip on the San Andreas fault. In contrast, the 1989 Loma Prieta earthquake exhibited subequal components of strike slip and reverse faulting. This result, together with other seismic and geologic data, may indicate that the two earthquakes occurred on two different fault planes.

In order to understand fully the tectonic setting of the 17 October 1989 M_s 7.1 Loma Prieta earthquake and the implications of this event for earthquake hazard assessment it is important to understand its relation to the most recent large earthquake on this part of the San Andreas fault system, the great 1906 San Francisco earthquake. In this report, we recalculate the horizontal surface displacements accompanying the 1906 earthquake as determined by historical triangulation measurements. We contrast the 1906 deformation in the Loma Prieta region with that occurring in the October 1989 earthquake and discuss the implications of these results for earthquake recurrence estimates and for future earthquake hazards in the Santa Cruz Mountains.

Much of what is known about the 1906 earthquake has come from analysis of geodetic survey measurements[1]. The surveys consisted of horizontal angle measurements, made with a theodolite, taped baselines, and astronomic azimuth sightings. Pre-earthquake surveys were conducted in the San Francisco Bay region in the 1850s and 1880s; the region was resurveyed following the earthquake in 1906 and 1907 (1906–7).

In 1908 Hayford and Baldwin[2] published displacement vectors found by taking the difference of coordinates derived from the 1906–7 survey and coordinates derived from earlier measurements. Their results showed that displacements of several meters parallel to the San Andreas fault extended many kilometers from the fault trace. At the northern (Point Reyes) and southern (Monterey Bay) ends of the network, however, many of the displacement vectors derived by Hayford and Baldwin are opposite to the right-lateral motion observed across the San Andreas fault.

We suggest that the anomalous displacements reflect computa-

(continued)

tional limitations rather than measurement errors. In the 1908 calculations it was assumed that Mount Diablo, Mocho, and Santa Ana, located 20 to 40 km east of the San Andreas fault (Fig. 1), did not move during the earthquake. Errors accumulate with distance from these arbitrarily fixed stations, and any true motion of these sites would bias the calculated displacements. Computational limitations prevented simultaneous inversion of all the data and also prevented determination of confidence intervals.

Given the considerable advances in computing in the last 80 years, we can adopt a different strategy. We reanalyzed the data using only repeated horizontal-angle and astronomic azimuth measurements, foregoing the less accurate distance determinations. The relative displacements of three survey points (between the post-earthquake and pre-earthquake surveys) can be simply related to the change in the enclosed angle[3]. We then used a generalized matrix inverse to solve for the displacements in terms of the measured angle and azimuth changes. The result is not unique because the data do not constrain some components (null vectors) of the displacement field. The null vectors are constrained by prior information, and this process yields a so-called "model coordinate solution"[4]. Our prior model was a simple elastic dislocation model of the earthquake[5].

In order to eliminate effects of the 1868 earthquake (located on the Hayward fault) and the 1865 earthquake (thought to be located in the Santa Cruz Mountains) we first solved for the displacements of the ten stations[6] that were surveyed in the 1880s and again in 1906–7. These stations were connected by 20 angle changes and three astronomic azimuth changes. In this solution there are three null vectors corresponding to rigid body translation of the entire network in the two horizontal directions and uniform areal dilatation. Rigid body rotation of the network is controlled by the astronomic azimuths. We then computed the displacements of the remaining stations using angle changes from the 1850s to 1906–7. Because these angle changes do not form a well-connected network, we fixed the displacements of the ten 1880s stations to the values determined by the analysis of the 1880s to 1906–7 data[7]. The displacements of the 1850s stations are consequently less well determined (Fig. 1).

In contrast to the results of Hayford and Baldwin (2), the recalculated displacements (Fig.1) are entirely consistent with right-lateral strike slip in the 1906 earthquake. Relative displacements vary considerably along the strike of the fault decreasing from 5 to 6 m on the Point Reyes Peninsula to 3 to 4 m on the San Francisco Peninsula. The station at Loma Prieta, in the middle of what was to become the rupture zone of the 1989 earthquake, displaced slightly more than 1 m parallel

(continued)

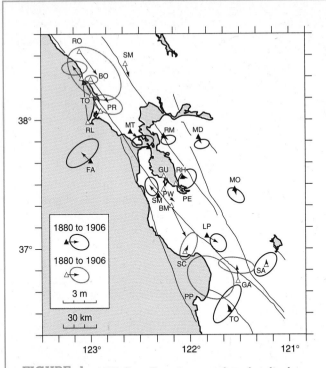

FIGURE 1 1906 San Francisco earthquake displacements. Displacements from the 1880s to 1906–7 and the 1850s to 1906–7 are distinguished by different symbols. Ellipses represent 95% confidence intervals. FA, Farallon lighthouse; MT, Mount Tamalpais; RM, Rocky Mound; MD, Mount Diablo; RH, Red Hill; SM, Sierra Morena; MO, Mocho; LP, Loma Prieta; SA, Santa Ana; TO, Mount Toro; PE, Pulgas East; PW, Pulgas West; GU, Guano; PR, Point Reyes. Other stations also referenced by two lettercodes [see (2)].

to the trace of the San Andreas fault. This motion indicates that more than 2 m of strike slip occurred on this part of the fault in 1906[8].

Horizontal displacements during the 1989 Loma Prieta earthquake have been measured with a variety of techniques. Most of the information has come from laser electronic distance measurements (EDM)[9]. Changes in distance constrain horizontal displacements up to rigid body translations and rotations of the network (4). Global Positioning System (GPS) measurements between Loma Prieta and stations Eagle, Allison, Mount Hamilton, and Fort Ord constrain the relative displacement vectors between these sites and thus the rotational component of the displacement field. The displacement of the Fort Ord site relative to stations remote from the epicentral region has been determined by Very Long Baseline Interferometry (VLBI)[10], constrain-

(continued)

ing the translational components of the displacement field. A least-squares estimate of the horizontal displacements during the 1989 earthquake from the available EDM, GPS, and VLBI measurements is shown in Fig. 2[11].

It is obvious that the displacement of Loma Prieta was markedly different in magnitude and orientation during the 1989 and 1906 earthquakes (Fig. 2). In 1989 Loma Prieta moved 0.19 m, 0.15 m parallel to the fault and 0.11 m perpendicular to the fault. This oblique motion is reflected in results from uniform-slip elastic dislocation models, which yield a ratio of strike-slip to dip-slip motion of 1.4 (9). In contrast, during the 1906 earthquake, Loma Prieta was displaced 1 m parallel to the fault. The difference in the magnitude of the displacement is because there was more shallow slip in 1906 than in 1989. The difference in orientation means that sense of slip in the two earthquakes must have been distinctly different.

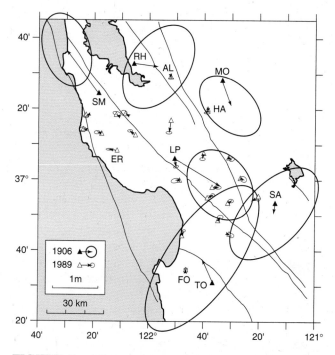

FIGURE 2 1989 Loma Prieta earthquake displacements (small error ellipses, 95% confidence intervals) compared to 1880s to 1906–7 displacements (large error ellipses). The 1989 displacements are from a joint solution with currently available EDM measurements, and GPS, and VLBI vectors. SM, Sierra Morena; RH, Red Hill; AL, Allison; MO, Mocho; HA, Mount Hamilton; LP, Loma Prieta; ER, Eagle Rock; SA, Santa Ana; FO, Fort Ord; TO, Mount Toro.

(continued)

Elastic dislocation models indicate that the 1906 displacements can be adequately fit by pure strike-slip movement on either a vertical fault or a fault dipping 70° to the southwest, as in the 1989 Loma Prieta earthquake. Allowing both strike and dip slip adds a small component of normal slip, opposite to the observed reverse slip in the 1989 earthquake, but does not significantly improve the data fit. Furthermore, if the ratio of strike slip to reverse dip slip in the Loma Prieta segment is set equal to that observed in the 1989 earthquake, the fit to the 1906 data is significantly degraded[12]. The observation that the displacement of Loma Prieta was so different in the two earthquakes means that the slip directions must have been different in the two events.

This interpretation is further supported by the difference in the long-term rates of horizontal and vertical displacement. At the latitude of Loma Prieta the San Andreas fault accommodates ~20 mm/yr of the horizontal motion between the North American and Pacific plates[13]. The rate of vertical motion, as evidenced by the rate of uplift of marine terraces along the Santa Cruz coastline, is roughly 0.5 mm/yr[14]. Even though the Loma Prieta segment of the fault is misoriented with respect to the relative plate motion vector, horizontal displacement rates exceed vertical rates by more than an order of magnitude. As noted by others, this means that earthquakes such as the 1989 Loma Prieta event with subequal amounts of strike slip and reverse slip must be relatively infrequent[15] and that predominantly strike-slip earthquakes, such as the 1906 event, must account for the bulk of the horizontal plate motion.

If the 1906 and 1989 earthquakes occurred on the same fault plane, then the difference in the orientation of the two slip vectors presents a problem. As long as the slip occurred in the direction of the resolved shear stress acting on the fault, the shear stress would have to have built up in a horizontal sense before 1906 and then rotated, so that between 1906 and 1989 stress accumulation on the fault was oriented at ~35° from the horizontal. Although such a rotation is not impossible, the observation that fault slip rates are nearly constant over thousands of years[16] implies that the loading process is nearly steady state.

Although models of the 1906 earthquake with pure strike slip in the upper 4 to 5 km and either no slip or oblique right reverse slip below this depth are consistent with the geodetic data, they are inconsistent with the geology if this slip pattern is repeated over any length of time. As discussed above, the long-term slip must be dominantly right-lateral strike slip at all depths. Furthermore, a change in slip vector from strike slip to oblique slip at a depth of 5 km results in an incompatible strain field unless another fault takes up the reverse slip component at shallower depths.

(continued)

The different mechanisms in 1906 and 1989 present no problem, however, if the two earthquakes occurred on two separate faults with different dips. Recently, Olson[17] found evidence in the microseismicity for a vertical fault distinct from the southwest-dipping Loma Prieta rupture plane. She relocated earthquakes in the Loma Prieta region for the 10 years before the 1989 earthquake. Surprisingly, the data show no evidence of a 70° southwest-dipping plane. Instead, the seismicity weakly outlines a vertical plane extending beneath the mapped trace of the San Andreas fault. This plane may represent the fault that ruptured in 1906.

A vertical strike-slip fault (the San Andreas) and a 70° dipping fault (the Loma Prieta rupture) with the same strike can both be driven by a temporally and spatially uniform stress. There are a family of applied stress states consistent with the observed rakes in the two earthquakes. Maximum compression directions can range from 64° from the fault if the intermediate (vertical) and minimum principal stresses are equal, to <90° if the intermediate and maximum principal stresses are equal[18].

The geodetic, geologic, and seismic data, suggest that the 1906 earthquake resulted from pure strike slip on a vertical San Andreas fault, whereas the 1989 Loma Prieta earthquake resulted from oblique slip on a separate, southwest dipping fault. If correct, this conjecture has extremely important implications for the evaluation of recurrence models and for present earthquake hazard. Before the Loma Prieta earthquake, several groups estimated earthquake recurrence times for this region using the "time predictable model"[19]. In this model the earthquake repeat time is estimated by the slip in the most recent event divided by the long-term slip rate on the fault. There has been some controversy over whether it is more appropriate to use reported surface offsets of 0.4 to 1.5 m, or the geodetic slip estimates of ~2.5 m for slip in 1906. Using the smaller surface offsets yields a shorter repeat time, ~75 years, versus ~125 years using the geodetic estimates. As shown above, Loma Prieta was displaced 1 m parallel to the fault in 1906. Because of the proximity of this site to the fault, there must have been somewhat more than 2 m of slip at shallow to intermediate depths in 1906. Thus, our analysis supports the earlier interpretations of the geodetic data (5).

More importantly, if the Loma Prieta earthquake occurred on a separate fault, then it is not at all clear that simple recurrence models are appropriate. Although one could argue that strain energy is released in some volume of the earth's crust and therefore it does not matter which fault the slip takes place on, we consider that the earlier forecasts should be reevaluated if indeed the Loma Prieta earthquake

(continued)

did not occur on the predicted fault. Finally, if the vertical San Andreas fault did not slip in 1989, we should not dismiss the potential for a future earthquake on this structure[20]. Although the 1989 earthquake must have decreased shear stress on a vertical San Andreas fault at some depths, it presumably concentrated stress at shallow to intermediate depths.

In summary, geodetic, seismic, and geologic data are consistent with the notion that the 1906 earthquake resulted from horizontal slip on a vertical fault, whereas the 1989 earthquake resulted from oblique slip on a separate southwest-dipping fault. If correct, this interpretation implies that earlier recurrence estimates should be reassessed and that the present earthquake hazard in the Santa Cruz Mountains is not negligible.

References and Notes

1. H. F. Reid, in *The California Earthquake of April 18, 1906, Report of the State Earthquake Investigation Commission*, A. C. Lawson, Ed. (Carnegie Institution of Washington, Washington, DC, 1910), vol. II, pp. 3–55; W. Thatcher, *J. Geophys. Res.* 84, 4862 (1975).

2. J. F. Hayford and A. L. Baldwin, in *The California Earthquake of April 18, 1906, Report of the State Earthquake Investigation Commission*, A. C. Lawson, Ed. (Carnegie Institute of Washington, Washington, DC, 1908), vol. I, pp. 114–145.

3. The change in horizontal angle ($\angle AOB$) $\delta\Theta$ is

$$\delta\theta = \frac{\sin(\alpha_1)}{R_1}(u_1^O - u_1^A) + \frac{\cos(\alpha_1)}{R_1}$$

$$(u_2^O - u_2^A) - \frac{\sin(\alpha_2)}{R_2}(u_1^O - u_1^B) + \frac{\cos(\alpha_2)}{R_2}(u_2^O - u_2^B)$$

where u_j^I is the displacement of station I in the j direction, and α_j and R_j are the angle (from x_1) and length of the vector from the vertex station (O) to station A and B, respectively, and $\Theta = \alpha_1 - \alpha_2$.

4. P. Segall and M. V. Matthews, *J. Geophys. Res.* 93, 14,954 (1988).

5. The a priori model is essentially equivalent to that given by W. Thatcher and M. Lisowski, *ibid.* 92, 4771 (1987).

6. Farallon lighthouse, Mount Tamalpais, Rocky Mound, Mount Diablo, Red Hill, Sierra Morena, Mocho, Loma Prieta, Santa Ana, and Mount Toro.

7. There are 38 angle changes between the 1850s and 1906–7 employed to determine the displacements of 13 additional stations. There is one additional null vector in this solution corresponding to dilatation of the lines between Red Hill and stations

(continued)

Pulgas East Base, Pulgas West Base, and Guano. In all of these calculations the angle changes are assumed to be independent.

8. For a simple screw dislocation in an elastic half-space with slip s from the earth's surface to depth d, the displacement decays with distance from the fault x according to $u = (s/\pi)\tan^{-1}(d/x)$. For slip to a depth of 10 km, this gives a slip of 2.9 m for 1.2 m of slip at Loma Prieta ($x \approx 3$ km). A more precise three-dimensional dislocation calculation yields a slip of 2.5 ± 0.4 m from the surface to 10 km.

9. M. Lisowski, W. H. Prescott, J. C. Savage, M. J. Johnston, *Geophys. Res. Lett.* 17, 1437 (1990).

10. T. A. Clark *et al.*, *ibid.*, p. 1215.

11. The EDM measurements were made to station Eagle; the GPS occupations were at Eagle Un, ~40 m away. In the calculation the two markers were treated as one. On Loma Prieta the GPS occupations and the south-directed EDM lines were from LP1, whereas the EDM lines to the north were from Loma USE. We have corrected all measurements to LP1. For this solution there are 59 observations, 46 unknowns (2 for each of 23 stations), and no null space.

12. For strike slip on a vertical dislocation extending from the earth's surface to depths of 5 to 15 km, the mean-square lack of fit [the residual sum of squares corrected for misclosure; see (4)] is 1.4. For strike slip on a plane dipping 70° to the southwest the mean-square lack of fit is 1.5. Both models provide acceptable fits to the data. W. Thatcher and G. Marshall [*Eos* 17, 554 (1990)] presented similar results. In contrast, if the slip vector is constrained so that the strike-slip component is 1.4 times the dip-slip component, the mean-square lack of fit is 2.8, a significantly poorer fit. With 18 linearly independent observations and three parameters, models with mean-square lack of fit greater than 1.7 are rejectable at the 95% confidence level. The variance in the triangulation measurements estimated from the network misclosure is completely consistent with the a priori estimate of the data variance.

13. Near Loma Prieta a total of 33 to 38 mm/yr of slip is partitioned between the San Andreas and Calaveras faults. Estimates of slip on the San Andreas fault range from 13 mm/yr [W. H. Prescott, M. Lisowski, J. C. Savage, *J. Geophys. Res.* 86, 10,853 (1981)] to 26 mm/yr [M. Matsa'ura, D. D. Jackson, A. Cheng, *ibid.* 91, 12,661 (1986)].

14. W. C. Bradley and G. B. Griggs [*Geol. Soc. Am. Bull.* 87, 433 (1976)] estimated uplift rates of as much as 0.26 mm/yr. T. C. Hanks, R. C. Bucknam, K. R. Lajoie, and R. E. Wallace [*J. Geophys. Res.* 89, 5771

(continued)

(1984)] estimated an uplift rate of 0.35 mm/yr. All rates apply to the Santa Cruz coastline, the uplift rate at the fault trace is not well known.

15. R. S. Anderson, *Science* 249, 397 (1990); Valensise and Ward, in preparation.

16. The average rate of slip on the San Andreas fault in the Carrizo Plains has been 33.9 ± 2.9 mm/yr during the last 3,700 years and 35.8 + 5.4/−4.1 mm/yr during the last 13,250 yr [K. E. Sieh and R. H. Jahns, *Geol. Soc. Am. Bull.* 95, 883 (1984)]; the geodetically determined slip-rate somewhat northwest on the creeping segment of the fault has been 33 ± 1 mm/yr over the last 100 years [W. Thatcher, *J. Geophys. Res.* 84, 2283 (1979)].

17. J. Olson, *Geophys. Res. Lett.* 17, 1492 (1990).

18. As pointed out by T. Heaton (personal communication). Erect a coordinate system with x'_1 horizontal and normal to fault strike, x'_2 horizontal and parallel to fault strike, and x'_3 vertical. The condition that the vertical fault be pure strike-slip is $\sigma'_{13} = 0$. If we assume that one principal stress is near vertical, then σ'_{23} must also vanish. The horizontal and vertical shear stresses acting on the dipping fault are given by $\tau_h = \sigma'_{12}\sin(\delta)$ and $\tau_v = (\sigma'_{11} - \sigma'_{33})\sin(\delta)\cos(\delta)$, where δ is the fault dip. Consider the intermediate principal stress (σ_2) to be vertical and the minimum compression (σ_3) to be at an angle Θ clockwise from x'_1. The observed rake of the slip vector in the Loma Prieta earthquake requires that $\tau_h = \beta\tau_v$; $\beta \sim 1.4$ or $\phi = \sin^2(\Theta) - \sin(\Theta)\cos(\Theta)/\beta\cos(\delta)$ where the parameter $\phi \equiv (\sigma_2 - \sigma_3)/(\sigma_1 - \sigma_3)$ describes the shape of the stress ellipsoid. Note that $\phi \geq 0$, as required by definition, for $\Theta \geq \tan^{-1}[1/\beta\cos(\delta)] \approx 64°$. The direction of maximum compression, consistent with the slip vector in the two earthquakes, ranges from $64° \leq \Theta < 90°$. The minimum value of Θ (N16°E) occurs when $\sigma_2 = \sigma_3$, the maximum (N42°E) when $\sigma_2 = \sigma_1$.

19. A. G. Lindh, *U.S. Geol. Surv. Open-File Rep. 83-63* (1983); L. R. Sykes and S. P. Nishenko, *J. Geophys. Res.* 89, 5905 (1984); C. H. Scholz, *Geophys. Res. Lett.* 12, 17 (1985).

20. H. Kanamori (personal communication) pointed out that the two 1985 Nahani earthquakes, M_s 6.6 (October 1985) and M_s 6.9 (December 1985) had epicenters only 2 to 3 km apart, coincident aftershock zones, and similar focal mechanisms; these common features suggest that they either ruptured the same plane or closely spaced parallel planes; R. J. Wetmiller *et al.*, *Bull. Seismol. Soc. Am.* 78, 590 (1988); G. L. Choy and J. Boatwright, *ibid.*, p. 1627.

(continued)

21. We thank R. Kovach, T. Heaton, and H. Kanamori for valuable discussions; D. Richards, C. Williams, and Y. Du for assistance in the numerical calculations; J. Savage, W. Thatcher, D. Oppenheimer, G. Beroza, B. Ellsworth, and K. Lajoie for comments. This work was supported by the U.S. Geological Survey and by National Science Foundation grant EAR 90-03575 to Stanford University.

23 July 1990; accepted 19 October 1990.

P. Segall, Geophysics Department, Stanford University, Stanford, CA 74305, and U.S. Geological Survey, Menlo Park, CA 94025.
M. Lisowski, U.S. Geological Survey, Menlo Park, CA 94025.
   Science, v. 250, 30 Nov. 1990: 1241–1244.

■ FOR YOUR NOTEBOOK

1. Conduct a fire-safety check of your home or residence hall; record detailed descriptions of each hazard, its location, and the date.

2. Read several empirical research reports in such journals as *BioScience, Science, Mechanical Engineering*, and the *Journal of Geophysical Research*. Summarize each report, then compare your summaries with the abstracts that very likely accompany each report. Don't peek at the abstracts beforehand!

3. Rewrite the abstract in the report presented in For Discussion, topic 2, so that a layperson would understand it.

4. Choose one or more of the following subject areas for empirical research and suggest at least two specific topics from this area that would be appropriate for a 15-page report.

   Example: Subject: Animals in Zoos

   > Topic 1: Helping endangered mammals breed in captivity

   > Topic 2: Determining ideal zoo habitats for African mammals

   a. Active volcanoes
   b. Antarctic exploration
   c. Science labs on campus
   d. Learning a foreign language
   e. Renovating old machines or furniture
   f. Landscaping
   g. Vitamins and other food supplements

■ WRITING PROJECTS

1. You have been asked to analyze a soil sample from a farm to determine which pesticide contaminants, if any, are present. You set up your equipment and test for each of the five pesticides known to have been used on this farm over the past thirty years. Knowing the maximum allowable concentration for each pesticide, you set up the following table:

Pesticide Tested	Maximum Concentration Permitted	Actual Concentration Detected
Aldrin	050 ppm	15 ppm
Endrin	300 ppm	450 ppm
Xylogin	300 ppm	250 ppm
Nitrin	1500 ppm	4100 ppm
DDT	010 ppm	002 ppm

   Next you do some library research on the toxic effects each of the above pesticides can have on the environment, and on animals and humans who happen to ingest them; see, for example, *Pesticide Alert: A Guide to Pesticides in Fruits and Vegetables,* by Lawrie Mott and Karen Snyder (San Francisco: NRDC/Sierra Club Books, 1987); also consider contacting the National Coalition for Misuse of Pesticides, 530 Seventh St., S.E., Washington, DC 20003; (202) 543-5450.

   Prepare a report in which you articulate the specific problem, provide sufficient background information about pesticide contamination of farm soils, present and discuss your findings, reach conclusions, and establish clear recommendations for action.

2. Prepare a report on one of the following investigations:
   a. The nutritional quality of selected brands of dog (cat) food.
   b. Sunscreen protectiveness (relative to skin type).
   c. Effective methods of foreign-language study.
   d. Effects of one-on-one tutoring (in a particular subject, such as calculus, chemistry, or first-year composition).
   e. How a daycare center affects a preschool child's social behavior (or learning ability, or emotional behavior).
   f. Medicinal values of plants or herbs in your region.

16

Reporting
Feasibility
Research

If we install antitheft detection equipment in our store, will it virtually eliminate shoplifting as claimed?

Is the use of microwave radiation to kill termites as effective as the use of traditional pesticides?

Does using highest-octane gasoline improve fuel efficiency substantially enough to make it a good investment?

Can computer simulations effectively replace the use of live animals in some biological experiments?

Answers to such questions are obtained by means of feasibility research; and the document which discusses the results of that research and arrives at recommendations is known as a feasibility report. Feasibility research is conducted to determine whether something is capable of working, of being accomplished, of achieving desired or hypothesized results. It involves a study of options, of available alternatives; it helps readers recognize what the available alternatives are, helps them decide which alternative is best for their particular needs, and why.

THE IMPORTANCE OF FEASIBILITY RESEARCH

Investigating and reporting feasibility are important activities in business, industry, the sciences, and government. Even though one never can be absolutely certain about the effects of any change in policy, methodology, or equipment, a careful feasibility study will improve the likelihood that the changes decided upon are good ones. If you wish to upgrade the computer systems in your business, for example, you will want to make sure that the upgrades will meet your needs and neither fall short of them nor exceed them. Will you upgrade from inkjet to laser printers? The price tag from one model to another seemingly identical model can vary by hundreds of dollars; how will you know if the cheaper model will be sufficient for your needs? You either conduct your own feasibility study to find out, or you look for existing feasibility reports, such as those published regularly in computer magazines (*PC Magazine, Macworld, Byte,* etc.). They often appear in the guise of feature articles, in order to appeal to the general reader. Are you trying to make up your mind about the best type of laptop computer to purchase for your personal needs? Is task speed a major criterion? Before investing $2500, you would probably want to examine a feasibility report that would include information like that in Figure 16.1.

Feasibility studies are important to disciplines in which capability or potential-impact investigations, such as in urban redevelopment or environmental planning, are routinely conducted. How might the proposed sports arena

Laptop Speed Check

Times are in seconds. Shorter bars are better.

■ Green indicates fastest result in test.

Processor Tasks

We ran tests that show how the different processors perform. Our tests included recalculating a spreadsheet, finding and replacing a word, and sorting a database (in RAM).

Drive-Access Tasks

We ran tests that show how each computer performs tasks that rely heavily on the hard drive. Our tests included opening files and sorting a database (on disk).

Math Tasks

We ran tests that depend heavily on a math coprocessor to achieve reasonable performance, including recalculations involving scientific functions.

Display Tasks

We ran tests that show how each computer performs tasks that depend heavily on screen redraw. These included scrolling tests in Microsoft Word and Excel.

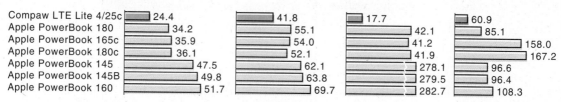

	Processor Tasks	Drive-Access Tasks	Math Tasks	Display Tasks
Compaw LTE Lite 4/25c	24.4	41.8	17.7	60.9
Apple PowerBook 180	34.2	55.1	42.1	85.1
Apple PowerBook 165c	35.9	54.0	41.2	158.0
Apple PowerBook 180c	36.1	52.1	41.9	167.2
Apple PowerBook 145	47.5	62.1	278.1	96.6
Apple PowerBook 145B	49.8	63.8	279.5	96.4
Apple PowerBook 160	51.7	69.7	282.7	108.3

The machines are listed from overall fastest to overall slowest, top to bottom. The Compaq LTE Lite 4/25c, running on an Intel 486 processor, beats the PowerBook 180c and all other PowerBooks by a wide margin. All computers were tested with 4MB of RAM; color laptops were tested in 256-color mode. The Compaq machine used a 120MB hard drive (which partly accounts for its superior performance on drive-access tasks), the 145B used a 40 MB hard drive; all others used an 80MB hard drive.

FIGURE 16.1 Product comparison: laptop speed.

affect local commerce? traffic? nearby residential areas? public safety? Will the use of a new genetically altered strain of E-coli bacteria be able to digest oil spills? Could side effects result from the use of this mutated organism? How would proposed housing projects along a twenty-mile stretch of pristine Oregon seacoast pose a threat to that ecosystem? Investigations into such matters culminate in the writing and publication of what are sometimes called environmental impact statements.

The report on pages 369–379 will give you some idea of the kind of research that is necessary for preparing an environmental feasibility report. The topic is environmental reclamation, and the purpose is to determine whether the biodiversity of a coastal salt marsh can be restored.

Analysis of "Restoring Diversity in Salt Marshes"

Notice, in the introduction (the first two paragraphs on page 369), how the author, Joy Zedler, describes the problem ("development has reduced the area of coastal wetlands and endangered certain wetland-dependent species"); provides relevant background information (development there continues; decline continues despite the California Coastal Act, and why; the magnitude of the problem); and states the scope of the report.

In the discussion or collected data section ("Restoration Plans"), Professor Zedler zeroes in on the specific wetland sites along the California coast where wildlife has been disrupted by industry and where plans are being made for restoration. She subdivides this section to emphasize the key elements in the investigation: restoration projects that have resulted in reduction of wetland areas, and restoration projects that modify wetland habitat. She also includes visual aids (a map and three photographs) to reinforce understanding.

The concluding portion of the report consists of four sections: (1) conclusions regarding wetland-restoration planning; (2) actual implementation of the plans; (3) a look ahead; and (4) acknowledgments. As you can see, Professor Zedler's report structure varies somewhat from the norm. Many writers would have made sections (2) and (3) subsections of (1). Also, the standard "front matter" of a report is missing—only because Professor Zedler's report was published in book form. Had it been submitted separately to a company executive, the front matter would have been included.

THE PROCESS OF REPORTING FEASIBILITY

Place yourself in this situation: A friend of yours is about to begin a carpet-cleaning service and is concerned about providing the most effective cleaning results using the most cost-efficient equipment and chemicals. He or she hires you, an experienced market researcher, to prepare a report that would discuss the feasibility of using relatively inexpensive carpet cleaning equipment to do professional-level work. What steps should you take to conduct this study, and to write the report? What follows is just one of many possible approaches. If you find it a bit too structured for your taste, then work out a comfortable variation for yourself. For example, you might prefer to do Step Four immediately after Step Two; that is, plunge directly into a draft as soon as you have your data—or even begin drafting parts of the report during the research. The guidelines must work for you, not the other way around!

Step One: Work out a plan of investigation. Even before you start surveying supply outlets for the types of equipment available for carpet cleaning, sit down and prepare a list of things you need to do. Your list might look like this:

> Determine what will be the upper and lower price limit of "low-priced equipment."
> Find out who supplies low-priced equipment.
> Determine which models are the most efficient within this range.

Supplier	Brand	Special Features	Test Notes	Cost
Argus Equipment Co.	(1) DeepKleen	Removable bristle heads (3) for differing carpet grades; jet spray.	Handles inconsistently. Leaves slight streaks.	$355
Elliott & Sons	(2) Steemer	Jet spray; guaranteed to penetrate; swivel wheels for easy maneuvering.	Handles with ease. Effective on shag; easy to change bristle heads.	405
Crown Co.	(3) Miracle Clean	Jet spray.	Awkward to move; ineffective on deep shag; spray is weak.	250

Table 16.1 Data Sheet for Feasibility Study

Arrange to test the equipment; observe the differences in performance and ease of use among these machines. Do the differences in performance justify the difference in price?

Find out if any of the suppliers offer "package deals" (such as free carpet-cleaning shampoo with purchase), or quantity discounts.

Tabulate all data in a notebook so that it will be easy to compare one supplier's equipment and prices with another's.

The first four items on the list suggest a considerable amount of footwork, of making contact with people in the business, is necessary. Be sure to obtain the names and telephone numbers of everyone you contact; you need to ask follow-up questions.

Step Two: Conduct the investigation. List in hand, you contact the suppliers, arrange to inspect the equipment, and record all relevant data. Your finished data sheet might look like the one in Table 16.1.

Once again, ask questions! The experts know more than you do, and they will almost always supply you with the information you need. Be candid about the purpose of your research.

ORGANIZING AND WRITING THE FEASIBILITY REPORT

As soon as you feel that you have more or less completed your investigation, it is time to begin planning out the report. Of course, once you get the outline or draft going, you may discover a gap in your research—actually, this happens a lot. The act of writing can often makes you more aware of what you need to research than you may be while actually doing the research.

Step Three: Prepare an outline for the report. First, you set up the basic structural elements:

I. INTRODUCTION
   A. Purpose
   B. Background Information
      1. Explanation of method of investigation
      2. Problems, limitations
   C. Scope of the Report
II. COLLECTED DATA
   A. Testing
      1. Model X
      2. Model Y
      3. Model Z
   B. Analysis
III. CONCLUSIONS
IV. RECOMMENDATIONS

Next, you jot down a few notes for each element; for example:

I. A. Purpose.
      To determine the most effective, reliable, and economical steam-cleaning equipment for the small-business person.
   B. Background information.
      Three good steam-cleaning machines avail. cheap; two retailers carry them; will need to compare prices. These are relatively unpublicized brands; will need to be tested.

Step Four: Begin the draft of your report. One way to begin the draft is simply to expand your outline, as follows:

I. INTRODUCTION
   A. Purpose: The purpose of this report is to determine the feasibility of using relatively inexpensive carpet-cleaning equipment to do a professional quality job.
   B. Background Information.

1. Method of investigation: The investigator examined and tested the two most popular low-priced steam cleaners in use, studied their respective warranties, and determined which supplier offered the best price.

2. Method of testing: Each steam cleaner test was conducted by the investigator, using carpet swatches of typical neap thicknesses made available by the suppliers. The same carpet shampoo (Formula 500, industrial strength) was used for all tests.

C. Scope of the report: This report first describes the two top carpet cleaners' major features and test results. Next, it analyzes the costs of the equipment. Finally, it evaluates the findings and recommends which machine and which brand of chemicals to purchase.

II. COLLECTED DATA

A. Steam Cleaners Selected for Testing: The following steam cleaners of comparable price reputation were chosen for testing:

1. DeepKleen (Argus Equipment Co., Inc.)
2. Steemer (Elliott & Sons Mfg.)
3. Miracle Clean (Crown Co.)

B. Cost Comparisons.

1. DeepKleen: $355 (full accessories) at Jones Hardware ($7.00 less than at Price-Cut)
2. Steemer: $405 (full accessories) at Jones Hardware ($11.00 less than at Price-Cut)
3. Miracle Clean: $250 (accessories extra) at Price-Cut ($4.00 less than at Jones Hardware)

C. Test Results.

1. DeepKleen: Handles easily with nonshag carpeting; not as easily with shag; jet spray relatively weak. Some slight streaking detected after cleaning.
2. Steemer: Handles with remarkable ease on all types of carpeting; strong, penetrating jet spray. Excellent cleaning results.
3. Miracle Clean: Doesn't clean as thoroughly as the others; hard to maneuver; weak spray.

III. CONCLUSIONS

A. The best performing steam cleaner was the Elliott & Sons Steemer. It cleaned all of the samples with equal thoroughness, ease, and speed. It has the longest warranty: three years.

B. DeepKleen and Miracle Clean worked best on the nonshag samples; they both left slight stains on the shag samples. Miracle Clean was more difficult to maneuver than the other two. Both of these machines carry only one-year warranties.

IV. RECOMMENDATIONS

Based on the testing described above, Steemer is the best choice. It is the most expensive of the three, but its overall performance, along with the three-year warranty, justifies the additional expense.

If you have prepared an elaborate outline like the preceding one, drafting the report may be little more than removing the framework and adding some complete sentences. However, some employers will prefer to receive strictly practical documents like feasibility and progress reports in "bare bones" outline format; the more telegraphic the better.

Step Five: Edit the report for accuracy, style, appropriate format. Never slight this stage. Every document you submit in the workplace must be error free. One misspelled word or typo, one grammatical or punctuation error, can cause your readers to wonder about your accuracy and observational skills in more substantive areas. This may seem unfair, but it is the way of the world. Edit with great care!

Also, be sure to format your final draft properly. Identify each segment of your introduction and collected data with subheads. See "Formatting the Report," below. Consult Appendix C for models of these report elements.

Outlining, as you will recall from Chapter 5, is a heuristic device: doing it brings particular aspects of your topic into focus that otherwise might have been overlooked. Writers commonly add, delete, and rearrange material in their outlines—a process which continues into the drafting stage.

If you have prepared an elaborate outline, drafting the report may be little more than removing the framework and adding some complete sentences. However, for many persons, drafting anything, even the most tightly structured report, can be a frustrating and messy business.

FORMATTING THE REPORT

Your first draft concentrated on the "guts" of the report. Now, in addition to revising to include additional necessary information and delete any unnecessary information, and to strengthen readability, you take care to format the document in a manner that facilitates easy reference, paying special attention to informa-tive, logically sequenced subheads, use of conspicuous textual markers such as bullets and boxes, and a clear interconnection between text and visual aids.

Finally, you prepare the "front matter": a letter of transmittal; a title page; a table of contents (if the report has several subdivisions and is more than four pages long); a table of figures and tables (if there is more than one); a glossary (if there are more than two technical terms central to the report); and an infor-mative abstract. Refer to Appendix C for the appropriate models.

DRAFTING A FEASIBILITY REPORT: A CHECKLIST

☐ Do I have a clear understanding of the problem?

☐ Have I determined which sites to visit in order to carry out my research?

☐ Are the alternatives I present accurately represented?

☐ Have I contacted the appropriate experts for background information?

☐ Did I include appropriate tables or figures to facilitate understanding or provide evidence?

☐ Have I carefully represented the test results?

☐ Are my conclusions definitive, based soundly on the collected evidence?

☐ Are my recommendations based soundly on my conclusions?

■ CHAPTER SUMMARY

Study of feasibility is a form of analytical research that determines degree of possibility or capability for one or more of the available alternatives. Readers of feasibility reports want answers to two basic questions: (1) What are the available alternatives; (2) Which alternative is best for our needs, and why? Feasibility studies are important for business, environmental, and urban plan-ning. To prepare a good feasibility study, one first works out a plan of investi-gation, then conducts the investigation (which would likely include testing, observing, gathering testimonials through interviews and published reports, comparing, evaluating); then plans the report structure. Outlines are especially useful in the drafting of feasibility reports.

■ FOR DISCUSSION

1. Suggest a problem in the context of your major that could be the basis for a feasibility study.

2. What specific tasks would need to be done for each of the following proposed feasibility studies?
   a. Should the city of Fair Oaks transform a nearby swamp into a shopping mall?
   b. Which anti-computer-virus program is best for our system?
   c. Which brand and grade of exterior latex paint would be best suited for our wood-frame house?
   d. What type of microwave oven would be best suited for our family?
   e. What kind of herbicide would be most effective against the weeds invading my vegetable garden?

■ FOR YOUR NOTEBOOK

1. List potential feasibility studies that could be conducted at various sites on your campus, such as the football or track field, the library, the media center, or the student union.

2. Jot down possible alternatives to your existing

   methods of study
   note taking during lectures
   diet
   exercise regimen
   savings or investment program

   Which of the available alternatives would be an improvement over the existing one? Why?

■ WRITING PROJECTS

1. Write an abstract for the feasibility report, "Restoring Diversity in Salt Marshes: Can We Do It?"

2. Prepare a feasibility report that answers the following question: What are the most environmentally safe alternatives to existing mercury and cadmium household batteries? The excerpts from an environmental impact report on municipal solid waste (pages 380–389) should prove helpful.

3. Conduct a feasibility study and write a report on one of the topics listed in For Discussion, topic 2.

RESTORING DIVERSITY IN SALT MARSHES
Can We Do It?

Joy B. Zedler
Professor of Biology, San Diego State University,
San Diego, California

Along the U.S. coastline, development has reduced the area of coastal wetlands and endangered certain wetland-dependent species. Despite the threats to biodiversity, development of wetland habitat is still permitted by regulatory agencies if project damages can be mitigated by improving degraded wetlands or creating new wetlands from uplands. For example, the California Coastal Act allows one-fourth of a degraded wetland to be destroyed if the remaining three-fourths is enhanced. The expectation is that increased habitat quality will compensate for decreased quantity.

The concept sounds reasonable, but biodiversity is continuing to decline. Why? First, the process allows a loss of habitat acreage. Second, there is no assurance that wetland ecosystems can be manipulated to fulfill restoration promises. The magnitude of the problem is well illustrated by examples from southern California, where more than 75% of the coastal wetland acreage has already been destroyed, where wetland-dependent species have become endangered with extinction, and, where coastal development pressures rank highest in the nation. This [report] reviews several restoration plans and implementation projects and suggests measures needed to reverse the trend of declining diversity.

RESTORATION PLANS

Several large development projects in southern California wetlands have recently been approved by the California Coastal Commission (see Figure 1). Three federally endangered species are affected by such projects: the California least tern (*Sterna albifrons browni*), light-footed clapper rail (*Rallus longirostris levipes*), and salt marsh bird's beak (*Cordylanthus maritimus* spp. *maritimus;* see Figure 2). There would also be an impact on the Belding's Savannah sparrow

E. O. Wilson, ed., *Biodiversity*. Washington, DC: National Academy Press, 1986: 317–325.

(continued)

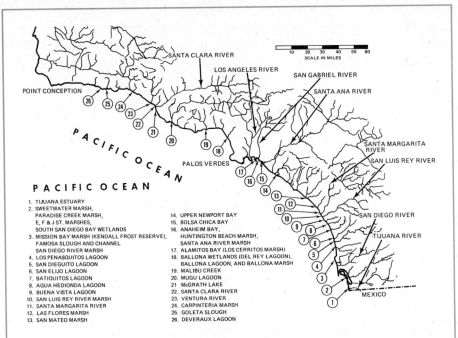

FIGURE 1 Sites of some coastal development projects in southern California. In all, there are 26 coastal wetlands between Point Conception and the Mexico-U.S. border.

Map legend:

1. TIJUANA ESTUARY
2. SWEETWATER MARSH,
   PARADISE CREEK MARSH,
   E, F & J ST. MARSHES,
   SOUTH SAN DIEGO BAY WETLANDS
3. MISSION BAY MARSH (KENDALL FROST RESERVE),
   FAMOSA SLOUGH AND CHANNEL
   SAN DIEGO RIVER MARSH
4. LOS PENASQUITOS LAGOON
5. SAN DIEGUITO LAGOON
6. SAN ELIJO LAGOON
7. BATIQUITOS LAGOON
8. AGUA HEDIONDA LAGOON
9. BUENA VISTA LAGOON
10. SAN LUIS REY RIVER MARSH
11. SANTA MARGARITA RIVER
12. LAS FLORES MARSH
13. SAN MATEO MARSH
14. UPPER NEWPORT BAY
15. BOLSA CHICA BAY
16. ANAHEIM BAY,
    HUNTINGTON BEACH MARSH,
    SANTA ANA RIVER MARSH
17. ALAMITOS BAY (LOS CERRITOS MARSH)
18. BALLONA WETLANDS (DEL REY LAGOON),
    BALLONA LAGOON, AND BALLONA MARSH
19. MALIBU CREEK
20. MUGU LAGOON
21. McGRATH LAKE
22. SANTA CLARA RIVER
23. VENTURA RIVER
24. CARPINTERIA MARSH
25. GOLETA SLOUGH
26. DEVERAUX LAGOON

(*Passerculus sandwichensis beldingi*), which is listed as endangered by the state, and on several plant species of regional concern (Ferren, 1985).

Projects That Show Losses in Wetland Area

At Bolsa Chica Wetland, more than 1,200 acres (480 hectares) of lagoonal wetland will be reduced to 951 acres (366 hectares) of restored wetland (California State Coastal Conservancy, 1984). Mitigation plans are not final, but the draft concept includes cutting an ocean inlet to serve a new marina. Inland from the marina are sites for restored wetlands with controlled tidal flushing. Uplands designated as "environmentally sensitive habitat areas" that lie within the lowland area and that will be destroyed during development are to be relocated adjacent to the restored wetland in a bluff-edge (linear) park. The draft concept plan accommodates development, but does not ensure maintenance of biodiversity. The restoration activities are based on the assumption that

(continued)

FIGURE 2   The salt marsh bird's beak grows near the upper wetland edge. As an annual plant, its seeds germinate after winter rainfall to maintain the population; as a hemiparasite, its seedlings grow roots that can attach to those of other plants, thereby increasing its supplies of water and nutrients. Photo by J. Zedler.

habitat values can be created and moved about at will.

In Los Angeles Harbor, about 400 acres (160 hectares) of shallow water fisheries habitat will be filled to construct new port facilities. At this project site, there is no habitat available to be restored—all the wetlands have been filled or dredged. Thus, off-site mitigation has been approved. Batiquitos Lagoon, more than 80 miles (130 kilometers) south of Los Angeles, will be dredged to create deep-water habitat and increase tidal flushing. According to plans (California State Coastal Conservancy, 1986), the net loss of aquatic habitat in Los Angeles will be mitigated by altering (not increasing) habitat elsewhere. The dredging of Batiquitos Lagoon will remove sediments and, at least temporarily, solve the occasional problems of algal blooms (odors and fish kills after sewage spills). However, maximizing tidal flushing at Batiquitos Lagoon (to replace fisheries habitat in Los Angeles Harbor) will destroy existing salt marsh habitat (Figure 3) and reduce the area of shallow water and mudflat habitat. The mitigation plan contains two strikes against biodiversity—the loss of area and the loss of existing functional wetland types.

(continued)

At Aqua Hedionda Lagoon, about 14 acres (5.6 hectares) of wetland were filled to build a four-lane road. The mitigation plan (U.S. Army Corps of Engineers, 1985) promised to enhance diversity and increase the functional capacity of the lagoon. Brackish-water ponds were planned for a wetland transitional area (itself a rare habitat type); a 2-acre (0.8-hectare) dredge spoil island was to be built for bird nesting; and a 7-acre (2.8-hectare) debris basin was proposed within a riparian area to reduce sedimentation into the lagoon. Flaws in the plan became clear when construction of the brackish ponds began. Pits were dug to a depth of 6 feet (1.8 meters) without encountering groundwater. Areas that were modified included transition habitat, pickleweed marsh (Figure 4), brackish marsh, and riparian habitat. The wetland lost both acreage and habitat quality.

All these projects show a net loss in wetland habitat area. Proponents argue that the lost areas are already degraded. However, they could be enhanced to maintain biodiversity. The fact that four wet-

FIGURE 3 Salt flats may appear to have low habitat value, but many unusual insects, some of them threatened with extinction, are found only in these open areas. In the winter, runoff and high tides inundate the areas, and they become highly productive ecosystems. What appears to be barren in summer is heavily used by shorebirds and dabbling ducks in winter, as migrants visit the flats and feed on the abundant insects and algae. Photo by J. Zedler.

(continued)

FIGURE 4 Pickleweed marsh may seem monotonous, but close inspection will reveal a variety of insects, invertebrates, and dozens of species of microscopic algae. Individuals and trails of the California horn snail (*Centhidea californica*) are visible in the tidal pool. Photo by J. Zedler.

land-dependent species have become endangered in Southern California while coastal wetlands have shrunk by 75% indicates a cause-effect relationship. There is some minimum area required to support regional biodiversity and a limit to the number of species that can be packed into individual wetlands. Populations are dynamic; some migrate and use several wetlands, whereas others experience local declines and must reinvade from another refuge. The need to maximize area available for wetland species is indicated by population declines that have followed human disturbance and environmental catastrophes.

Several species may be lost simultaneously if a wetland experiences multiple catastrophes. At Tijuana Estuary, the combination of destabilized dune sands (following long-term trampling), the winter storms of 1983, dune washovers, and channel sedimentation led to closure of the ocean inlet in April 1984. The drought of 1984 coincided with an 8-month nontidal period. The population of endangered light-footed clapper rails dropped from about 40 pairs to 0 and did not fully recover after tidal flushing was restored (16 pairs were present in 1987). In

(continued)

addition, there were major declines of three salt-marsh plant species—cordgrass (*Spartina foliosa*), annual pickleweed (*Salicornia bigelovii*), and sea-blite (*Suaeda esteroa*)—and none has recovered to pre-1983 levels. This salt marsh has shifted from the region's most diverse to a species-poor wetland (Zedler and Nordby, 1986).

Maintenance of the region's resources through years of wet and dry periods, with and without closure to tidal flushing, requires that each habitat type be maintained at several different wetlands so there will be refuges during periods of environmental extremes. Further losses in habitat area cannot be justified.

Projects That Replace Functional Wetland Habitat with Modified Wetland Habitat

Some restoration projects retain acreage but exchange one type of habitat for another. In these cases, functional wetland habitats may be destroyed in order to create some other habitat type. Following are some examples.

The City of Chula Vista's Bayfront Development Plan calls for several developments near and in the last major salt marsh within San Diego Bay (90% of the Bay's wetland has already been developed). The plan includes a multistory hotel and a nature center to be built on an island that is surrounded by Sweetwater Marsh and San Diego Bay. Residential and commercial buildings would surround the marsh. To provide access, three roads are to be built over the wetland. The plan will also require modification of the wetland for the construction of debris basins. Wetland restoration is planned to mitigate impacts. The U.S. Fish and Wildlife Service has concluded that portions of the project jeopardize the following endangered species: California least tern, light-footed clapper rail, California brown pelican (*Pelicanus occidentalis*), and salt marsh bird's beak.

At Los Cerritos Wetland near Los Angeles, a complicated plan (California State Coastal Conservancy, 1982) proposes development of some wetland in exchange for an equal area of wetland creation. In all, 129 acres (51.6 hectares) of wetland will be retained. Some dikes that now prevent tidal flushing will be breached; some new areas will be graded to allow tidal flow. The restored wetlands will be divided into four segments and surrounded by high-density urban uses. Buffers between the wetland and development are as narrow as 25 feet (7.5 meters) for much of the project. A main concern is that existing wetland habitat will be lost and that the artificially created replacements cannot guarantee maintenance of biodiversity.

(continued)

At Upper Newport Bay, a sediment control plan within a California State Ecological Reserve has received wide political support, in part because dredging in the upper bay reduces sedimentation in the lower bay's marina. Sedimentation in the upper bay is a long-term threat to the marsh habitat, but sudden changes in hydrology may have a negative impact on the habitat of endangered species. Upper Newport Bay has the highest density of light-footed clapper rails in the state of California and some of the region's most robust cordgrass vegetation. Shallow-water and transitional habitats are being traded for deeper channels, and the value of the new habitats to biodiversity is uncertain.

At Buena Vista Lagoon, sediment control measures were also taken. Shallow-water areas were deepened, and dredge spoils were placed alongside them in the wetland. The dredge spoil islands became hypersaline, bricklike substrates that have not developed the desired vegetation or significantly improved the status of the least tern population.

CONCLUSIONS CONCERNING RESTORATION PLANNING

Several observations on the status of restoration plans can be made:

- Many large projects result in a loss of wetland acreage.

- Mitigation measures for lost habitat often involve changing one type of wetland habitat into another, rather than creating wetland from upland habitat.

- Proposed projects are planned and reviewed individually rather than with a regional perspective. While cumulative impacts may be considered, there is no regional coordination to set priorities and guide decision making. There is no way to ensure that the wetlands with the greatest potential for maintaining clapper rails will be managed for clapper rails.

- There is no single source of information on restoration projects, no center that keeps records on changes in biodiversity to ensure that resource agencies are aware of changes in individual wetlands, no comprehensive monitoring programs to assess changes in biodiversity (although some endangered species are censused annually), and no mechanism to require suitable and comparable methods for the few monitoring programs that have been planned.

(continued)

IMPLEMENTATION OF RESTORATION PROJECTS

To enhance, restore, or create wetland habitat requires manipulation of the physical environment (especially the topography and the degree and timing of fresh- and seawater influence) as well as the biota (e.g., by introducing target species and eliminating undesirable ones). Research in this area is just beginning; most of the work has been done on a trial-and-error basis, and the evaluation criteria are not yet standard.

Assessing Success

Restoration success must be measured in time scales that relate to the species being managed and to the periodicity of extreme environmental conditions characteristic of the region. Successful creation of clapper rail habitat cannot be measured by censusing mortality of cordgrass a few weeks after transplantation. Rather, such projects need to be followed at least until clapper rails establish breeding populations. Measures of restoration success must be done within spatial scales that relate to whole ecosystems. The degree to which breaching a dike and restoring tidal flushing can enhance a lagoon must be measured beyond channel biota and water quality, because there will also be substantial impacts on intertidal marshes. Likewise, dredging to improve fish diversity cannot be considered successful if endangered birds become extinct in the process. In short, restoration success must be measured at the ecosystem level and with long-term evaluation. To date, this has not been done.

Summary of Trials

All the projects described above incorporate some element of habitat creation or restoration, for which there are no guaranteed benefits to threatened species. Many projects have been designed to restore wetlands and mitigate losses, but in no case have ecosystem functions been duplicated, nor have endangered species been rescued from the threat of extinction. Projects to reduce sedimentation have as one goal the creation of fish and benthic invertebrate habitat (sometimes to provide food for the California least tern). Marsh restoration projects have focused on vegetation used by target bird species (cordgrass for clapper rails, pickleweed for Belding's Savannah sparrows).

While some wetland plant species can be transplanted successfully and others will invade voluntarily given suitable conditions

(continued)

(Zedler, 1984), there are only a few cases where the marsh ecosystem has been monitored for several years (Broome et al., 1986; Homziak et al., 1982) and no example of a threatened species that has been increased as desired. Attempts to restore wetlands in southern California have generally failed to attract target species. In a few cases, the California least tern has nested on dredge spoil islands—but not always where its use was planned. One briefly successful site was an 80-acre (132-hectare) island in south San Diego Bay, which was planned for salt marsh and fish habitat.

Conclusions Concerning Implementation

Four observations on the status of wetland habitat restoration in southern California can be made.

- Selected plant species (e.g., cordgrass) can be transplanted successfully.

- No plant or animal populations have been taken off the endangered list as a result of restoration projects.

- Wetland restoration assessments have not been made for entire ecosystems but have been limited to one or a few target species.

- No studies have been conducted to determine the minimum wetland area or configuration of multiple wetlands required to maintain regional biodiversity.

Therefore, it is premature to conclude that an artificial tidal wetland can develop and replace the functions of a natural one. Furthermore, there is no evidence that restoration of degraded wetland habitat can compensate for lost habitat area.

PROSPECTS FOR THE FUTURE

To restore biodiversity in the nation's coastal wetlands, we must understand the factors controlling these ecosystems and develop the ability to modify them to meet desired management goals. We must make substantial advances in ecotechnology—the scientifically sound manipulation of ecosystems to maintain natural diversity and achieve specific management objectives. The field is relatively new in ecology. Only one journal, *Restoration and Management Notes*, and a few books focus on the topic. Although most of the work in this area concerns disturbed ecosystems, all ecosystems need some

(continued)

management to maintain their natural hydrology as well as air and water quality.

Ecosystems of greatest concern tend to be those whose areas have been reduced and whose species are threatened with extinction. Rare species are difficult to study, because the conditions that allowed them to thrive no longer exist. Manipulative experimentation, required to establish cause-effect relationships, cannot always be done without threatening the endangered populations even further. Bringing animals or plants (even seeds) into the laboratory may reduce field populations to levels that jeopardize population recruitment. Thus, maintenance of biodiversity must be based on an understanding of the factors that control the ecosystems in which rare species persist—the type of long-term, ecosystem-level research now funded by the National Science Foundation, the National Oceanic and Atmospheric Administration through its National Sea Grant College Program, and other agencies. A new research emphasis could allow major advances to be made in wetland ecotechnology. I recommend manipulative experimentation, first in replicate mesocosms (medium-size artificial ecosystems), followed by experimental restoration at the ecosystem level. This approach was adopted by the U.S. Environmental Protection Agency in their research plan for the nation's wetlands (Zedler and Kentula, 1985).

Ecosystem-level experiments have not been incorporated into wetland restoration projects. The contention that artificial or restored wetlands can maintain biodiversity must be tested. Every restoration project can include experimentation in its design, e.g., to provide different tidal flows; to test different hydroperiods, salinities, and nutrient inputs; to use different transplantation regimes; or to vary the width of buffer zones, with treatments appropriately replicated. Detailed, long-term evaluation of the experiments will document success or failure to maintain natural diversity. In either event, we will learn whether it can be done and why it succeeds or fails. The present practice of poorly planned, unreplicated, undocumented *trials* leads mainly to *errors* whose causes cannot be identified. Only as our understanding of factors controlling wetland ecosystems improves can we ensure the restoration and maintenance of biodiversity.

ACKNOWLEDGMENTS

Research on wetland restoration was funded in part by NOAA, National Sea Grant College Program, Department of Commerce,

(continued)

under grant number NA80AA-D-00120, project number R/CZ-51, through the California Sea Grant College Program, and in part by the California State Resources Agency.

REFERENCES

Broome, S. W., E. D. Seneca, and W. W. Woodhouse, Jr. 1986. Long-term growth and development of transplants of the salt-marsh grass *Spartina alterniflora*. *Estuaries* 9:63–74.

California State Coastal Conservancy. 1982. *Los Cerritos Wetlands: Alternative Wetland Restoration Plans Report*. State of California—Resources Agency, State Coastal Conservancy, Oakland. 49 pp. + appendixes.

California State Coastal Conservancy. 1984. *Staff Presentation for Public Hearing: Bolsa Chica Habitat Conservation Plan*. State of California—Resources Agency, State Coastal Conservancy, Oakland. 31 pp. + exhibits.

California State Coastal Conservancy. 1986. *Batiquitos Lagoon Enhancement Plan Draft*. State of California—Resources Agency, State Coastal Conservancy, Oakland. 183 pp. + appendixes.

Ferren, W., Jr. 1985. *Carpinteria Salt Marsh*. Publication #4, Herbarium, University of California, Santa Barbara, Calif. 300 pp.

Homziak, J., M. S. Fonseca, and W. J. Kenworthy. 1982. Macrobenthic community structure in a transplanted eelgrass (*Zostera marina*) meadow. *Mar. Ecol. Prog. Ser.* 9:211–221.

U.S. Army Corps of Engineers. 1985. Public Notice of Application for Permit, Application No. 85–137–AA. U.S. Army Corps of Engineers, Los Angeles District. 13 pp.

Zedler, J. B. 1984. *Salt Marsh Restoration: A Guidebook for Southern California*. Report No. T-CSGCP-009. California Sea Grant, La Jolla, Calif. 46 pp.

Zedler, J. B. In press. Salt marsh restoration: Lessons from California. In J. Cairns, ed. *Rehabilitating Damaged Ecosystems*. CRC Press, Boca Raton, Fla.

Zedler, J. B., and M. Kentula. 1985. *Wetland Research Plan*. Corvallis Environmental Laboratory, U.S. Environmental Protection Agency, Corvallis, Oreg. 118 pp.

Zedler, J. B., and C. S. Nordby. 1986. *The Ecology of Tijuana Estuary: An Estuarine Profile*. U.S. Fish Wildl. Serv. Biol. Rep. 85(7.5). 104 pp.

Reducing Battery Waste

CONSUMER USE

The average household used 32 batteries a year in 1988. But some heavy-use households buy up to 92 batteries a year. These households are typically young married couples, with children ages 6 to 12 or 13 to 20.[1] End-uses for household batteries were estimated in 1987 as follows: portable audio units, 30 percent; toys and games, 21 percent; flashlights and lanterns, 17 percent; photo, 12 percent; remote controls, 5 percent; clocks, 4 percent; calculators, 3 percent; smoke detectors, 3 percent; and other, 5 percent.[2]

QUANTITY OF THE PRODUCT

U.S. battery sales for 1990 were expected to be over 3 billion. Sales of alkalines were projected at 1,928,000,000 and carbon-zincs at 710,000,000.[3] Household sales of nickel cadmium rechargeables were estimated at 174,000,000.[4] Sales figures for button cells were not identified. . . .

QUANTITY OF BATTERIES DISCARDED

A report prepared for EPA by Franklin Associates estimated that battery discards into MSW [Municipal Solid Waste] in 1986 included 930 tons of cadmium, accounting for 52 percent of cadmium in MSW. Franklin projected discards of cadmium in batteries to increase to 2,035 tons (and 76 percent of cadmium in MSW) by the year 2000 (Figure 1).

Not all of the mercury and cadmium used in batteries is used in household batteries. Much of it is used in specialized military, industrial, and scientific applications, which are outside the scope of this paper. However, these other types of batteries can still have environmental impacts.

Mercury and cadmium are not the only metals used in batteries. One study estimated that 4,400 tons of nickel, 18,500 tons of zinc, and 37,400 tons of manganese dioxide were used in household batteries in 1987.[5]

From WWF, *Getting at the Source: Strategies for Reducing Municipal Solid Waste*, Washington, DC: Island Press, 1991: 111–116.

(continued)

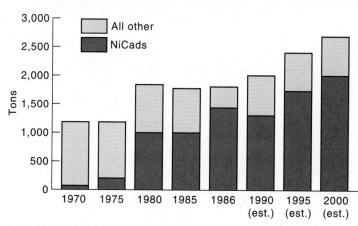

Source: Franklin Associates.

FIGURE 1   Cadmium Discarded in Municipal Solid Waste

FUNCTION OF THE PRODUCT
OR MATERIAL OF CONCERN

Mercury and cadmium are the materials of primary concern in batteries. Figure 2 presents estimates of the levels and amounts of mercury and cadmium in household batteries. (Note that these estimates are from different sources and are not necessarily comparable.)

Battery manufacturers and product designers face a complex set of trade-offs between such factors as size constraints, shelf life (how long the battery lasts when not in use), voltage, current, capacity, power density (capacity-to-weight and -size ratios), frequency of use, drain rate, cost, response to temperature changes, shape of the discharge curve, change in discharge with use, and cycle-life (for rechargeables). Choices of which materials to use in a battery, and what battery to use in a product, are made on the basis of these trade-offs.

Mercury-oxide button cells have relatively steady discharge characteristics, a high capacity-to-volume ratio, good high-temperature characteristics, and good resistance to shock, vibration, or acceleration.[6] These qualities make them suited for use in numerous products such as hearing aids, calculators, and watches. The mercury-oxide acts as the positive electrode, and thus the level cannot be reduced without reducing the capacity of the battery. Mercury-oxide batteries used 116.4 tons of mercury in 1988, 46.8 tons of which were in button cells used by consumers.[7] But, since mercury-oxides are often recycled, they probably contribute a relatively lower share of the mercury from

(continued)

FIGURE 2 Mercury and Cadmium in Household Batteries

Battery Type	Typical Mercury or Cadmium Weight per Cell (%)	Consumer Sales	Estimated Quantity of Mercury or Cadmium (Tons)
Mercury-oxide	35–50% mercury .3–.45% cadmium	Not available	32.6 tons mercury
Silver-oxide	About 1% mercury .3–.45% cadmium	Not available	0.003 tons mercury
Zinc-air	About 2% mercury 1.1–1.22% cadmium	Not available	1.05 tons mercury
Alkaline-manganese	<1% mercury .25–.54% cadmium	1,928,000,000	256.8 tons mercury 16.5 tons cadmium
Carbon-zinc	<0.01% mercury .366–.6% mercury	710,000,000	3.52 tons mercury 20.6 tons cadmium
Lithium	None	Not available	None
Nickel-cadmium rechargeables	18–22% cadmium	174,000,000	1,200 tons cadmium (in appliances) 262.5 tons cadmium (sold separately)

Note: These estimates are from different sources and may not be comparable.

Source: Carnegie-Mellon; Minnesota Pollution Control Agency; National Electrical Manufacturers Association.

batteries entering the environment.

The U.S. market for mercury-oxides totals $195 million, of which $100 million is in hearing aids and $60 million in watches. Mercury-oxide batteries currently have about 40 percent of the button-cell market for hearing aids. The number used in hearing aids is growing at about 4 percent a year, but the share is rapidly declining as they are being replaced by zinc-air batteries.[8] There will probably be a shift away from mercury-oxides to lithium and zinc-air batteries in other uses as well.

The other types of batteries that contain mercury have much lower levels, of a few percent or a fraction of a percent. Mercury is used to coat the negative electrode and acts to prevent the formation of bubbles of hydrogen gas. Such gassing lowers the voltage available from the battery and shortens its shelf life. Gas buildup may also cause leakage and corrosion by breaking the battery's safety vents (which are designed to prevent an explosion due to excessive gas pressure).

(continued)

Attention has centered on the use of mercury in alkaline batteries because of their high sales volume.

Nickel-cadmium batteries are the main source of cadmium from batteries. (Cadmium also occurs as an impurity in the zinc used in primary batteries). NiCads are 18 to 22 percent cadmium. NiCads' advantages include long cycle-life, good low-temperature and high-rate performance capability, and long shelf-life in any state of charge.[9] Because cadmium is used as the negative electrode in NiCads, the amount cannot be reduced without reducing battery capacity.

The National Electrical Manufacturers Association (NEMA) estimates that 75 to 80 percent of NiCad batteries used by consumers are sealed into appliances. The remaining 20 percent of NiCad batteries, purchased by consumers in blister packs in drug and hardware stores, represents only 1 percent of household batteries that are purchased and disposed of. This share is likely to increase in the future. More appliances will be built with rechargeable batteries that are removable, in order to facilitate recycling.

TOXICITY

Household batteries are an issue in MSW disposal because of the toxicity of the metals they contain. Household batteries are made of a variety of materials, including zinc, manganese dioxide, mercury, cadmium, silver, and lithium. Concern has focused on the use of mercury and cadmium, because they are toxic at low levels and because batteries are major sources of these elements in the waste stream.

Mercury can damage the kidneys and the central nervous system (producing fatigue, rash, cramps, fever, personality changes, memory loss, and delirium) and can cause fetal damage. Human exposure to mercury is most likely to occur through fish consumption, since the bioaccumulation factor in fish is from 10,000 to 100,000 the level in ambient water. According to one estimate, mercury levels of two parts per trillion in lake water may result in fish consumption advisories.[10] Thus, people who consume large quantities of fish are at increased risk. Fetuses are the most sensitive population, even though their mothers may show no adverse effects.

Cadmium is a probable human carcinogen and can cause lung and kidney disease. The ingestion of food is the prime pathway, since cadmium is bioconcentrated in grains and cereal products, and animals that feed on them, as a result of fertilizer use and disposal of sewage sludge. Cadmium residues in the general population are reaching

(continued)

maximum safe levels, and thus all exposure pathways are of importance when controlling for exposure. Populations at risk include smokers, people living near smelters and incinerators, those who eat organ meats, fish, grains, and vegetables from cadmium-enriched soil and water, and those using acidic drinking water. Various factors can affect sensitivity. Besides the human health risk, both mercury and cadmium may have other environmental effects, such as on aquatic organisms.

Use of other metals (including zinc, manganese, and nickel) raises potential human health and environmental concerns. However, knowledge of the health effects of these metals is based largely on occupational exposure at high levels. Data on the effects of the low exposure levels (such as would be attributed to the use of these metals in batteries) are very sparse, so these metals will not be discussed further.

The effects of battery disposal are not well known at this time. Exposure could come from incinerator emissions, incinerator ash, or landfill leachate. At this point, the risks from any of these sources are uncertain, but there is some information available.

Mercury is difficult to control in incinerator emissions because of its volatility. Pollution control equipment can remove 45 to 85 percent of the mercury in the flue gas, depending on the technology.[11] As would be expected, given that batteries have historically been the largest consumers of mercury, several foreign and domestic studies have shown that removing household batteries from the waste stream reduces mercury emissions from incinerators by 60 to 80 percent.[12] However, the studies do not appear to have addressed the actual risk from these emission levels. Studies have estimated that direct inhalation of mercury emissions from incinerators creates low risks to humans. However, precipitation will deposit atmospheric mercury into lakes, where it will bioaccumulate in fish and may create a risk to humans.

Incineration also produces incinerator ash. If mercury batteries are incinerated, some of the mercury will be captured in the incinerator ash. The rate of capture depends on the type of pollution control device being used. Because cadmium is not as volatile, nearly all of the cadmium is likely to be captured in the ash.[13] Tests of incinerator ash have found that samples often exceed EPA's extraction procedure (EP) toxicity test for cadmium. Depending on the final regulatory status of incinerator ash this could limit beneficial uses of the ash or force it to be disposed of as a hazardous waste. Actual leachate collected from ash monofills may fail drinking water standards but generally does not exceed the EP toxicity levels.

Industry studies done in the past have shown that discharged alkaline and carbon zincs sometimes passed the EP toxicity test and

(continued)

sometimes failed it.[14] An independent study of NiCads found that they exceeded the levels in the EP toxicity test when their casings were damaged.[15] Two studies have shown that disposing of batteries in landfills does not appear to present a threat in the short term, because the metals are not likely to escape from the batteries very quickly.[16] The long-term fate is less certain. Whether mercury and cadmium eventually enter groundwater would depend on whether landfills are designed adequately to contain the leachate and on the degree to which the metals bind to the soil (which in turn depends on the type of soil and the properties of the leachate).

INTENDED LIFE OF PRODUCT AND TRENDS IN AVERAGE LIFE SPAN

A battery's life span depends on the type of product that it is being used in. For example, a battery that lasts 100 hours when used in a low-drain application for 4 hours a day may only last 20 hours when used continuously in a high-drain application.[17] In general, battery life span has improved in recent years.

Batteries may be disposed of before the end of their useful life. Franklin Associates estimated that consumers discard half of all NiCads sold in blister packs within the first three months of purchase, because they become frustrated with recharging them.[18] (The remaining batteries are estimated to last up to four years.) NiCads sealed into consumer products will also be discarded when the appliance is thrown away, whether or not they are still working.

Manufacturers have continually been able to improve the performance of the various battery types. Eventually, though, the limits will be reached, and producers will have to switch to different chemistries to improve performance or reduce toxicity.

ECONOMIC LIFE CYCLE AND TRENDS

Some battery types are at the innovation stage (lithium); others are mature (alkalines); still others are being replaced by alternatives (carbon-zinc, mercury-oxide).

(continued)

MAJOR ALTERNATIVES

Mercury Alternatives

For *alkalines* and *carbon-zincs*, the alternative is to substitute a different material than mercury to control gassing. Manufacturers are continuing work toward reducing mercury levels. They are using the experience they have gained in Europe to introduce low-mercury batteries in the United States, with the goal of producing mercury-free batteries. The industry is also searching for substitutes that perform the same function as mercury, including indium, thallium, and organic compounds.[19] Information on the full range of substitutes is proprietary, so it is difficult to estimate the overall effect on waste-stream toxicity. In addition, there is less data on the environmental effects of the substitutes than on the material they are replacing.

The options for *mercury-oxides* involve switching to different battery chemistries that require little or no mercury. Alternatives include silver-oxide, zinc-air, and lithium.

Silver-oxide batteries can substitute for mercury-oxides in a number of applications. Their performance advantages include higher voltage, greater capacity, and superior low-temperature capacity. However, they can be considerably more expensive than mercury-oxides.[20]

Zinc-air button cells are replacing mercury-oxides in many applications. Zinc-air cells use oxygen taken from the air as the active cathode material. The air intakes are covered with a seal, and the cell is activated by removing the seal. Once started, this reaction cannot be stopped. This makes zinc-air batteries unsuitable for products used only intermittently. They work best in frequently used products with moderate drain rates, such as hearing aids. They are not well suited for low-drain applications such as watches, because they would deteriorate before using their full capacity.[21] They can sell for up to twice as much as equivalent mercury-oxide button cells, but they can last twice as long in some applications.

Lithium batteries can substitute for mercuric-oxide button cells and small alkalines. There are a number of lithium chemistries available or under development. Lithium batteries have high energy density and a long shelf-life (up to 10 years). They have been produced in various sizes, most commonly in button cells, AA, and AAA. One use has been in computers to provide current to memory chips if the power fails. They are also well suited for high-drain applications such as automated 35-mm cameras. Manufacturers are working on developing lithium batteries in popular consumer sizes, although safety issues may limit their size.

(continued)

Most lithium metals are highly reactive with air and water. Lithium batteries have safety features to prevent them from catching fire or exploding if short circuited, and their safety record has been good.[22] Questions have been raised, however, about whether they may present problems in collection and recycling programs. (Some programs deactivate them before disposal.[23]) Manufacturers are currently working to develop lithium batteries without lithium metal, which should be inherently safer. Lithium batteries are projected to increase in popularity, and their environmental implications deserve further study.

Cadmium Alternatives

The options for reducing cadmium use also involve switching to different battery types. One alternative is to increase the usage of sealed *lead-acid* batteries, which are already used in a limited number of household applications. This is not source reduction, however, since lead is also toxic. The advantage of lead is in the ease of recycling. There are no domestic recyclers of NiCads at this time, although some material is sent overseas for reclamation. By contrast, an infrastructure for lead recycling does exist (since most lead-acid car batteries are recycled). However, there are questions about the environmental records of secondary lead smelters. And lead-acid batteries cannot be recharged as many times as NiCads, so they are not as effective at conserving resources.

Rechargeable lithium batteries were seen at one time as having the potential to replace NiCads in some applications. Unlike lithium primary batteries, lithium rechargeables had serious safety problems, and production was halted. New, safer types of lithium rechargeables are under development, but safety will continue to be an even bigger issue in lithium rechargeables than in lithium primary batteries. Even when the safety problems are solved, rechargeable lithiums will probably be limited to expensive products such as lap-top computers, because they require special circuitry to control their charging and discharging.

The battery industry sees *nickel metal hydride* rechargeables, currently being developed, as the best option to reduce cadmium use. These batteries will replace cadmium with various rare earth metals. One metal hydride battery under development uses a vanadium-zirconium-titanium-nickel-chromium alloy as the negative electrode, nickel hydroxide as the positive electrode, and potassium hydroxide as the electrolyte.[24] Other possible elements include lanthanum, cerium, cobalt, and neodymium.[25]

(continued)

Nickel hydrides may be introduced for applications requiring low to medium drain rates (such as lap-top computers and VCR cameras) by 1991. Nickel hydrides have not yet attained the high drain rates required for use in appliances such as power tools. It is likely to be several years before nickel hydrides are commercially available for high drain applications.

Another set of alternatives for both mercury and cadmium is to use nonbattery alternatives such as solar cells and super-capacitors. However, those alternatives will not be addressed here.

1. "Innovative Battery Promotions Charge Drug Store Sales," *Drug Store News*, June 20, 1988, p.116; Kanner, Bernice, "Market for Batteries Is Now Supercharged," *Drug Store News*, February 1, 1988, p. 26.

2. "Batteries Are Still Making Waves Through Niche Marketing," *Drug Store News*, January 8, 1990, p. 21.

3. Information provided by the National Electrical Manufacturers Association.

4. Minnesota Pollution Control Agency, Karen Arnold, Nancy Misra, and Randall Hukriede, *Household Batteries in Minnesota: Interim Report of the Household Battery Recycling and Disposal Study* (St. Paul: Minnesota Pollution Control Agency, 1990), appendix A.

5. Carnegie Mellon University, Department of Engineering & Public Policy, School of Urban and Public Affairs, and Department of Social and Decision Sciences, *Household Batteries: Is There a Need for Change in Regulation and Disposal Procedure?* (Pittsburgh: Carnegie Mellon University, 1989), p. 18.

6. David Linden, ed., *Handbook of Batteries and Fuel Cells* (New York: McGraw-Hill, 1984), pp. 8–25.

7. Information provided by the National Electrical Manufacturers Association.

8. Ibid.

9. Linden, *Handbook of Batteries and Fuel Cells*, p. 18-1.

10. Minnesota Pollution Control Agency, Division of Water Quality, *Assessment of Mercury Contamination in Selected Minnesota Lakes and Streams: Report to the Legislative Commission on Minnesota Resources*, Executive Summary (St. Paul: Minnesota Pollution Control Agency, 1989), p. 3.

11. Minnesota Pollution Control Agency, *Household Batteries in Minnesota*, p. 20.

(continued)

12. Ibid., pp. 47–60.

13. Ibid., p. 20.

14. Ibid., p. 42.

15. Carnegie Mellon University, *Household Batteriesy*, p. 113.

16. Minnesota Pollution Control Agency, *Household Batteries in Minnesota*, p.28.

17. "Dry-cell Batteries," *Consumer Reports*, November 1987, p. 705.

18. Franklin Associates Ltd., *Characterization of Products Containing Lead and Cadmium in Municipal Waste in the United States, 1970 to 2000*, prepared for U.S. Environmental Protection Agency, Office of Solid Waste, EPA/530-SW-89-015A (Washington, D.C.: U.S. Environmental Protection Agency, January 1989), p. 157.

19. Minnesota Pollution Control Agency, *Household Batteries in Minnesota*, p. 44.

20. Science Application Inc., *Economic Impact Analysis of Effluent Limitations and Standards for the Battery Manufacturing Industry*, EPA 440/2-84-003 (Washington, D.C.: Environmental Protection Agency, February 1984), prepared for U.S. EPA, Office of Analysis and Evaluation, p. 4–12; Linden, *Handbook of Batteries and Fuel Cells*, p. 9-1.

21. Trudy Bell, "Choosing the Best Battery for Portable Equipment," *IEEE Spectrum*, March 1988, pp. 31–33.

22. Timothy Somheli, "Charging the Industry," *Appliance*, February 1990, p. 49; Douglas Bahniuk, "New Lithium Cells Charge Up Consumer Electronics," *Machine Design*, February 21, 1989, p.104; Dana Gardner, "New Look for Lithium Batteries," *Design News*, November 6, 1989, p. 97.

23. David Cohen, "Environmentally Sound Disposition of Household Batteries," presented to the Legislative Commission on Solid Waste Management Conference on Solid Waste Management and Materials Policy, New York, N.Y., February 2, 1990, p. 3; Carnegie Mellon, *Household Batteries*, p. 86.

24. "Hydride Batteries Charge Up," *Chemical Week*, May 25, 1988, p. 9.

25. Minnesota Pollution Control Agency, *Household Batteries in Minnesota*, p. 44.

17

Writing Instructions and Guidelines

The great majority of technical-writing projects involve the designing, writing, and updating of assembly/installation and operation/maintenance manuals—either for technicians inside and outside the workplace or for consumers (such as purchasers of computer systems).

It would be difficult to overstate the importance of instructional writing in industry. Without instructions, the machinery of the world could not be built, operated, or maintained. And instructions must also be written *well*. A multimillion-dollar investment in a new video digitizer, let us say, could fail because the instructions for operating it were poorly written, confusing the users. Safety guidelines for construction workers and power-equipment users could result in serious injury or death if not written with utmost clarity and precision.

To be able to write instructions with clarity and precision, you need to become familiar with the product you are writing about. This involves four things: (1) knowing the product's operating principles; (2) acquiring hands-on experience operating the product; (3) knowing the product's component parts; and (4) knowing how to care for the product.

Professional technical writers routinely acquire this "foundation knowledge" as a first step in writing instructional documents.

There are four types of procedures for which instructions are written:

1. assembly
2. operation
3. maintenance and repair
4. calculation

This chapter discusses methods of composing instructions for each type.

INSTRUCTIONS FOR ASSEMBLY

To write effective assembly instructions, the writer must think carefully about *every* step that is involved in putting parts together. This may not be as easy as it sounds: the writer, as expert in the assembly task, can all too easily overlook steps that may seem "obvious."

The best way to begin planning the document, then, is to consider your readers: technicians? If so, what kind? Members of a particular profession? Which one? Ordinary consumers (i.e., do-it-yourselfers)? Once you've decided on the general type of readership, you need to ask specific questions, such as,

"I say we read the directions
before we get any deeper."

A DOG AND HIS BOY, © Rick Stromoski

How familiar will this audience be with the terminology associated with assembling the product? Here are other specific questions to consider:

- Do I need to mention the tools required for assembly?
- Do I need to describe and/or illustrate each tool?
- Should I adopt a formal or informal style?
- If informal, should my approach be straightforward or humorous?
- Should I include an introduction or conclusion or both?
- How should the document be formatted?

Of course, any given audience type can be accommodated in more than one way, but the idea is to choose the strategies that would seem to be most suitable to the kinds of readers you envision. For completely inexperienced readers, it might be a good idea to illustrate everything—not only each stage of the procedure but each tool and each part. Images of proper positioning (of one part with another in an assembly process, say), including the way the tools or parts should be held, can also be beneficial to the beginner. Certainly, a clear introduction and a conclusion that reminds the reader to double-check every stage of the procedure would be welcome. Humor somehow works well with beginners or the "all thumbs" set—those who rush to purchase books like *DOS for the Complete Idiot* the first day it appears.

Let's consider the following set of instructions for assembling a bookcase. (Figure 17.1) How would you characterize the overall tone of the document? What rationale underlies the writer's choice of style and format? How clear are the instructions themselves?

The writer of these bookcase-assembly instructions has opted for an informal, even humorous approach. The idea, of course, is to give the impression that the assembly process is both easy and fun. Is this a good strategy for the project in question? Would it be appropriate for more complex assembly documents, such as for building a canoe or a birdhouse? Clearly, the nature or complexity of certain projects, such as building a canoe, would make this kind of approach inappropriate.

In addition to an engaging introductory paragraph to better involve the target audience, this document makes good use of another fundamental of assembly instructions: ability to keep the target audience engaged. When preparing your own assembly document, it is a good idea to plan along these lines:

- identification and description of parts, materials, tools
- preparation for assembly
- careful delineation of each step
- arrangement of steps in a clear, logical sequence
- judicious use of visual aids

Identification and description of parts, materials, tools. Notice that the author encourages the assembler to check over the parts before beginning. By doing so, the assembler quickly learns the name of each part, thereby making the instructions easier to follow: the letters in the left column correspond to the letters placed next to the images of the parts, to make identification even easier. Quantities are given for each part, so that the assembler can be certain that all the parts have been included. If not, missing (or damaged) parts may be ordered, using the part numbers provided in the righthand column. Materials and tools accompanying the assembly package (3mm Allen wrench, glue) are also labeled and identified; tools the assembler must provide (sponge, hammer) are also referred to.

Preparation for assembly. In addition to including a parts list, assembly instructions should also have a "get ready" section, such as the "Super ideas for

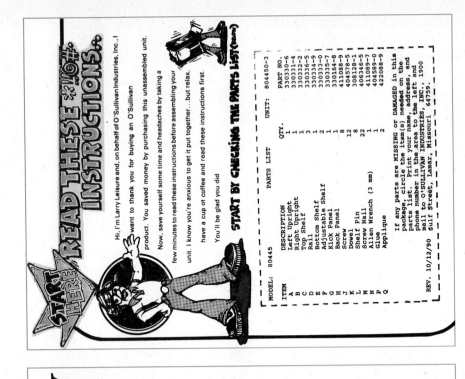

FIGURE 17.1 Assembly instructions: bookcase.

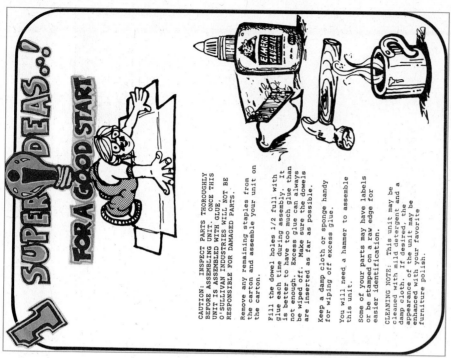

SUPER IDEAS...! FOR A GOOD START

CAUTION: INSPECT PARTS THOROUGHLY BEFORE ASSEMBLING UNIT. ONCE THIS UNIT IS ASSEMBLED WITH GLUE, O'SULLIVAN INDUSTRIES WILL NOT BE RESPONSIBLE FOR DAMAGED PARTS.

Remove any remaining staples from the carton and assemble your unit on the carton.

Fill the dowel holes 1/2 full with glue each time during assembly. It is better to have too much glue than not enough. Excess glue can always be wiped off. Make sure the dowels are inserted as far as possible.

Keep a damp cloth or sponge handy for wiping off excess glue.

You will need a hammer to assemble this unit.

Some of your parts may have labels or be stamped on a raw edge for easier identification.

CLEANING NOTE: This unit may be cleaned with mild detergent and a damp cloth. If desired, the appearance of the unit may be enhanced with your favorite furniture polish.

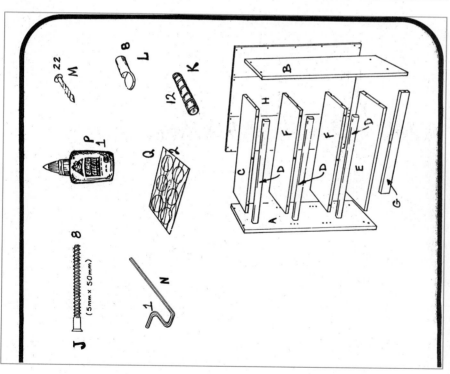

FIGURE 17.1 (continued)

2 THEN

Put it together

NOTE: RAW EDGES UP

WOODGRAIN SIDE

(5mm x 50mm)

J

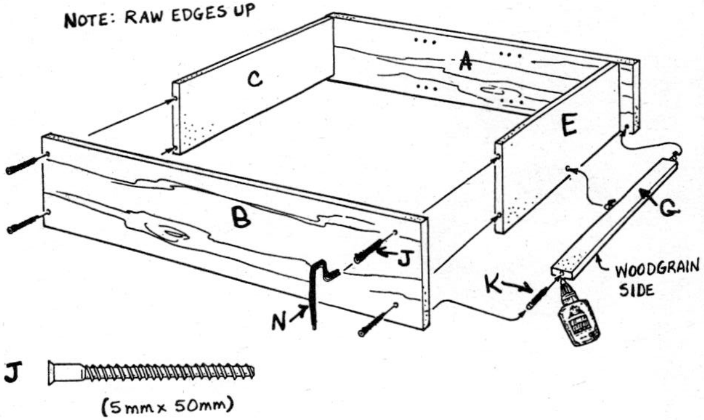

Install dowels "K" in the kick panel "G." Press the kick panel onto the bottom shelf "E" as shown. Fasten the shelves "C" and "E" to the uprights "A" and "B" with the screws "J" using the Allen wrench "N" as shown. Tighten each screw until the head of the screw is flush with the upright. DO NOT OVERTIGHTEN.

2 LET'S GET STARTED...

TOP

K

D

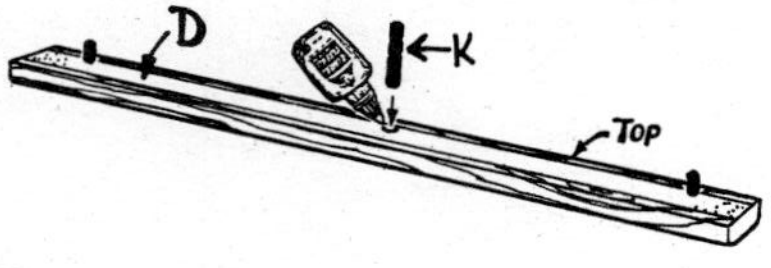

Install the dowels "K" in all three rails "D" as shown.

FIGURE 17.1 (continued)

5 THE FINAL STEP

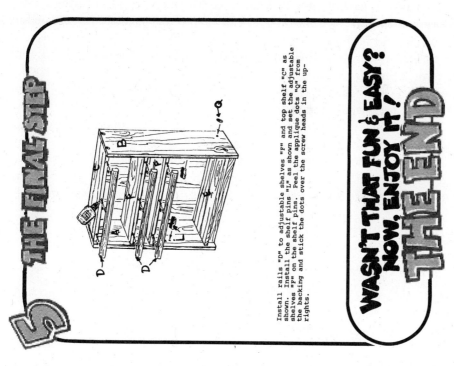

Install rails "D" to adjustable shelves "F" and top shelf "C" as
shown. Install the shelf pins "L" as shown and set the adjustable
shelves "F" on the shelf pins. Peel the applique dots "Q" from
the backing and stick the dots over the screw heads in the uprights.

WASN'T THAT FUN & EASY?
NOW, ENJOY IT!
THE END

4 NOW SQUARE-UP THE UNIT & PUT ON THE BACK PANEL...

Align the back panel "H" with the uprights "A" and "B" and
shelves "C" and "E" as shown. Push or pull one side of the unit,
if necessary, to get the back panel aligned straight and evenly on
all sides. Fasten with screw nails "M" as shown.

FIGURE 17.1 (continued)

a good start" page from the example. Here the emphasis is on common sense rather than technical know-how: inspect parts for possible damage before assembling them; use the carton as the work surface; use the glue abundantly because too much glue is better than not enough; keep a damp cloth handy; and so forth. Remember that your readers are typically "weekend" do-it-yourselfers who always appreciate being reminded of these pointers that are second nature to more experienced assemblers.

Careful delineation of each step. Once you have properly tended to the preliminaries—clear identification of parts, getting ready—the actual step-by-step instructions can be written in a concise, straightforward manner without having to include a lot of parenthetical reminders. Notice in the example that the first step of assembly (Step 3) actually consists of three short substeps:

    a. install dowels in kick panel;

    b. press kick panel onto bottom shelf;

    c. fasten shelves onto uprights, using the Allen wrench.

One might wonder if these might have been presented as three separate steps. Possibly; but the benefit derived from placing them under one step is that of coherence: in this case, the substeps all contribute to the assembly of the book-case's frame.

Arrangement of steps in a clear, logical sequence. The challenge for writers here is deciding what the assembler should do first. Sometimes the sequence will be obvious (the top, bottom, and sides of the bookcase must be assembled before the back can be attached); but other times it will not be. In the bookcase assembly, the writer makes gluing the dowels into the rails the *first* step (Step 2) of the assembly process, and installing the rails to the adjustable shelves the *last* step (Step 5) so that the glue would have a sufficient chance to dry. Step 2 could have been incorporated into Step 5 for a more experienced audience.

Judicious use of visual aids. Even the most experienced assemblers appreciate illustrations that visualize the procedure. Any ambiguity generated by the instructions is usually removed (and the clearest of instructions can contain ambiguities); also, the procedure is more easily and quickly processed when instructions and their visual counterparts reinforce each other.

INSTRUCTIONS FOR OPERATION

Operation instructions constitute the most common form of instructional writing. All work that is conducted with tools and machines, computer hardware and software, laboratory and medical equipment, requires written instructions, usually in the form of operational or "user" manuals or instruction sheets.

With any kind of operational instructions, the writer has got to proceed cautiously. Nothing can be omitted or stated in ambiguous language, or else the reader could get hurt! Even when everything is presented clearly and accurately, it is necessary to add cautionary reminders, made conspicuous by placing them in a box, perhaps accompanied by a caution symbol.

Here we will look at two common varieties of operational instructions: for using machinery (in this case, an electric sander), and for software (in this case a set of options from a paint program).

Operational Instructions for Machinery

Any machine, be it an electric toothbrush or an automobile, is accompanied by operational instructions (also maintenance instructions and safety guidelines). Very often, the instructions must include the use of secondary equipment or materials—toothpaste for the electric toothbrush, drill bits for an electric drill—and in the following example, sandpaper for an electric sander (Figure 17.2.)

The writer of these instructions has made use of a number of rhetorical techniques that enable the reader to follow the instructions without difficulty:

1. Key points are discussed separately, under their own subheads: "Sandpaper Selection," "Installing Sandpaper," "General," and "Extension Cords."

2. Emphatic and cautionary language is used to reinforce main points: "Selecting the correct size and type sandpaper *is an extremely important step* in achieving a high quality sanded finish"; "*WARNING: CHECK EXTENSION CORDS BEFORE EACH USE. . . .*" [emphasis added]

3. Explanations are given for cautionary statements. Understanding why one must or must not do something increases the likelihood that the precautions will be heeded. Example: "Do not force the tool or apply downward pressure. *The weight of the tool supplies adequate pressure for effective sanding. Excessive pressure will cause overheating. . . .*" [emphasis added]

4. Sandpaper installation is broken down into a linear series of easy-to-follow steps.

5. A diagram is used to help the user to see how to insert the sandpaper.

Despite the care the author took to make these instructions clear, accurate, and emphatic, they are not as readable as they could be. Can you see why? The prose is too clipped, too impersonal. Notice how the direct or indirect articles ("the" or "a/an") are omitted in many instances. This makes the information harder to process, especially for a beginner.

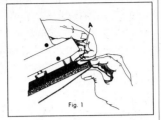
FIGURE 17.2 Operating instructions: sander.

Operational Instructions for Software

The incredible proliferation of computer software since the birth of the personal computer (late 1970s)—word processing, spreadsheets, paint and graphics, education, desktop publishing, database, utilities, entertainment—has resulted in an equal proliferation of user documentation: manuals that accompany the program diskette, as well as handbooks published separately, by independent publishers.

A writer of user manuals must, of course, be completely familiar with every facet of the program in question; that means having to interact with the program designer. The process usually works something like this: the designer of the program (say a word processing program) prepares a script that describes each capability in detail (margin settings, cutting and pasting, font size/selection, etc). A manual writer then uses the script to prepare a rough draft of the operational instructions. Meanwhile, computer programmers are busy translating the script into language the computer can understand. Much collaboration between program designer and computer programmer occurs before the program takes on anything resembling final shape. The next stage involves a lot of testing at the user end, so that awkward or complex commands are simplified, and any "bugs" in the program eliminated. All of this takes a lot of time and a lot of creative problem-solving by writers, designers, programmers, testers—all working together, drawing from each other's expertise.

One good way to get a sense of what it's like to write a computer-program operation manual is to start with one small segment. Let's say you are part of a team of writers working on a user manual for a paint-and-draw program. Your job is to draft instructions for using the basic geometrical figures the program will provide: the rectangle, rounded rectangle, oval, circle, polygon.

General principles: User first needs to be aware of elementary procedures such as "constraining."

How to select options from the tool palette (the menu of icons for brush, spray, lettering, predetermined shapes, etc.).

Basic techniques of painting shapes (painting from center, using patterns and borders).

Step-by-step instructions for working with each shape.

First you may want to do some freewriting or brainstorming (see Chapter 5) to come up with a rough sense of what the document should include. Then, after sharing your plan with others, you sit down to write a first draft. And after more sharing and drafting, you come up with a final draft such as the one in Figure 17.3.

The writers of the Superpaint™ document in Figure 17.3 quite likely mapped out their task in the following manner:

- *Give an overview of these paint techniques for the user:*
  "Painting" in the context of this computer program means creating (and erasing) patterns of dots, using the "tools" provided by the program.
- *Describe the tools, materials, and performance options available to the user:*
  Constraining; painting shapes (whole objects, such as rectangles and ovals, starting from the center; or painting borders and interior patterns).
- *Explain how the user may accomplish these techniques:*
  Select certain options with the mouse; use other options by pressing combinations of keys. (Note that the writers use screen diagrams to help the user visualize these techniques.)

The most important thing to understand about the Paint layer is that you will use it to create patterns of dots. Once you paint something, the only way to alter it (besides using the Undo command) is to erase dots or add dots.

CONSTRAIN-ING

Constraining is used to restrict a shape you are painting to some particularly useful subset of the tool's capabilities. For example, when painting an unconstrained line, you can select any pair of endpoints. When painting a constrained line, though, SuperPaint will restrict it to certain angles (e.g., vertical, horizontal, 45 degrees). Once you have selected a tool, constraints are invoked by pressing and holding one or more keys (usually Shift, but sometimes both Shift and Option) while you press the mouse button to use the tool. Generally, you must hold the constraint key(s) throughout that use of the tool.

The description of each tool includes all the constraint variations.

PAINTING SHAPES

Paint from Center

Normally, you will paint a shape by specifying the locations of its corners. Sometimes, that is not the most convenient way. That's why SuperPaint includes the Paint From Center option. You can enter this mode by choosing **Paint From Center** in the **Options** menu, or by double-clicking the Rectangle, Rounded Rectangle, Oval, or Circle. The affected shape tools reflect this mode by displaying a small crosshair. While in this mode, pressing the mouse button specifies the center of the shape, while releasing the mouse button specifies the outside location.

To restore the initial shape-drawing mode, choose **Paint from Corner** in the **Options** menu, or double-click one of the four shape tools.

Patterns and Borders

Remember that the patterns that are used to paint the borders and interiors of shapes are taken from the pattern palette. To paint a transparent shape, set the fill pattern to "None." For more information on patterns, see the discussion in Part 1.

You can also control the thickness of the border with the Line Size selector. See the discussion of the Line tool, below.

Rectangle — This tool lets you paint rectangular shapes. When you select it, the cursor becomes a solid crosshair. The thickness of the crosshair indicates which line thickness is currently selected. In Paint from Corner mode, move the cursor to a starting point, press the mouse button to select one corner, then drag the rectangle's opposite corner in any direction, as in Figure 2.1A.

Figure 2.1

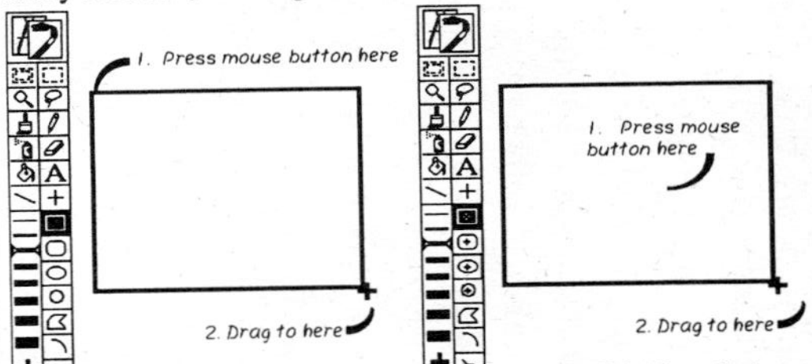

1. Press mouse button here
2. Drag to here
A. **Paint from Corner**

1. press mouse button here
2. Drag to here
B. **Paint from Center**

In Paint from Center mode, press the mouse button on the center point, then drag the rectangle's corner wherever you want it (see Figure 2.1B).

Constraints — The Shift key restricts this tool to the creation of squares.

Rounded Rectangle — This tool works just like the Rectangle tool, except it paints shapes with rounded corners.

Constraints — The Shift key lets you create squares with rounded corners.

Oval — The Oval tool lets you draw ovals. You actually specify a rectangle, within which SuperPaint paints an oval. Paint from Center works just as with the Rectangle tool.

Constraints — The Shift key restricts this tool to creating circles.

Circle — The Circle tool works precisely like a constrained Oval. With it, you can draw only circles.

Constraints — None.

FIGURE 17.3 Excerpt from a software user manual.

Polygon

The Polygon tool lets you paint a shape with as many sides as you want. Click once to set the starting point. Move the pointer and click at the second corner, then the third, etc. Double-click the last corner to complete the polygon, or click outside the window.

SuperPaint will automatically paint the final edge and fill the polygon with the current fill pattern. If the fill pattern selected is "None," when you double-click to finish the polygon, the cursor location becomes the final point. Unlike a filled polygon, a line will not be drawn from that point to the beginning point.

The polygon's border is painted according to the current line size.

Constraints The Shift key constrains the sides of the polygon to angles of 45 and 90 degrees.

Holding Shift and Option constrains polygon sides to angles of 30, 60 and 90 degrees.

Either constraint remains in effect only so long as you hold the key(s) down. You can even change (or eliminate) constraints in the midst of adding an edge to the polygon. If you press the Shift key while painting an edge, it immediately jumps to the nearest permitted angle. If you release the Shift key, it starts following the cursor again. Likewise for the Shift/Option constraint.

Freehand

The Freehand tool allows you to paint any shape you want. Move the cursor to the starting point, then drag it to outline the shape you want. Release the mouse button to define the ending point. If you have selected a fill pattern other than "None," SuperPaint automatically connects the starting and ending points.

Figure 2.2

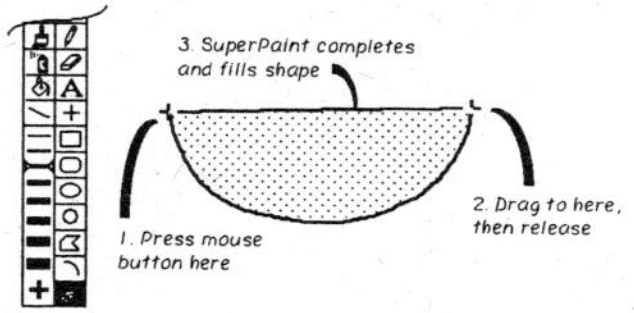

3. SuperPaint completes
and fills shape

2. Drag to here,
then release

1. Press mouse
button here

Constraints None.

14

Overview of the task; tools to be used, how the tools are to be used for completing the task: these are the stages writers of instructions should work through when designing effective, reader-oriented documents.

INSTRUCTIONS FOR MAINTENANCE AND REPAIR

Owner's manuals for machinery and tools almost always include instructions for maintaining optimum performance of the product. There may also be instructions for troubleshooting and for simple repairs (instructions for more complex repairs would often constitute a separate manual).

Maintenance When Servicing Use Only Identical Replacement Parts

Only the components shown on parts list, page six, are intended to be replaced by the customer. All other parts represent an important part of the double insulated system and are not intended to be replaced.

Avoid using solvents when cleaning plastic parts. Most plastics are susceptible to various types of commercial solvents and may be damaged by their use. Use clean cloths to remove dirt, carbon dust, etc. **WARNING: DO NOT AT ANY TIME LET BRAKE FLUIDS, GASOLINE, PENETRATING OILS, ETC. COME IN CONTACT WITH PLASTIC PARTS. THEY CONTAIN CHEMICALS THAT CAN DAMAGE, WEAKEN, AND/OR DESTROY PLASTICS.**

When electric tools are used on fiberglass boats, sports cars, wallboard, spackling compounds, or plaster, it has been found that they are subject to accelerated wear and possible premature failure, as the chips and grindings are highly abrasive to bearings, brushes, commutator, etc. Consequently, it is not recommended that this tool be used for extended work on any fiberglass material, wallboard, spackling compounds, or plaster. During any use on these materials it is extremely important that the tool is cleaned frequently by blowing with an air jet. **ALWAYS WEAR SAFETY GLASSES OR EYE SHIELDS BEFORE BEGINNING POWER TOOL OPERATION OR BLOWING DUST.**

MAINTENANCE TAPE A and TAPE B

To ensure the continued high performance of your unit, clean the heads, pinch rollers and capstans periodically whenever dust or a reddish-brown oxide has accumulated. Failure to clean these parts can result in inferior sound quality, distortion of recorded sound, deterioration of reproduction of high frequencies and inconsistent tape speed.

1. Open both cassette compartment lids by pressing both STOP/EJECT buttons.
2. Press both PLAY buttons.
3. Moisten a cotton swab (1) with head cleaner or methylated spirit and apply it to the faces of the heads (2 and 3), rubbing gently until all traces of dust or oxide are removed. Also, clean the surfaces of the pinch roller (4) and capstan (5) (Fig. 6).
4. Dry, clean and polish the faces of the heads with a piece of soft cloth.

TAPE A and TAPE B

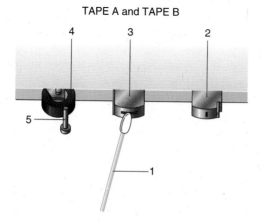

Sanyo MW 700
AM/FM Stereo Radio/Tape Recorder.

Sanyo Electric Co., Ltd. Osaka, Japan.

FIGURE 17.4   Maintenance instructions: tape recorder.

Maintenance Instructions

Most mechanisms require maintenance (upkeep) to assure proper, safe operation and long life, and there is almost always a section of the user's manual that covers maintenance. Here is the one for the power sander:

Relatively little needs to be done to maintain optimum performance of the sander; however, because wrong cleaning practices can harm the tool, "maintenance" in this context must also include what *not* to do.

Sometimes maintenance of a mechanism is best explained by a conventional paragraph giving a general explanation of the maintenance procedures, followed by a set of steps for actually performing the task. For an example, look at the maintenance instructions for a tape recorder (Figure 17.4).

Troubleshooting

The better an operator understands the workings of a machine, the more likely he or she can determine the causes of problems that are bound to occur, sooner or later. The technical writer translates the expertise of the experienced user into accurate yet easy-to-understand language for the inexperienced user. One effective way of achieving this is through the use of a three-column troubleshooting table: in Column A, the symptoms exhibited by the ill-functioning mechanism; in Column B, possible causes; in Column C, solutions to the problem. Study the troubleshooting table for a dishwasher (Figure 17.5).

Troubleshooting guides are essentially verbal tables. The possible problems are listed under the stub heading, "Problem." The two column headings are, logically, "Caused by" and "Solution," respectively. Any notations should appear at the bottom of the table.

Repair Instructions

Like maintenance instructions, repair instructions may consist of a brief overview of the task, followed by a series of steps to take to effect the repair, along with appropriate cautionary statements. For relatively simple repairs, such as patching a bicycle tube, the instructions may consist only of the set of steps, and appear on the packaging itself:

---

1. Use no gasoline. 2. Clean thoroughly with buffer. 3. Apply good coat of cement, allowing it to dry thoroughly before applying patch. 4. Remove blue plastic backing by bending at slit and peeling off. 5. Do not cement patch or touch prepared surface. Apply patch to tube, pressing well into place. 6. Printed plastic may be removed if desired. 7. Repaired article may be used immediately.

Although these are simple instructions for a simple product, the writing reveals considerable skill. In an allotted space of only $1\frac{1}{2}" \times 3"$ the author of those instructions has managed:

Common Dishwashing Problems and Solutions

PROBLEM	CAUSED BY	SOLUTION
Dishwasher will not start	Dishwasher is not receiving electrical power	Check house fuse or circuit breaker.
Dishes not washing clean	Water is not hot enough	Water temperature should be at least 140°F. Set water heater thermostat to higher setting. Run water at sink until hot before starting dishwasher and/or use Water Heat option.
	Improper loading	Make sure dishes are loaded so spray reaches all surfaces and items drain properly. Do not overload. Do not nest items.
	Spray arm not rotating freely	Check spray arms to make sure they rotate freely after loading. Be sure a utensil has not prevented turning.
	Not enough detergent or improper detergent	Use more dishwasher detergent. Use detergent with highest available phosphorous content, especially if you have hard water.
	"Old" detergent	Use only fresh dishwasher detergent. Store tightly closed in a cool, dry place. Discard old, lumpy detergent. Do not fill detergent dispensers until ready to wash.
	Water pressure may be low if dishwasher is not filling properly	If water pressure is low, do not use water for other purposes while dishwasher is running to assure correct fill levels.
Small particles deposited on items	Spray arm not rotating freely	Check spray arms to make sure they rotate freely after loading. Be sure a utensil has not prevented turning.
	"Old" detergent	Use only fresh dishwasher detergent. Store tightly closed in a cool, dry place. Discard old, lumpy detergent. Do not fill detergent dispensers until ready to wash.
	Improper loading	Make sure dishes are loaded so spray reaches all surfaces and items drain properly. Do not overload. Do not nest items.
Spotting and filming	Hard water	Fill detergent dispensers to capacity. Use dishwasher detergent with the highest available phosphorous content. May be necessary to install a water softener.
	Water is not hot enough	Water temperature should be at least 140°F. Set water heater thermostat to higher setting. Run water at sink until hot before starting dishwasher and/or use Water Heat option.
	Not enough detergent or improper detergent	Use more dishwasher Cetergent. Use detergent with highest available phosphorous content, especially if you have hard water.
	"Old" detergent	Use only fresh dishwasher detergent. Store tightly closed in a cool, dry place. Discard old, lumpy detergent. Do not fill detergent dispensers until ready to wash.
	Improper loading	Make sure dishes are loaded so spray reaches all surfaces and items drain properly. Do not overload. Do not nest items.
	No rinse agent	Does the rinse agent dispenser need filling? See instructions on page 9
	Heat Dry Off Option used	Drying without heat may cause some spotting on glasses and silverware.
Etching—permanent filming which cannot be removed	Too much detergent in soft water	Check for a rainbow hue on glasses. A rainbow hue is the first sign of etching (corrosion of glass). Reduce the amount of dishwasher detergent. Use a detergent with a lower phosphorous content.
	Inadequate rinsing	If water pressure is low, do not use water for other purposes while the dishwasher is running, to assure correct water fill levels. Make sure dishes and glassware are loaded properly to assure adequate rinsing and draining. Do not overload.
Dishes not dry	Water is not hot enough	Water temperature should be at least 140°F (60°C). Set water heater thermostat to a higher setting. Run water at sink until hot before starting and/or use WATER HEAT Option.
	Improper loading	Make sure dishes and glassware are loaded so spray reaches all surfaces and items drain properly. Do not overload. Do not nest items.
	No rinse agent	Fill rinse aid dispenser.
	Heat Dry Off	Allow more time when using HEAT DRY OFF Option, or use heated drying. Plastic items may need towel drying.
Cycle takes too long	Water is not hot enough.	Water temperature should be at least 140°F. Set water heater thermostat to higher setting. Run water at sink until hot before starting dishwasher and/or use Water Heat option.
Water left in bottom of dishwasher		Allow dishwasher to complete cycle.
	Clogged drain air gap	Some plumbing codes require the use of a drain air gap between an undercounter dishwasher and the drain system of the house. The air gap is usually located above the sink or on the top of the counter near the dishwasher to prevent the possibility of water backing up from the drain into the dishwasher due to a plugged drain. The drain air gap is not a part of the dishwasher and is not covered by the dishwasher warranty. The drain air gap should be kept clean to insure proper draining of the dishwasher.
	Drain hose kinked.	Reposition hose to eliminate kink.
Dishes and interior of dishwasher yellow or brown	Iron or manganese in water	Temporary Remedy: 1. Set empty dishwasher for Light Wash. 2. Start dishwasher and unlatch door to stop the cycle when it has filled for washing. 3. Add ¼ to ½ cup (60-120 ml) of citric acid crystals (usually available at drugstores). 4. Close and latch door to complete cycle. Permanent Solution: Install an iron or manganese filter in home water supply.
Film build-up on lower front of tub	Some detergent did not dissolve.	Use fresh dishwasher detergent. Use detergent with highest available phosphorous content, especially if you have hard water. Water temperature should be at least 140°F. Set water heater thermostat to higher setting. Run water at sink until hot before starting dishwasher and/or use Water Heat option. Remove with dishwashing detergent and warm water.

NOTE: To remove spots and film from dishes, try a vinegar rinse.
1. Wash and rinse as usual using the Heat Dry Off Option.
2. Remove all metal items.
3. Put 2 cups (500 ml) white vinegar in a container on the bottom rack.
4. Run dishwasher through complete washing cycle. Vinegar will splash out during washing.

13

FIGURE 17.5 Troubleshooting guide: dishwasher.

to incorporate cautionary remarks ("Use no gasoline"; "Do not cement patch or touch prepared surface");

to emphasize especially important points with carefully chosen expressions ("Clean thoroughly"; "allowing it [cement] to dry thoroughly"; "pressing well into place");

to mention options ("Printed plastic may be removed if desired").

Note how mentioning this option also removes any ambiguity the user may have about whether the plastic ought to be removed.

As might be expected of instructions this condensed, some ambiguity exists anyway. How much cement is a "good" amount? How long a drying period is "thoroughly"? The author in effect seems to be saying that using one's common sense to interpret these ambiguous expressions will suffice for such a relatively simple procedure as patching an inner tube.

INSTRUCTIONS FOR CALCULATION

Our daily lives are filled with number-crunching moments: maintaining budgets of all sorts (personal, family, vacation, expense account, managerial/administrative); calculating income tax; converting one standard of measure to another (such as metric to English measure); as well as solving mathematical problems. The more complex the problem, the more we rely on clear instructions to guide us through the steps to their solution. Even relatively simple problems in math can be tricky to present clearly enough for beginners to follow along. Here, for example, is a set of instructions for solving an elementary problem in calculus:

Problem: A photocopy service has a fixed cost of $2000 per month (for rent, depreciation of equipment, etc.) and variable costs of $.04 for each page it reproduces for customers. Express its total cost as a (linear) function of the number of pages copied per month.

To solve the problem:

1. Let x represent the number of pages copied per month.
2. The variable cost may then be expressed as .04x.
3. Thus, total cost = fixed cost + variable cost.
4. Expressed as a linear function of the number of pages copied per month: $f(x) = 2000 + .04x$.

The copy-service retailer could then easily compute his or her total cost by multiplying x (equal, say, to 6000 copies for a given month) by 4¢ and adding the $2000 overhead cost to arrive at a total-cost estimate of $2240 for that month.

Adapted from Goldstein, et al., *Brief Calculus and Its Applications*, 4th ed. Englewood Cliffs: Prentice-Hall, 1987: 21, 22.

As with other forms of instruction writing, it is very easy to overlook a necessary step because it seems so obvious. But such steps are anything but obvious to those first learning the problem-solving technique. Always remember to identify what each element in the equation represents, and to do so in "list" fashion (listing one element at a time), as in the preceding example.

GUIDELINES

Guidelines may be regarded as a type of activity-instruction writing that enables the reader to perform a particular task, but unlike instructions—steps for what *must* be done exactly as written—guidelines present advice on what *ought* to be done. Many guidelines are set up as checklists arranged by category, sometimes phrased as questions, and accompanied by rationales. "Poison-Proof Your Home" (Figure 17.6) is an example of such a set of guidelines.

WRITING INSTRUCTIONS AND GUIDELINES: A CHECKLIST

The following list of questions should be referred to during the initial drafting stages of your procedural document.

1. *Assembly Instructions*
   - ☐ Have I determined who my primary readers will be and slanted the document to accommodate them?
   - ☐ Did I include a list of tools and materials needed for assembly?
   - ☐ Is my approach (informal, formal; humorous, straightforward) appropriate for the assembly task presented?

2. *Operational, Repair, and Maintenance Instructions*
   - ☐ Did I include enough preliminary information?
   - ☐ Are all the required steps included and in logical sequence?
   - ☐ Did I insert cautionary statements where necessary?
   - ☐ Did I prepare a troubleshooting guide?

3. *Calculation Instructions*
   - ☐ Have I broken down the task into clear, easy-to-follow steps?
   - ☐ Have I identified all elements in the equation(s)?

4. *Guidelines*
   - ☐ Are my guidelines clear and sensible?
   - ☐ Have I included all relevant suggestions for the task at hand?

Poison-Proof Your Home

HOME SAFETY

Here's an easy checklist for possible dangers in your home. Every *"no"* answer indicates a hazard you should fix quickly.

Poison-proofing the kitchen	Yes	No
1. Do all harmful products in your kitchen cabinets have child-resistant caps? *Furniture polishes, drain cleaners and some oven cleaners should have safety packaging.*	☐	☐
2. Are all potentially harmful products in their original containers? *Labels on the original containers often list ingredients, helping experts treat poisonings. But if products are stored in containers like drinking glasses or pop bottles, someone may think they're food and swallow them.*	☐	☐
3. Are harmful products stored away from food? *If they aren't, someone may accidentally confuse a food with a poison and swallow the poison.*	☐	☐
4. Have all potentially harmful products been placed out of reach of children? *The best way to prevent poisoning is to make sure that poisons are impossible for children to find and get at. Locking all cabinets that hold dangerous products is the best poison prevention.*	☐	☐
5. Have medicines been taken off kitchen counters and window sills? *These storage places may be convenient, but they put danger within easy reach of children.*	☐	☐
Poison-proofing the bathroom		
1. Do aspirin, medicines and other potentially harmful products have child-resistant closures? *Aspirin and most prescription drugs come with caps small children can't open. Check to see if yours have them.*	☐	☐
2. Have you thrown out all out-of-date prescriptions? *As medicines get older, the chemicals inside them can change.*	☐	☐
Poison-proofing the garage		
1. Did you know that almost everything in your garage or storage area that can be swallowed is a poison? *Swallowing such everyday substances as gasoline, charcoal lighter fluid, paint thinner and remover, antifreeze and turpentine can be life-threatening.*	☐	☐
2. Do all these poisons have child-resistant caps?	☐	☐
3. Are they stored in the original containers?	☐	☐
4. Are the original labels on the containers?	☐	☐
5. Are all these harmful products locked up and out of sight and reach?	☐	☐

Be sure to keep a bottle of syrup of ipecac in your home, but don't use it unless someone from a poison control center recommends it. Check the front pages of your phone directory for the phone number of the poison control center in your area.

March 1992

FIGURE 17.6 A checklist.

CHAPTER SUMMARY

Instruction writing is the most common form of technical writing, comprising assembly; operation (for mechanisms; for software); maintenance, repair, troubleshooting; and calculation instruction. Guidelines are similar to instructions in that they present a sequence of steps, but they are recommendations or suggestions rather than requirements.

Considerable stylistic possibilities are available to the writer of instructions, depending on the nature and complexity of the task to be performed and the audience to be reached. An informal or even humorous approach can be appropriate for relatively simple instructions aimed at an audience of beginners whose confidence in assembling furniture or operating machinery is not very high.

Readers of instructional documents are best served when writers give them an overview of the task; present a clear description of the tools, materials, and performance options available; then explain how the procedures may be accomplished, using visual aids where feasible.

Instructions are often one segment of an owner's manual (or pamphlet) that accompanies the product. The document typically includes a parts list (with accompanying descriptions); safety precautions; preparation for assembly or operation; maintenance; and troubleshooting instructions.

■ FOR DISCUSSION

1. Bring to class a set of assembly or operation instructions. Form small groups and discuss the strengths and problems in each document. Keep notes on passages that everyone agrees are strong or that need reworking; also note those passages that get mixed reviews. Share the highlights of your findings with the rest of the class.

2. Working in pairs, obtain two operation manuals for a similar product, such as for two different CD players or VCRs. Evaluate each document according to clarity, thoroughness, safety precautions, and maintenance. Report your findings to the class.

3. Locate a set of instructions that you have had difficulty in understanding or following. In class, form small groups and see if others in the group have the same difficulty you had. Finally, collaborate on improving the instructions to your mutual satisfaction.

4. Bring a challenging mathematical problem to class. Form small groups and together develop a set of instructions that would help a beginner solve the problem with ease.

5. Study the maintenance instructions for a power mower (page 412). Pinpoint the rhetorical elements that make these instructions successful. Are there any elements that could be strengthened, in your opinion? How helpful are the visuals? How clear is the writing?

6. You will get a lot of laughs out of the *Star Trek* satire on page 414, which also makes a clear point about a common problem with written instructions. Discuss this problem, and why it is so prevalent.

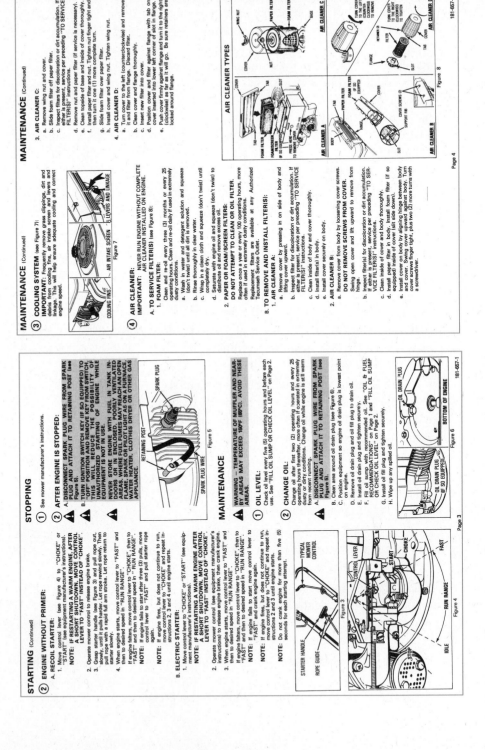

FOR YOUR NOTEBOOK

1. Compare two or more sets of instructions for similar products, such as two sets of instructions for two different brands (but same type) of answering machine or coffeemaker. Comment on the differences in clarity, concision, thoroughness of detail, effectiveness of visual aids.

2. Practice rewriting instructions to improve their clarity. Also practice rewriting instructions for an audience other than the one intended.

3. Write a set of guidelines for conducting a physics, chemistry, or biology experiment that you originally had trouble performing.

4. Analyze the strategies a writer might have used for developing a segment of a software user manual that you are familiar with. Review the discussion of the Superpaint™ program on pages 399–402 to get an idea of what to look for.

WRITING PROJECTS

1. Prepare a set of assembly instructions, complete with sections for tools/materials needed, visual aids, and precautions, for one of the following:
   a. bicycle or exercycle
   b. CD or videotape storage cabinet
   c. skateboard
   d. doghouse
   Slant the document for a teenage audience.

2. Prepare a set of operational instructions, complete with a general description of the product, precautions, and appropriate visual aids, for one of the following:
   a. popcorn popper
   b. cappuccino maker
   c. microwave oven
   d. stepladder
   Slant the document for an adult audience.

3. Prepare a troubleshooting guide for one of the products in the preceding topic 2.

4. As a follow-up on For Discussion, topic 4, write a set of instructions for calculating a tricky mathematical problem. Your audience should consist of students who are having difficulty with this kind of math.

SHOUTS AND MURMURS

STARDATE 12:00 12:00 12:00

BY CHRISTOPHER BUCKLEY

"I watch science-fiction movies. . . .I like to watch them on tapes, so I can examine them closely. There's only one problem: I still can't figure out my VCR." —William Shatner, in *TV Guide*.

CAPTAIN KIRK: Captain's log, stardate 7412.6 . . . hello? The red light still isn't going on. Testing, 1-2-3-4. Chekov, it's not recording.

CHEKOV: I know, Keptin. Perhaps a negative function with the clock-timer.

UHURA: Captain, I'm getting indications of a Klingon presence.

KIRK: Mr. Spock?

SPOCK: I confirm at least six Imperial Klingon warships, Captain, and heading toward our position at Warp 7.

KIRK: No, the Captain's log. Why won't it record?

SPOCK: Might I suggest, Captain, that we first remove ourselves to a more secure sector and then address the matter of your log? That would be the. . . logical approach.

KIRK: There's nothing logical about the instruction manual. Chekov?

CHEKOV: Keptin?

KIRK: Try this. "With the Rec-On day flashing, press the 5 key."

CHEKOV: I did already, Keptin. *Still* negative function.

SULU: Captain, I'm having difficulty holding course.

KIRK: Shut down engines. Chekov, "Press the number for the day. For Sunday, press the 1 key, for Monday, the 2 key, and so on."

CHEKOV: Affirmative, Keptin. Still negative function. Perhaps ve should go back to page 15, vere it said to press Rec-Off time and enter two digits for hour.

SPOCK: Captain, the Klingons are arming their photon torpedoes.

KIRK: Engineering.

SCOTTY: Aye, Captain?

KIRK: Mr. Scott, we've got a malfunction in the log. We're going to need full deflector power while we get it fixed.

SCOTTY: I canna guarantee it, Captain. The systems are overloaded as it is.

CHEKOV: Keptin, the flashing 12:00 disappeared!

KIRK: Good Work, Chekov!

CHEKOV: Den it came right back.

KIRK: Damn it. Analysis, Mr. Spock.

SPOCK: It would appear, Captain, that this instruction manual that you and Mr. Chekov have been attempting to decipher was written in Taiwan.

KIRK: Taiwan?

SPOCK: A small island in the Pacific Rim Sector, formerly inhabited by a determined people who believed that the adductor muscles in giant clams, *Tridacna gigas*, conferred sexual potency. In the later twentieth century, they became purveyors of early video equipment to what was then the United States. They were able to successfully emasculate the entire U.S. male population by means of impenetrable instruction manuals. It was this that eventually led to the Great Conflict.

KIRK: But this is 7412.6. How did a Taiwanese instruction manual get aboard the Enterprise?

SPOCK: It is possible that a Taiwanese computer virus was able to infiltrate Star Fleet Instruction Manual Command and subtly alter the books so that not even university-trained humans could understand them.

KIRK: It's diabolical.

SPOCK: On the contrary, it is perfectly logical. Their strategy was based on an ancient form of Oriental persuasion known as water torture.

In this case, instead of water a digital rendering of the hour of twelve o'clock is flashed repeatedly and will not disappear until the unit is correctly programmed.

KIRK: And for that you need a manual you can understand.

SPOCK: Precisely. Unless . . .

KIRK: Spit it out, Spock.

SPOCK: You have Star Log Plus. A small device that permitted the Americans to bypass the instruction manuals and program their units so that they would not end up with six hours of electronic snow instead of "Masterpiece Theatre" or, more likely, "American Gladiators."

KIRK: Could you make one of these things, Spock?

SPOCK: It would take more than the one minute and twenty seconds that we have until we are within range of the Klingon weapons.

DR. MCCOY: Jim, you know I hate to agree with Spock, but he's right. We've got to get out of here. There are hundreds of people on this ship, young people, with homes and families and futures, and pets—little hamsters on treadmills, Jim. You can't sacrifice them just because you can't figure out how to program your damn log!

KIRK: I know my responsibilities, Bones. Spock, would it be possible to beam the flashing 12:00 into the Klingons' control panel?

SPOCK: Theoretically, yes.

KIRK: Do it.

UHURA: Captain, I'm picking up a Klingon transmission.

KIRK: Put it on the screen.

KLINGONS: QI'yaH, majegh!

KIRK: Translation, Spock.

SPOCK: It appears to have worked, Captain. They are surrendering.

KIRK: Take us home, Mr. Sulu. Mr. Chekov, try pressing the OTR button twice. ♦

18

Communicating Orally in the Workplace

In industry and business, information is routinely presented orally. For example, a project manager, having spent the last several days completing a written report, calls a meeting in order to debrief his or her project team. The next day the manager must meet with clients, executives, and stockholders and give a much different presentation, one that addresses the particular concerns of this group. For them, the focus will probably shift from the technical, design aspects of the project to the marketing aspects.

How does one go about preparing effective oral presentations, including improvement of delivery and voice control? How does one make good use of visuals during presentation? How can non-native speakers of English improve their speech? These are the questions addressed in this chapter.

THE PURPOSE OF ORAL PRESENTATIONS

Reports that require a thoroughly detailed investigation of their subject are put in writing so that every facet of the investigation may be referred to whenever needed. Written documents become part of the permanent record, available to anyone at any time.

In view of this, what importance could oral presentations possibly have? Wouldn't it suffice simply to distribute copies or summaries of the document to interested persons?

These are always possible options, but oral presentations are important for reasons other than transmitting information per se. Perhaps the most important role of the oral presentation is that of achieving dynamic engagement with the audience. An effective talk will not only clarify and highlight essential ideas and data; it will also involve the audience in the issues raised, thus generating **feedback.** The underlying premise here is that no report is ever "complete," no matter how thorough a job the writer has done to gather and analyze the data. Every analysis generates new concerns; an oral presentation enables the audience to interact with the author-speaker in ways that the author alone could not have anticipated.

Also, oral presentations are (or should be) multimedia events to some degree. In presenting technical information, especially to a heterogenous group, one does not merely stand behind a lectern and read. One reinforces statements with visual illustrations, just as one does in the written report itself—but with this difference: the visuals are much more colorful and varied than in a written document. One can include a larger quantity of visuals, as well as refer to any of them more than once, in different contexts. As you can see, a well-thought-out oral presentation is a *show,* not just a lecture. That added dimension of entertainment goes a long way toward generating comprehension and involvement.

PRESENTATION TOOLS

To give an effective talk, it is important that you familiarize yourself with presentation tools and make efficient use of them.

We live in a computer and video society. Long gone are the days when an audience would sit attentively through long paper-read lectures. To speak engagingly on virtually any topic, you must supplement your talk with audio or visual aids, demonstrations, handouts, or a combination of these. In addition to the old-fashioned (but never outdated) chalkboard, the following presentation tools are commonly used:

- An overhead projector.
- $8^1/_2 \times 11$ transparencies for use with the overhead projector: blanks for making illustrations with a marker (or different-colored markers) during the presentation, or precopied/predrawn transparencies.
- A carousel slide projector for conventional slide-viewing, or a computer-linked overhead projector for projecting computer-generated images.
- Flip-charts.
- A videocassette recorder and monitor.
- An audiocassette recorder.

ORAL PRESENTATION TECHNIQUES

Like any skill, giving good oral presentations is based on certain techniques that, when studied and practiced, can be mastered without difficulty. The techniques are these:

1. *Carefully integrate words with images.* Just as in a written report, a visual aid in a talk should reinforce the concept being discussed, and vice versa.
2. *Speak directly to individuals, not to a faceless mass.* When facing a large group of people, it is all too easy to speak at them rather than to them; the audience members do not feel as involved in the talk when they feel that they are not being addressed directly. In short, you must engage different people's faces as you speak.
3. *Keep the microphone approximately one inch from your mouth.* Inexperienced speakers often keep shifting the distance between mouth and microphone, and this causes their amplified voices to modulate annoyingly.
4. *When using an overhead projector or flip-chart . . .*

- Orient yourself to the audience, not to the visuals. Learn the contents of each visual well enough to discuss it while glancing at it only briefly and occasionally.

- Organize your transparencies in the exact order of showing; taking time to shuffle through transparencies during your presentation will disrupt the continuity of your talk.

- Make sure each image is large enough for everyone in the audience to see clearly; always double-check the focus.

- Keep the image up long enough for the audience to process, but don't forget about it.

- Switch the projector off when not showing an image. Bright, blank screens or unreferred-to images are distracting.

- Do not remove a visual too soon.

5. *Be mindful of pacing.* Speak neither too slowly nor too swiftly. It is also advisable to vary your pace somewhat during your talk; avoid speaking in a monotone.

6. *Be mindful of timing,* especially if you're to speak with others on a panel. Inexperienced speakers tend to prepare a longer presentation than their time frame permits; and that cuts into the time allotments of your fellow speakers. The best way to avoid this all-too-common problem is to time your talk in rehearsal, and make whatever adjustments are necessary.

An oral presentation must be thoroughly prepared if it is to go over well. If you are uncomfortable before an audience, then you must prepare that much harder; do several rehearsals before friends. The more time you spend getting ready, the more confident you will feel and the more coherent and interesting your presentation will be. There is no such thing as being overprepared.

Clearly, giving a talk is not like acting in a play, in which the actors may envision a "fourth wall" instead of an audience. Everything about a good speech reflects audience awareness: the shorter and simpler sentences, the emphases, the intonations and pauses, the repetitions and asides. This is all part of what is known as audience interaction.

We all have heard speakers whose first words were, "Can you hear me okay?" Corny as that question might sound, it does in fact signal the speaker's desire to be interactive with his or her audience. A much more functional kind of interaction, of course, is the question-and-answer period; but even before this, during the presentation itself, speaker-audience interaction is created by eye contact, movement, and personableness.

Eye contact: This involves more than the superficial meeting the glance of one person at a time; it means addressing audience members as if they were your best friends, sitting in your living room. "Many speakers," writes Dale Carnegie, the veteran advice-giver for success in public life, "seem to be delivering a soliloquy. There is no . . . give and take between the audience and the speaker."[1] The real force behind good eye contact, in other words, is good *attitude* toward your audience.

Movement: Move around erratically and you will distract your audience from your speech and give them vertigo; stay motionless and you may tend to lean onto the lectern or stiffen up like a robot. Occasional movements, like taking a few steps toward your listeners, is a kind of gesture, reinforcing the involvement you create with eye contact. Movement also communicates energy and enthusiasm; people listen up, feel more involved with what you have to say, when you come across as energetic and enthusiastic.

Personableness: This hard-to-define quality is a cumulative impression an audience receives from a speaker, a feeling that the speaker enjoys addressing the audience, takes delight in sharing information with them. It is achieved by positive facial expressions: smiling, not scowling ("Your face," writes James G. Gray, Jr., "is your ambassador to the world"[2]) by being at ease (or at least seeming to be at ease); by being friendly in the literal sense (treating the audience as if they were your friends); and by dressing in good taste.

SCRIPTING AN ORAL PRESENTATION

When you write down what you are going to say to your audience, it is important to think about how you are going to say it and what you are going to do with your physical self while you are saying it; this is why every oral presentation you give should be thought of as a **script.** To produce a successful script for yourself, keep these three principles in mind:

1. Know your audience's needs
2. Plan a forceful and engaging delivery
3. Make sure your audience is following along.

---

[1]Dale Carnegie, *How to Develop Self-Confidence and Influence People by Public Speaking,* rev. ed. New York: Pocket Books, 1956: 101.
[2]James G. Gray, Jr., *Strategies and Skills of Technical Presentations: A Guide for Professionals in Business and Industry.* New York: Quorum Books, 1986: 94.

Know Your Audience's Needs

One big advantage that a speaker has over a writer is that the former knows in advance what type of audience he or she will be addressing. Whether specialists, generalists, preprofessionals, executives, technicians, or a mixed group, every audience type requires a different slant, and knowing ahead of time how to slant your talk will pay off. Should you devote much time to defining technical concepts? How much background information regarding the topic of discussion does your audience already know? Not to think out these matters can result in boring or confusing your audience.

Plan a Forceful and Engaging Delivery

To achieve an engaging style of delivery, it is best not to read from a paper; if you must do so, then spend at least as much time looking up from the paper and speaking extemporaneously as reading from it. One way to make this work is to write prompts in the margin—akin to stage directions in a playscript (see Figure 18.1). This scripting technique can also be used to remind yourself when to show visuals or play a tape.

A more effective presentation, though, can be achieved with notecards: 4×6 index cards are fine, although some prefer 3×5s—they're less conspicuous and easier to slip into a pocket.

What you place on a notecard should be thought of as a series of prompts; different colored inks, perhaps, for different types of prompts (see Figure 18.2). You do not want to read verbatim from notecards any more than you want to read verbatim from a paper. Of course, you must come to know your topic so well that a quick glance at the prompt—a phrase, a short sentence, sometimes just a single word—will bring what you need to say instantly to mind.

Make Sure Your Audience Is Following Along

The most attentive audiences can sometimes lose the thread of your argument for a number of reasons: maybe you've been digressing; maybe the auditorium is too stuffy; maybe you've been speaking in a monotone. You need to be alert to subtle forms of feedback. Signs of inattention include yawns, coughs, glancing at watches; frowning, scowling, head scratching or glassy-eyed gazing into space. Be prepared to slow or speed up your pacing, as the case may be, or to ask the audience, "Would you like me to go over that again?" or words to that effect. Of course, a speaker should always allow for a question-answer period at the end. This part of the presentation, which so frequently gets shortchanged

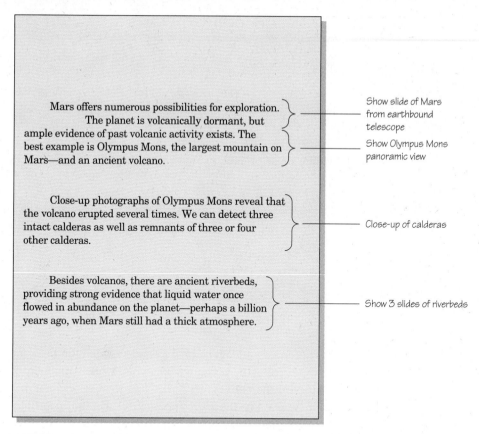

Mars offers numerous possibilities for exploration. The planet is volcanically dormant, but ample evidence of past volcanic activity exists. The best example is Olympus Mons, the largest mountain on Mars—and an ancient volcano.

— Show slide of Mars from earthbound telescope

— Show Olympus Mons panoramic view

Close-up photographs of Olympus Mons reveal that the volcano erupted several times. We can detect three intact calderas as well as remnants of three or four other calderas.

— Close-up of calderas

Besides volcanos, there are ancient riverbeds, providing strong evidence that liquid water once flowed in abundance on the planet—perhaps a billion years ago, when Mars still had a thick atmosphere.

— Show 3 slides of riverbeds

FIGURE 18.1   "Scripted" page from an oral presentation with slides.

(mainly because of not adhering to time limits!) offers the best opportunity for richly informative audience-speaker interaction.

SUGGESTIONS FOR DIALECT SPEAKERS OR NON-NATIVE SPEAKERS OF ENGLISH

Because so much information gets communicated orally in the workplace, either informally or via formal oral presentations, it is important that one's elocution or "accent" not interfere with "clarity" from the point of view of one who is not familiar with the dialect or accent. Alas, it is a fact that many people claim to be confused more out of an unwillingness to tolerate a foreign accent than an inability to understand it. Be that as it may, it is a good idea, for strictly practical "job survival" reasons, to work on your elocution if your customary speech

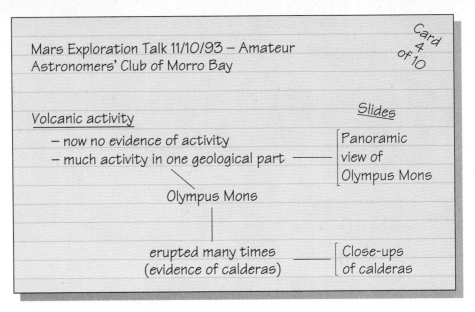

FIGURE 18.2 Sample notecard with directions for showing visuals.

differs markedly from that of your colleagues or if you are among the millions of immigrant Americans, visitors, or exchange students whose manner of speech has been problematic for the natives.

The suggestions that follow are those of Jonathan C. Berton, a professional Accent Improvement Trainer who directs the San Francisco Bay Area Language and Accent Specialists.

1. **Reduce your rate of speech and slow down.** Many foreign-born individuals are accustomed to speaking quite rapidly in their native language. Slowing down when you speak English will often help native Americans to understand your accented speech.
2. **Do not omit sounds and syllables at the end of the words.** Many speakers of Asian languages such as Chinese often are unaware that they have dropped a sound at the end of a word or sentence. Pay special attention to all of the sounds, especially in longer words.
3. **Use the dictionary to find appropriate vowel and consonant symbols along with accent marks for word stress.** Knowing how to use the dictionary to help you pronounce words can be a valuable tool.
4. **Do not be embarrassed to ask native Americans to check your pronunciation.** Many native speakers are glad to assist if you take the first step and ask for help.
5. **Avoid monotone voice quality.** Many Asian languages such as Vietnamese tend to have a choppy intonation pattern that does not sound natural when applied to spoken English.
6. **Use volume and pitch to emphasize important words and ideas.** Intonation is the music and rhythm of a language. English uses varied pitch and loudness to stress syllables and words.

7. **Write down difficult technical words which you have trouble pronouncing.** It is frustrating to constantly have to repeat words or wonder whether you have pronounced the word correctly. Simply ask a co-worker or friend to tape record the words for you.

8. **Check if your company will pay for you to take a class to improve spoken English.** Many large companies are anxious to support education and language training for their employees.

These suggestions should help you focus on some of the important areas of spoken English. Most foreign-born engineers and workers who are having difficulty being understood in English are aware that they must improve in order to succeed in their careers. . . . It is imperative that workers have clear English skills for successful job interviewing and promotion.

From: Jonathan C. Berton, M.A., "Improve Your English Skills For Better Communication," *High Technology Careers*, Aug.–Sept., 1993: 40.

PREPARING TO GIVE AN ORAL PRESENTATION: A CHECKLIST

- ☐ Have I planned my talk with the audience firmly in mind?
- ☐ Does my talk reflect the particular needs of my audience?
- ☐ Are my visuals useful? clearly interconnected with the text of my talk?
- ☐ Have I scripted notes for impromptu elaborations of key points?
- ☐ Does my talk fall within the requisite time frame?
- ☐ Do I know the substance of my talk well enough to elaborate extempore on key points? To respond to questions from the audience?
- ☐ Have I rehearsed my speech?

■ CHAPTER SUMMARY

In industry and business, reporting information is as much an oral pursuit as a written one; developing one's skills in oral presentation can thus be a valuable asset. To prepare an effective talk, one must understand the needs of the particular audience and be conscious of the audience's presence at every stage of the presentation. This means being alert for subtle kinds of feedback, such as yawning or frowning, and being prepared to go over parts of the talk that may not have been understood. A good speaker also needs to be adept at using audiovisual equipment, and to integrate visuals effectively with discussion. One efficient way to prepare for such integration of presentation tools with the talk itself is to "script" the talk: write out directions for what tools to deploy alongside the appropriate statements.

■ FOR DISCUSSION

1. What are some of the common problems in oral presentations that you have encountered? Why were they problems for you?
2. Why are visual aids important in an oral presentation? How can they be misused?
3. If a campus lecture, panel, or other type of oral presentation coincides with your study of this chapter, attend it, tape it, then play the tape to the class and discuss the strengths and shortcomings of the speaker(s).

■ FOR YOUR NOTEBOOK

1. Do a self-assessment of your speaking skills. What are your strengths and weaknesses? What, if anything, have you done to overcome them?
2. Listen to speeches on tape or live and take notes on the speaker's pacing, intonation, emphasis, and articulation.

■ ORAL PRESENTATION PROJECTS

1. Transform an old term paper or report into a 10- to 15-minute "script" for an oral presentation, using the scripting techniques suggested in this chapter.
2. Give an oral presentation that introduces your current term project to a lay audience. Use appropriate visuals.
3. Prepare a 10- to 15-minute oral presentation on a topic of your choice, and have someone videotape you while you are presenting it. Afterward, study the tape, take notes on what to improve upon, and then give the presentation a second time; again arrange to have it videotaped. How worthwhile was it to study your presentation on tape?
4. Prepare a script for a 10- to 15-minute oral presentation—drawing from the material in the following report—to explain how and/or why two Colorado River dams, Parker Dam and Davis Dam, were consolidated into a unified hydroelectric project.

Arizona: Mohave and Yuma Counties
California: San Bernardino County
Nevada: Clark County
Lower Colorado Region
Water and Power Resources Service
(Formerly Bureau of Reclamation)

In 1954, the Parker Dam Power Project and the Davis Dam Project were consolidated to form the Parker-Davis Project. The major works include Davis (originally named "Bullshead") Dam and Powerplant, Parker Dam and Powerplant, a high-voltage transmission system, and substations which sectionalize the long transmission lines. The original capacity of the Davis Powerplant was 225,000 kilowatts. In 1973, generator replacement of stator windings was initiated. Completed in 1976, the new windings increased the capacity of the powerplant to 240,000 kilowatts. The rated capacity of the Parker plant is 120,000 kilowatts. The transmission system includes 1,609.2 miles of high-voltage transmission lines and 31 substations. Parker Dam and Davis Dam are located on the Colorado River, 155 miles and 67 miles, respectively, downstream of Hoover Dam.

Lake Havasu, formed by Parker Dam, provides a forebay and desilting basin from which the Metropolitan Water District pumps water into its Colorado River Aqueduct. Parker Dam Powerplant was added to provide low-cost electrical energy to Arizona and southern California. Davis Dam provides reregulation of the Colorado River below Hoover Dam and facilitates water delivery beyond the boundary of the United States as required by treaty with Mexico. The Davis Dam portion of the project also provides for production and transmission of electrical energy, contributes to flood control, irrigation and municipal water supplies, navigation improvement, recreation, and wild waterfowl protection and related conservation purposes.

Parker Dam, Powerplant, and Reservoir

Parker Dam is a concrete arch structure with a structural height of 320 feet, and a volume of 380,000 cubic yards. At its crest, the dam is 856 feet long and is controlled by five 50-foot-square gates.

Lake Havasu backs up behind the dam for 45 miles and covers

(continued)

over 25,000 acres. The total capacity of the reservoir is 648,000 acre-feet. The Metropolitan Water District's W. P. Whitsett Intake Pumping Plant for the Colorado River Aqueduct is located on the shore of Lake Havasu about 2 miles upstream from the dam. The aqueduct begins at the intake pumping plant and extends 242 miles to its terminus at Lake Mathews near Riverside, Calif. About half of the power generated at the Parker Powerplant is reserved by the district for pumping water along the aqueduct. The Bureau of Reclamation retains the other half of the power output. The contract limits the use of active storage in Lake Havasu to the uppermost 180,000 acre-feet.

The Parker Powerplant includes a penstock gate structure, four penstock tunnels, and a powerplant building housing four hydroelectric generating units. Each of the four tunnels and the penstocks conveying river water from the forebay at the left end of the dam to the turbines is 22 feet in diameter and has a water capacity of 5,575 cubic feet per second.

Davis Dam, Powerplant, and Reservoir

Davis Dam spans the Colorado River in Pyramid Canyon 67 miles downstream from Hoover Dam and 88 miles upstream from Parker Dam. The Mexican Treaty of 1944 required the United States to construct Davis Dam for regulation of water to be delivered to Mexico. The reservoir formed by the dam, Lake Mohave, is used for that purpose through integrated operations of Hoover and Davis Powerplants.

Davis Dam, rising 200 feet above the lowest point of the foundation and about 140 feet above the level of the river, is a zoned earthfill structure with concrete spillway, intake structure, and powerplant. It has a crest length of 1,600 feet, and a top width of 50 feet. Its reservoir, Lake Mohave, has a total storage capacity of 1,818,300 acre-feet, and at high-water stages extends 67 miles upstream to the tailrace of the Hoover Powerplant.

Almost 5 million cubic yards of rock and earth were excavated to form the diversion and forebay channel and foundations for the dam, spillway and intake structures, and powerplant. More than 3,642,000 cubic yards of earth and rockfill were required to form the dam, and about 600,000 cubic yards of concrete and 23 million pounds of reinforcing steel were placed in the spillway, powerplant, and other structures.

The semi-outdoor type Davis Powerplant is on the Arizona side of the river immediately downstream from the dam embankment. Water is delivered from the forebay to the powerplant through five 22-foot-diameter penstocks.

(continued)

Transmission System

The transmission system includes 31 substations with a total capacity of 2,113,083 kilovolt-amperes, and 51 transmission lines with a total length of 1,609.2 miles. The high-voltage switchyards near the powerplants are the takeoff points for a system of transmission lines and substations which interconnect the Davis, Hoover, and Parker Powerplants, and extend to load centers in central and southern Arizona, southern Nevada, and southern California.

DEVELOPMENT

Investigations

Parker Dam

Population growth of municipalities within the greater Los Angeles area in California created a domestic water demand in excess of the supply from the local streams and the more remote Owens Valley source. After intensive investigations, it was determined that sufficient water could be obtained from the Colorado River. The construction of Hoover Dam, by virtue of the resulting river regulation and power generation, made feasible a plan to construct a dam on the Colorado River below the mouth of the Bill Williams River. Surveys initiated by the Bureau of Reclamation on June 25, 1934, established the best site for the location of Parker Dam.

Davis Dam

The Reclamation Service investigated a possible damsite at the lower end of Pyramid Canyon, 67 miles below Hoover Dam, as early as 1902–03. Until Hoover Dam controlled the Colorado River, however, a dam at the Davis site was not practicable.

In 1930, the Bureau of Reclamation made further investigations and explorations of the site in Pyramid Canyon, which led to authorization of the Davis Dam Project.

Authorization

The Parker-Davis Project was formed by the consolidation of the Parker Dam Power Project and the Davis Dam Project under the terms of the

(continued)

act of May 28, 1954 (68 Stat. 143). The Parker Dam Power Project was authorized by the Rivers and Harbors Act of August 30, 1935 (49 Stat. 1028). The Davis Dam Project was found feasible and authorized April 26, 1941, by the Secretary of the Interior under provisions of the Reclamation Project Act of 1939 (53 Stat. 1187).

On October 1, 1977, in conformance with Public Law 95-91, the Department of Energy Organization Act of August 4, 1977, the power marketing function of the Bureau of Reclamation, including operation and maintenance of transmission lines and attendant facilities, were transferred to the Department of Energy.

Construction

With funds advanced by the Metropolitan Water District of southern California, contracts were awarded by the Bureau of Reclamation and excavation for the Parker Dam and Powerplant commenced in October 1934. The dam was substantially completed in September 1938. Construction of the powerplant, consisting of four units, began in July 1939. Concurrently with construction of the powerplant, transmission lines and substations of the project were constructed and put into operation. Because of the onset of World War II, certain features were constructed with temporary materials or were omitted until proper materials could be made available and installed. Postwar work included replacement of temporary wood supporting structures with permanent steel structures in the substations.

A contract for the construction of Davis Dam and appurtenant works was awarded in June 1942. Work was halted after the War Production Board revoked priority ratings required to obtain the necessary materials for construction. Construction resumed in April 1946, and was completed in 1953.

Operating Agencies

The dams, hydroelectric powerplants, and attendant facilities are operated and maintained by the Bureau of Reclamation. The Parker Dam and Davis Dam Field Division of the Parker-Davis Project and the Boulder Canyon Project (Hoover Dam) were combined in a single operating unit administered by the Lower Colorado Dams Project Office located at Hoover Dam. The marketing functions, including the operation and maintenance of the transmission lines and attendant facilities

(continued)

of the Parker-Davis Project, are administered by the Boulder City Area Office of the Western Area Power Administration.

BENEFITS

Municipal and Industrial Water

Parker Dam diverts about 1,080 cubic feet per second of water daily to the Colorado River Aqueduct for use in the metropolitan area of Los Angeles.

Hydroelectric Power

Davis, Hoover, and Parker Powerplants are interconnected. The electrical integration and interconnection of these Bureau of Reclamation powerplants provides maximum generation of power with efficient use of water resources. The highly developed agricultural base and the complex industrialization of the Pacific Southwest benefit greatly from Colorado River hydroelectric energy.

Flood Control

Just above Parker Dam, the Bill Williams River pours flash floods into Lake Havasu. These floods are trapped in the reservoir and the downstream lands are protected. Parker Dam and Davis Dam both reregulate water releases from Lake Mead through the Hoover Powerplant for use downstream.

Recreation and Fish and Wildlife

Lake Havasu and most of the large marsh area extending 10 miles above the reservoir are included in the Havasu National Wildlife Refuge. Cabin sites are available for lease. Principal activities are camping, picnicking, swimming, boating, and year-round fishing—primarily for large-mouth black bass, bluegills, and crappie. Migratory waterfowl hunting is permitted in season.

(continued)

Lake Mohave is included in and administered as part of the Lake Mead National Recreation Area. Several concessions operate in the area with cabins, camping and trailer parks, and boats for hire. Camping, picnicking, swimming, boating, and excellent year-round fishing are the major activities. From Hoover Dam downstream to Cottonwood Landing, where Lake Mohave begins to widen, rainbow trout fishing is good. Below Cottonwood Landing, bass, catfish, and bluegills predominate.

PROJECT DATA

Facilities in Operation

Storage dams 2
Powerplants 2
Transmission lines[1] . . . . 1,609.2 mi
Substations 31

[1]The power marketing function, including the operation and maintenance of transmission lines and attendant facilities, was transferred to Western Area Power Administration, Department of Energy, in 1977.

Power Generation

Fiscal Year	Davis Powerplant (kWh)	Parker Powerplant (kWh)	Total
1968	912,961,000	433,611,000	1,346,572,000
1969	915,507,000	437,581,000	1,353,088,000
1970	926,879,000	434,108,000	1,360,987,000
1971	949,674,000	446,645,724	1,396,319,724
1972	972,116,000	455,856,000	1,427,972,000
1973	914,652,000	435,800,794	1,350,452,794
1974	997,680,000	477,109,291	1,474,789,291
1975	959,710,000	475,751,232	1,435,461,232
1976	961,381,000	478,940,452	1,440,321,452
1977	941,940,000	438,740,000	1,380,680,000

(continued)

ENGINEERING DATA

Water Supply

COLORADO RIVER

(See Boulder Canyon Project for information
  on drainage area and discharge.)

Annual diversion at Parker Dam[2]
Maximum (1977)	1,280,000 acre-ft
Minimum (1946)	80,400 acre-ft
Average	740,500 acre-ft

---

[2]Pumped by the Metropolitan Water District of Southern California.

Storage Facilities

DAVIS DAM

Type: Zoned earthfill
Location: On the Colorado River, 32 mi west
  of Kingman, Ariz.
Construction period: 1942–50
Date of closure (first storage): January 1950
Reservoir, Lake Mohave:

Total capacity to El. 647	1,818,300 acre-ft
Active capacity, El. 533.39-647	1,810,000 acre-ft
Surface area	28,500 acres

Dimensions:

Structural height	200 ft
Hydraulic height	140 ft
Top width	50 ft
Maximum base width	1,400 ft
Crest length	1,600 ft
Crest elevation	655.0 ft
Total volume	3,642,000 yd^3

Spillway: Concrete ogee weir in end of forebay channel at east end of dam, controlled
  by three 50-ft-square fixed-wheel gates.

Elevation top of gates	647 ft
Crest elevation	597.0 ft
Capacity at El. 647	214,000 ft^3/s

Outlet works: Two openings, one on each side
  of spillway section, each controlled by one 22-

(continued)

by 19-ft radial gate.

Capacity at El. 610	43,400 ft³/s

Foundation: Badly fractured and faulted
   porphyric granite gneiss overlain by silt,
   sand, and gravel in river channel.

Special treatment: Cement grout curtain
   under dam; intensive intermediate-zone
   grouting under concrete structures.

PARKER DAM

Type: Concrete arch
Location: On the Colorado River 12 mi
   northeast of Parker, Ariz.
Construction period: 1934–38
Power plant constructed in 1939–42
Date of closure (first storage): July 16, 1938
Reservoir, Lake Havasu:

Total capacity to El. 450	648,000 acre-ft
Available capacity, El. 400–450	180,000 acre-ft
Surface area	20,400 acres

Dimensions:

Structural height	320 ft
Hydraulic height	75 ft
Top width	39 ft
Maximum base width	100 ft
Crest length	856 ft
Crest elevation	455.0 ft
Total volume	380,000 yd³

Spillway: Overflow section at center of dam
   controlled by five 50-ft-square Stoney gates.

Elevation top of gates	450.0 ft
Crest elevation	400.0 ft
Capacity at El. 455	400,000 ft³/s

Outlet works: Four 22-ft-diameter steel
   penstocks through right abutment, each
   controlled by one 22- by 35-ft fixed-wheel
   gate.

Capacity at El. 450	22,300 ft³/s

Foundation: A hard, firm porphyric gneiss
   with subordinate masses of granite cut by
   several aplitic dikes; clay seams and fractured rock in right abutment.

(continued)

Special treatment: Cement grout curtain near
axis of dam, supplemental grouting in
abutments.

Mass concrete: Natural aggregate from pit on
Bill Williams River 3.5 mi from dam; low-
heat portland cement; mixing and placing
temperatures controlled, artificial cooling
through embedded pipe system.

Maximum size aggregate	6 in
Average net water-cement ratio (by weight).	0.58
Cement content	1.09 bbl/yd^3

Power Facilities

PARKER POWERPLANT

Location: Parker Dam
Year of initial operation: 1942
Year last generator placed in operation: 1943

Nameplate capacity		120,000 kW
Number and capacity of generators	(4)	30,000 kW
Maximum head		78 ft

DAVIS POWERPLANT

Location: Davis Dam
Year of initial operation: 1951
Year last generator placed in operation: 1951

Nameplate capacity^3		240,000 kW
Number and capacity of generators	(5)	48,000 kW
Maximum head		136 ft

SUBSTATIONS[4]

Number in operation	31
Total capacity of transformers	2,113,083 kVA

TRANSMISSION LINES[4]

Total number of lines	51
Total circuit miles	1,609.2

Description	Voltage, kV	Conductors and supporting structures	Circuit miles	Year placed in service
Parker-Davis No. 1	230	795 Steel	69.9	1951
Davis-Mead	230	795 Steel	60.7	1951

(continued)

Description	Voltage, kV	Conductors and supporting structures	Circuit miles	Year placed in service
Davis-Davis Switchyards 1 through 5	230	CU 500 Steel	1.5	1951
Davis-Prescott	230	795 Steel	142.5	1951
Hoover-Basic (North Basic Line)	230	CU 500 Steel	15.0	1942
Mead-Basic	230	795 Steel	12.8	1942
Mead-Hoover States	230	CU 500 Steel 795 Steel	8.5	1942
Mesa-Coolidge	230	CU 500 Steel 795 Steel	39.9	1951
Parker-Gene (MWD)	230	795 Steel	1.7	1947
Prescott-Pinnacle Peak	230	795 Steel	74.9	1951
Blythe-Knob	161	477 Wood-H	64.4	1951
Gila-Knob	161	CU 300 Wood-H	20.2	1943
Gila-Wellton-Mohawk (P.P. No. 2)	161	397.5 Wood-H	12.7	1956
Parker Powerplant-Parker 161-kV Switchyard (Nos. 1 through 4)	161	CU 300 Steel	0.5	1942–43
Parker-Blythe No. 1	161	CU 300 Wood-H 477 Wood-H	64.6	1943–51
Parker-Blythe No. 2	161	954 Wood-H	63.9	1946
Parker-Gila	161	CU 300 Wood-H 477 Wood-H	116.5	1943
Parker 161-kV-Parker 230-kV (Trans. 5 & 6)	161	477 Steel	0.5	1947–54
Parker-Phoenix No. 1[5]	161	CU 300 Wood-H	136.9	1942
Parker-Phoenix No. 2	161	477 Wood-H	139.8	1946
Knob Tap-Drop 4 Tap	161	CU 300 Wood-H	28.6	1943
Coolidge-BIA Coolidge	115	CU 4/0 Wood-H	0.4	1950
Coolidge-Elect. District No. 2 (via Signal)	115	336.4 Wood-H	12.2	1965
Coolidge-Oracle	115	CU 4/0 Wood-H	44.7	1943
Coolidge-Saguaro	115	795 Wood-H 336.4 Wood-H	47.1	1948–65
Maricopa-Saguaro	115	336.4 Wood-H 795 Wood-H	58.5	1948–54
Oracle-Tucson	115	CU 4/0 Wood-P CU 4/0 Wood-P	25.0	1943

(continued)

Description	Voltage, kV	Conductors and supporting structures	Circuit miles	Year placed in service
Phoenix-Coolidge	115	CU 4/0 Wood-H 336.4 Wood-H	52.5	1943–48
Phoenix Maricopa	115	336.4 Wood-H CU 4/0 Wood-H	36.1	1943–48
Saguaro-Oracle	115	795 Wood-H	19.0	1954
Saguaro-Tucson	115	795 Wood-H 336.4 Wood-H	35.4	1948–54
Tucson-Cochise	115	336.4 Wood-H	79.7	1952
Davis Powerplant-Davis 69-kV Switchyard	69	3/0 Steel	0.2	1965
Davis-CPU Tap (Needles)[6]	69	2/0 Wood-H 4/0 Wood-H CU No. 2 Wood-H	12.2	1946–53
Davis-CUC Tap (Kingman)	69	2/0 Wood-H	27.3	1947
Parker Powerplant-Parker 69-kV (Transf. 3)	69	CU 250 Wood-H	0.1	1943
Parker Powerplant-Parker 69-kV (Transf. 4)	69	CU 250 Wood-H	0.1	1943
Parker-Bagdad[7]	69	CU No.2 Wood-H	64.3	1943
Gila-Yuma Tap	34.5	CU 2/0 Wood-P	9.8	1943
Parker (Indian Service)-Parker Dam Camp	34.5	CU No. 2 Wood-P	0.2	1943
Wellton-Mohawk-Wellton-Mohawk P.P. No. 1	34.5	266.8 Wood-P	4.9	1951
Wellton-Mohawk-Wellton-Mohawk P.P. No. 3	34.5	336.4 Wood-P	3.5	1951
Parker LV-Colorado (APS)	13.2	795 Wood-P	0.2	1963

[3]In 1976, the new winding of generators increased the capacity from 225,000 to 240,000 kW.

[4]See footnote 1.

[5]To be replaced and upgraded to 230 kV as part of the Granite Reef Aqueduct transmission system of the Central Arizona Project.

[6]To be transferred to the Bureau of Indian Affairs.

[7]Planet Tap to the Cyprus Tap section (57.1 mi) to be sold to the Mohave Electric Cooperative, Inc.

III

Scientific and Technical Journalism

CHAPTER

19

Writing Articles About Science and Technology

These days nearly everyone feels overwhelmed by the avalanche of information from countless specialized fields. New advances in medicine; new insights into health and nutrition; new dangers to the environment (and new ways of replenishing it); the latest developments in the aviation, space, and automotive technologies—all of these advances affect us all. When medical researchers disclose important differences between two types of cholesterol ("bad" and "good"), all of us need to be able to understand those differences: our health can depend on it. Similarly, improvements in telecommunications technology can significantly affect the way one operates a business or even manages one's personal finances.

No wonder, then, that millions of people untrained in technology or science—as well as those who are so trained (for everyone is a layperson outside of his or her field)—are always looking for lucid and engagingly written articles that enable them to understand this vital information. But how can specialized knowledge be rendered comprehensible to nonspecialists without seriously distorting "the facts"? To answer this question we need to identify the characteristics of such writing.

CHARACTERISTICS OF ARTICLES ON SCIENCE AND TECHNOLOGY

An article that presents scientific or technical information to a general audience can vary greatly in style according to its field and the writer's purpose, but generally has the following identifiable characteristics:

- Presents a strong opening that stimulates curiosity as well as reveals the central purpose of the article.
- Relates sufficient background information about the topic.
- Defines or explains unfamiliar terms and concepts; often employs analogies—examples from commonplace experience that would parallel the unfamiliar concept; often uses direct quotations from experts to reinforce explanations.
- Reveals a clear organizational scheme; is attractively formatted.
- Is engagingly but never simplistically written throughout; avoids jargon and excessive formality.

That last attribute may be the most challenging for writers who have been trained to communicate to fellow specialists or at least to those within their professional domain. As communication specialist Diane Dowdey states, "Writing about scientific subjects for a general audience can be one of the most challenging tasks facing technical and scientific writers." (275). Professor Dowdey discovered—in her in-depth study of scientists who have been acclaimed for their science writing—"that they all use rhetorical strategies which primarily focus on involving or addressing readers so that they are asked to become a participant in the communication process."[1]

Let's examine the kinds of articles that are most in demand; these include the process description ("how it works"); the procedure explanation ("how to do it"); the investigation; the new discovery; and the biographical profile.

WRITING AN ARTICLE THAT DESCRIBES A PROCESS ("HOW IT WORKS")

How does an electron microscope work? How do psychoanalysts interpret dreams? What new nonsurgical techniques are being used to treat glaucoma patients? How does air pollution damage our health? Descriptive articles typically answer questions like these.

This kind of writing has been around a long time: Aristotle, in the fourth century B.C., wrote treatises on physics and biology; Pliny the Elder (A.D. 23–79) wrote a full-fledged and engaging *Natural History,* in which he describes, among other things, the property of magnets and the natural history of human beings. In the seventeenth and eighteenth centuries, when the public became curious about how nature worked, modern science reopened the formal study of these and many other related topics.

Starting with the 1890s, when electricity and the internal combustion engine started changing the world, the public became mesmerized by the power of technology: locomotives, automobiles, dynamos, the telephone, and perhaps even more dramatic than the others—the electric light. Science books began to be written even for children. One wonderful example is *The Fairy-Land of Science,* published in 1893, which includes chapters with titles such as "Sunbeams, and the Work They Do"; "A Drop of Water on Its Travels"; "The Two Great Sculptors—Water and Ice"; "The History of a Piece of Coal"; and "The Life of a Primrose." Here is how the author, Arabella R. Buckley, begins the last-mentioned chapter:

> When the dreary days of winter and the early damp days of spring are passing away, and the warm bright sunshine has begun to pour down upon the grassy paths of the wood, who does not love to go out and bring home posies of violets, and bluebells, and primroses? . . . But tell me, did you ever stop to think, as you

---

[1]Diane Dowdey, "Rhetorical Techniques of Audience Adaptation in Popular Science Writing," *Journal of Technical Writing and Communication* 17:3 (1987): 275.

added flower after flower to your nosegay, how the plants which bear them have been building up their green leaves and their fragile buds during the last few weeks? If you had visited the same spot a month before, a few last year's leaves, withered and dead, would have been all that you would have found. . . .

We cannot learn all about this little [primrose], but we can learn enough to understand that it has a real separate life of its own, well worth knowing. For a plant is born, breathes, sleeps, feeds, and digests just as truly as an animal does, though in a different way.

The author then describes the nature of a primrose seed, its growth, and, lastly, explains the flower's reproductive organs:

First, look at the outside green covering, which we call the *calyx*. See how closely it fits in the bud, so that no insects can creep in to gnaw the flower, nor any harm come to it from cold or blight. Then, when the calyx opens, notice that the yellow leaves which form the crown or *corolla,* are each alternate with one of the calyx leaves, so that anything which got past the first covering would be stopped by the second.

—Arabella R. Buckley, *The Fairy-Land of Science.*
New York: D. Appleton and Co., 1893: 150–51; 163.

Adults have been no less curious than children about the world around them and, since the turn of the century, have had access to numerous mass circulating publications that explained just about anything they wanted to know. The article on pages 443–444, from a 1916 issue of *Electrical World,* is typical of the many short features, usually written by engineers, on some facet of a new technology, such as electrical street lamps.

The public's need to understand the scientific and technical underpinnings of everyday life is stronger than ever because of the rapid developments in computer, aerospace, and telecommunications technologies. Equally fascinating is the use of technology in the arts: sophisticated restoration techniques used to save five-hundred-year-old masterpieces like DaVinci's fresco, *The Last Supper;* incredible engineering feats to save ancient Egyptian monuments from the rising floodwaters of the Nile; or the subject of the article on pages 445–446, which answers the question, "What is being done to prevent thousands of valuable books from deteriorating?" The article originally appeared in a popular science magazine and was subsequently reprinted in the science section of a large metropolitan newspaper—a clear indication of its ability to reach a widespread audience.

Before reading this article, imagine that you have encountered it in an issue of its original publication—in other words, think of yourself as one of the author's intended readers. How immediately comprehensible is it to you? Next, reread the article, locating as many of the characteristics listed at the beginning of the preceding section as you can. Finally, compare your analysis with the one that follows the article.

A Convenient Method for Replacing Series Incandescent Street Lamps

Where series incandescent street lamps are installed at the end of long outriggers, attached to unstepped iron posts, it is customary to replace burned-out and broken lamps from a tower wagon or with a stepladder. The first method is quite satisfactory where there are hundreds of such posts in service. It frequently happens, however, that a lighting company has only a few of such posts to maintain, and these are often located several miles from the shop. In such cases the cost of operating a tower-wagon is prohibitive, and the use of a ladder is awkward and

FIGURE 1

introduces an element of danger. By means of the device shown in the accompanying illustrations the night inspector can quickly and easily replace a lamp from the ground.

The construction of the apparatus is shown in detail in Fig. 1. A pair of tongs is fitted with semicircular clasps that close around the lamp base, holding it rigid and allowing no lateral motion. These clasps press against the top of the base rim when the lamp is being removed. In returning the lamp to the receptacle a lug on each jaw fits under the rim and forces the lamp into place. About 3/4 in. play between the clasps and the lugs permits of quick adjustment of the jaws in grasping the lamp base. It will be noted that no part of the device touches the lamp itself, so that it is as easy to remove a broken lamp as one that is burned out or blackened. The tongs in the particular outfit

[1]Reprinted from the *Electrical World* for April 22, 1916, by permission of the publishers.

(continued)

shown have been rounded out to allow clearance for lamps of the shape now manufactured in sizes of 250 cp. and over. The tongs are opened and closed by means of a rod running the length of the stick upon which they are mounted. This rod is attached to an insulated operating lever at the lower end of the stick. A tension spring controls the operating lever so that lamp base can not be released even if the operator removes his hand from the handle. The apparatus was made at small cost by a local blacksmith, who used the stick and operating handle of an ordinary tree-trimmer in its construction. The writer has found the device a valuable time-and-trouble saver.

Homer Andrew Watt and Philip B. McDonald, *Composition of Technical Papers*, 2nd ed. New York: McGraw-Hill, 1925: 190–192.

FIGURE 2

NASA Comes to the Rescue of Aging Books

By Glenn Garelik
Discover Magazine

The marble rotunda of the Library of Congress imparts an air of permanence appropriate to the timeless value of its 80 million books, documents and manuscripts—the rarest collection of printed matter in the world. But in recent years the library has become a mausoleum for many of its treasures: Their pages are rapidly decomposing, turned to dust by the very chemicals used to make them.

Crumbling books have been an "excruciating problem" for decades, says Peter Sparks, the library's chief preservationist; their paper can turn to dust in as little as 25 years. But in a small laboratory not far from the rotunda, library chemists may have hit upon an effective method for inexpensively and permanently arresting the decay, with an unlikely pair of tools: a chamber that tests spacecraft, and a chemical that ignites on exposure to air.

Before the mid-19th century, paper was made from cotton or linen rags, and it could last for hundreds of years. But the growing demand for reading matter called for a cheaper source of paper. Wood pulp fit the bill. But untreated pulp-based paper is too absorbent to take a sharp imprint, so chemicals must be added to prevent the ink from running. These additives, especially alum, sooner or later combine with moisture in the paper to form sulfuric acid. The acid helps break down cellulose fibers, and the page begins crumbling.

This poses no threat to the library's Gutenberg Bible, which is printed on durable vellum (thin calfskin treated to take ink). But the priceless editions of Whitman, the first sketches of a telephone by Alexander Graham Bell, the letters of Freud and three-quarters of the library's 20 million books are in danger.

The library staff for years has immersed, sprayed or brushed rare books with an alkaline solution to neutralize the acids. But it's time-consuming and sometimes costs thousands of dollars a volume.

In 1974, library chemists George Kelly Jr. and John Williams began exploring ways to salvage books on a larger scale. They settled upon diethyl zinc gas, or DEZ, a substance used in making plastics. Infusing the books with DEZ would effectively neutralize the acid in paper, they thought.

But DEZ bursts into flames if exposed to air, and it explodes on contact with water. To use the chemical safely, they would have to apply it in a vacuum chamber.

(continued)

Then NASA came to the rescue. Scientists at the space agency's Goddard Space Flight Center, in Greenbelt, Md., offered to adapt one of their large vacuum chambers—originally designed to simulate the intense solar radiation of space. They estimated that it could handle 5,000 volumes at a time.

In October, thousands of books were shipped to Greenbelt. They were placed in the chamber, spines down, in ordinary milk crates, loosely packed and separated by chicken wire to give them maximum exposure to the gas.

To draw the natural moisture out of the books, the scientists removed the air from the chamber and heated it to 113 degrees Fahrenheit. Then they piped in the DEZ vapor. Six days later, when the books were impregnated with DEZ, they added water vapor and carbon dioxide. The water brought the moisture in the pages into equilibrium with the moisture content of the air; the carbon dioxide combined with the zinc in the paper to form zinc carbonate—a harmless alkali that should keep the paper from turning acid.

Kelly is predicting that the process should extend the lives of the books for as long as 500 years—and it can be repeated again and again. The library hopes to build a plant of its own by 1985—one that will be able to treat 15,000 to 20,000 books every two weeks for less than $5 a book.

¶1. Opening is at once intriguing and informative. What purpose is served by calling attention to the physical description of the Library of Congress? Note how the description leads directly to a statement of the problem in the last sentence.

The marble rotunda of the Library of Congress imparts an air of permanence appropriate to the timeless value of its 80 million books, documents and manuscripts—the rarest collection of printed matter in the world. But in recent years the library has become a mausoleum for many of its treasures: Their pages are rapidly decomposing, turned to dust by the very chemicals used to make them.

¶2. Information, derived from an interview, elaborates upon the problem.

Crumbling books have been an "excruciating problem" for decades, says Peter Sparks, the library's chief preservationist; their paper can turn to dust in as little as 25 years. But in a small laboratory not far from the rotunda, library chemists may have hit upon an effective method for inexpensively and permanently arresting the decay, with an unlikely pair of tools: a chamber that tests spacecraft, and a chemical that ignites on exposure to air.

Possible solution to the problem is hinted at. Note how reader interest is sustained with the phrase, "an unlikely pair of tools." Reader wonders: How can books be preserved by using a spacecraft testing chamber and a volatile chemical?

(continued)

¶3. Historical background (paper production) is given to enable lay readers to appreciate the context in which the problem exists. Answers a key question: Why are modern-day books decomposing?

Before the mid-19th century, paper was made from cotton or linen rags, and it could last for hundreds of years. But the growing demand for reading matter called for a cheaper source of paper. Wood pulp fit the bill. But untreated pulp-based paper is too absorbent to take a sharp imprint, so chemicals must be added to prevent the ink from running. These additives, especially alum, sooner or later combine with moisture in the paper to form sulfuric acid. The acid helps break down cellulose fibers, and the page begins crumbling.

¶4. Examples of vulnerable books are given to emphasize the severity of the problem.

This poses no threat to the library's Gutenberg Bible, which is printed on durable vellum (thin calfskin treated to take ink). But the priceless editions of Whitman, the first sketches of a telephone by Alexander Graham Bell, the letters of Freud and three-quarters of the library's 20 million books are in danger.

¶5–7. Why haven't conventional acid neutralizers worked? The author explains; this leads to discussion of a new kind of neutralizer and the obstacle preventing its use.

The library staff for years has immersed, sprayed or brushed rare books with an alkaline solution to neutralize the acids. But it's time-consuming and sometimes costs thousands of dollars a volume.

In 1974, library chemists George Kelly Jr. and John Williams began exploring ways to salvage books on a larger scale. They settled upon diethyl zinc gas, or DEZ, a substance used in making plastics. Infusing the books with DEZ would effectively neutralize the acid in paper, they thought.

Why do you suppose the author presented the DEZ information in two paragraphs instead of one?

But DEZ bursts into flames if exposed to air, and it explodes on contact with water. To use the chemical safely, they would have to apply it in a vacuum chamber.

(continued)

¶8. The solution to the problem: a vacuum chamber, provided by NASA.

Then NASA came to the rescue. Scientists at the space agency's Goddard Space Flight Center, in Greenbelt, Md., offered to adapt one of their large vacuum chambers—originally designed to simulate the intense solar radiation of space. They estimated that it could handle 5,000 volumes at a time.

¶9. The author explains a perplexing secondary problem: How is it possible to treat large quantities of packed books quickly and efficiently?

In October, thousands of books were shipped to Greenbelt. They were placed in the chamber, spines down, in ordinary milk crates, loosely packed and separated by chicken wire to give them maximum exposure to the gas.

¶10. Technical details of the treatment process are explained. Is the explanation sufficiently clear for lay readers? Why or why not?

To draw the natural moisture out of the books, the scientists removed the air from the chamber and heated it to 113 degrees Fahrenheit. Then they piped in the DEZ vapor. Six days later, when the books were impregnated with DEZ, they added water vapor and carbon dioxide. The water brought the moisture in the pages into equilibrium with the moisture content of the air; the carbon dioxide combined with the zinc in the paper to form zinc carbonate—a harmless alkali that should keep the paper from turning acid.

¶11. Conclusion: a look ahead.

Kelly is predicting that the process should extend the lives of the books for as long as 500 years—and it can be repeated again and again. The library hopes to build a plant of its own by 1985—one that will be able to treat 15,000 to 20,000 books every two weeks for less than $5 a book.

WRITING AN ARTICLE THAT EXPLAINS A PROCEDURE ("HOW TO DO IT")

A procedural or how-to article, instead of describing how something works or is done, *shows* readers how to do the task themselves. A recipe is an extreme example of a procedural piece—extreme because recipes aren't "articles"; they are purely directions without a larger context. A group of recipes, however, presented in the context of some theme, such as "Five Great Holiday Recipes from the Old South," would constitute a recipe-article.

Procedural articles introduce the task to be performed in an interesting, informative manner. An article on how to add a light fixture to your home, for example, begins thus:

> If you need to put a little more light into some of your rooms, and don't want the clutter of table and floor lamps, you can add ceiling and wall fixtures just about anywhere you want them. You'll need to plan your work carefully, of course, and an understanding of basic wiring is a must.

> —Merle Henkenius, "Adding Light Fixtures,"
> *Homeowner,* Jan./Feb. 1990: 58.

Notice that the author is accomplishing several things at once with his opening: he makes the task seem worthwhile (light fixtures eliminate clutter); he speaks directly to his readers in a friendly teacher-to-student manner ("*you* can add ceiling and wall fixtures" rather than "ceiling and wall fixtures can be added"); and finally, he limits his target audience by in effect stating a prerequisite: this feature is for any reader who already understands basic wiring techniques.

Most procedural features will contain the following elements:

- An interest-snagging introduction that identifies the task for the reader; may also include historical background if warranted.
- A description of preliminary concerns, such as (in the case of the light-fixture article) familiarity with basic wiring; permit requirements; and an understanding that the finished work will need to be checked by a certified inspector.
- A list and description of tools and supplies needed to complete the task.
- A step-by-step delineation of the task (steps either numbered and conspicuously formatted for easy follow-along, or introduced in a more conversational manner).
- Cautionary, rechecking, and troubleshooting reminders (not always included, but greatly appreciated by readers).

Let us examine a procedural article, opposite, that advises amateur photographers how to photograph aurora borealis—the northern lights. Once again, read the feature straight through to get a sense of the whole; then reread it analytically, looking for the characteristics typical of how-to features.

WRITING AN ARTICLE THAT INVESTIGATES A PROBLEM

In a technologically advanced society, technologically complex problems abound, and they affect every one of us. Writers perform an extraordinary service by calling attention to these problems and enabling the lay public to understand their natures and what dangers they pose to health and safety, to the environment, even to our pocketbooks.

Where to look for problems? Chances are good that if you look up from where you are sitting right now, you will find one. Do you feel a cold draft coming from a wall that you are close to? If so, the house or building may very well have an insulation problem. Also perhaps the lighting is too dim or the electrical outlets look as though they are not up to code.

Many problems, however, are not so readily observable—not unless you have a bit of the sleuth in you or have a healthy amount of suspiciousness about all those chemicals with unpronounceable names that appear in tiny print on the labels of countless products.

Such a problem made the front page of the *San Jose Mercury News* in October, 1991—not only the front page but the headline story. It appears on pages 455–458.

After you have read the article, go back and study it from a writer's perspective. Ask these questions of the piece:

1. Did the author (Scott Thurm) "hook" your interest quickly? How did he manage to do this?
2. What specific examples does Scott Thurm mention that convince you of his premise that household substances are hazardous?
3. Where does Thurm appear to have his readers in mind when discussing the more technical aspects of his subject?
4. How does Thurm manage to incorporate outside sources into his article without disrupting the readability of the piece?

Having a firm grasp of the rhetorical strategies at your disposal will help you write your own successful articles.

Aurora Borealis: Photographing the Northern Skies Ablaze

by John W. Warden

Nothing allows a photographer to paint with natural light on such a grand scale as the aurora borealis, otherwise known as the northern lights. Its magnificent displays and the photographic opportunities it provides are truly out of this world.

The aurora has intrigued humankind for nearly 2000 years. Early descriptions can be found in the Old Testament, in Medieval literature, and in mythology of the Lapps, Eskimos, and American Indians. Oral legends about the aurora have been passed down for generations. To some, the aurora was a pathway to heaven. To others, the aurora represented ghosts of past generations.

We now know, thanks to the advances of modern-day science, that this luminous phenomenon is caused by the interaction of the solar wind, a continuous ejection of plasma from the sun's surface, with the earth's magnetic field. The solar wind becomes trapped by the earth's magnetic pole, where the different gases contained in the solar wind are ionized. This results in a display of colorful light, otherwise known as the aurora borealis.

For photographers, the most significant aspect of this scientific understanding is that it is now possible to predict when auroras are most likely to occur. We can now know for a fact that within approximately 48 hours after a major solar flare, the aurora display will begin. During 1990, sun-spot activity should reach a peak in its cycle, and thus provide excellent opportunities for photographers to get out their cameras and point them to the skies. Note: Obviously, the closer you are to the North Pole, the better your view of the aurora will be. Nevertheless, on especially clear, cold nights, the aurora has been known to be visible as far south as Seattle, Washington, on the west coast, and the state of Maine on the East coast. [All photos in original article were taken in Alaska.]

PLANNING AHEAD

The key to achieving successful images of the northern lights is in planning ahead. This means looking or listening for solar-flare- or magnetic-storm-related announcements in local newspapers, radio, and tele-

(continued)

vision programs. You can also contact your weather bureau to determine any unusual circumstances that may lead you to believe the aurora will soon be appearing.

During the day, scout for locations that will provide a panoramic view of the horizon as well as interesting foreground elements, like trees, mountains, or buildings; a strong skyline can add visual structure to the nebulous forms of the northern lights. Try to avoid locations where power lines, traffic, or other distractions might conflict with the subject. In urban areas, scout for locations during the evening hours that will provide the darkest sky setting.

EQUIPMENT

The aurora is usually most visible on cold, crisp, and clear evenings, so take this into account as you make your preparations. Cold-weather gear, like warm footwear, gloves, and outerwear, is essential. Take along a sturdy tripod and cable release, since you will be working with long exposure times; and keep a spare set of batteries in a warm pocket, so that you'll be ready to replace the ones in your camera should the cold weather disable them. Also, since you'll be working in total darkness, it's a good idea to have a flashlight on hand to illuminate your camera's settings.

EXPOSURES, LENSES, AND FILM

Recommended exposure guidelines are listed below, based upon film speed and f-stops. While I've heard of photographers trying to take spot-metering readings off the aurora, this is not recommended. Also, forget about programmed autoexposure modes.

F-STOP	ISO 200	ISO 400
1.2	3 sec.	2 sec.
1.4	5 sec.	3 sec.
1.8	7 sec.	4 sec.
2	20 sec.	10 sec.
2.8	40 sec.	20 sec.
3.5	60 sec.	30 sec.

(continued)

Bear in mind, too, these are only guidelines. The actual exposure should vary with the intensity of the light. For example, for brighter auroras, less exposure time is required. It's a good idea to bracket your exposure times, just to play it safe. I recently witnessed the brightest, most intense auroras in my life, and came away with nothing but a bunch of overexposed photos as a result of going by the above chart, instead of bracketing exposure times.

In terms of lens selection, any focal length from normal to wide-angle will suffice. My personal preference is 28mm. Generally, the faster and wider the lens, the better. Avoid using long exposures beyond 40 seconds, as star streaks will appear. For best results, open your lens to its largest aperture for the shortest possible exposure time.

Regarding film selection, my personal preference is Fujichrome 100 or Ektachrome 100 HC, pushed one stop for an exposure index of 200. The color saturation in these films, in my estimation, is superior to that of Kodachrome 200. Furthermore, Kodachrome 64 is simply too slow, and the higher-rated ISO films are too grainy for my personal taste.

COMPOSITION

To create images that are more than just pictorial documentations, try to include foreground or background elements to add perspective, and thus interest, to the photo. Learn to anticipate the direction of the aurora by taking the time to do some keen observation.

As it moves across the sky, the aurora forms bands of light that often pulse in an on-off cycle, as if someone were flicking a switch in outer space. The key is to photograph the aurora during concentrated, high-peak light shows. Intense displays will often form, climax, dissipate, and then reform several times during the night. On a good night, be prepared to spend anywhere from 4–8 hours in the hopes of coming away with 1–2 rolls of film *at best*.

Try to avoid the inclusion of truly bright stars in the composition, as they will tend to dominate the photograph as pinpoints of light. The exception would be to include bright stars that make up recognizable constellations, like the Big Dipper, Orion, etc. Bright stars can be easily hidden by blocking them out with foreground elements, such as trees.

When using long exposures, it's critical to anticipate the direction the aurora is moving, since the light will move through the image frame as the shutter remains open. While you should avoid photographing the moon amidst the aurora, photographing the aurora on

(continued)

a full-moon night can be advantageous if the moon is behind you. In this instance, the moon can be used as a fill-light source to help illuminate snowcapped mountains and add definition to foreground subjects. Of course, to balance a bright moonlit sky, a strong showing on the part of the aurora is required.

By the way, when you change lenses in cold weather, using warm hands, watch that you don't fog the front or back element of the lens. You probably won't notice it at the time, but the effect of a fogged lens will be apparent on the film. Making sure your horizons are straight can also be a challenging task when shooting at low light levels.

The photographic rewards for your efforts may prove to be wonderful or disappointing. Yet, the experience of being among the stars, and watching the aurora dance across the sky, is inspirational and uplifting beyond any description. I have found no other photographic experience that even comes close to this sensation. It is truly an out-of-this-world experience.

Photographic, July 1990: 44–45.

WRITING AN ARTICLE THAT INTRODUCES A NEW DISCOVERY OR INVENTION

Few things in our culture undergo such rapid change as technology. New computer systems appear on the market before we are able to learn the systems they replace. Lists of substances that are bad for us or good for us change from week to week. But not all changes are frustrating; some are exciting—and the sooner we become familiar with them, the better prepared we will be to assimilate them into our lives.

Take nanotechnology, for example. What seemed hard to swallow even as science fiction not so long ago is now taking shape as reality: microscopic machines. It is a fascinating enough topic to have made the pages of one of the most popular magazines in the world, *Reader's Digest* (see pages 459–462).

Once again, you will see in this article an ideal example of reader accommodation: an attention-getting opening, lucid explanations, engaging prose style, coherent organization. In reviewing this article, see how many specific rhetorical strategies the author, Lowell Ponte, uses.

WRITING A BIOGRAPHICAL ARTICLE

Scientists and engineers are people like the rest of us, not mad wizards with unruly hair slouched over their test tubes or computer terminals, eager to take over

Pollution Danger Lurks Under Sink, in Garage

By Scott Thurm
Mercury News Staff Writer

Nine Bay Area dry cleaners were told in August to warn neighbors that breathing the air near their shops could pose a significant risk of cancer.

But there are no warnings about breathing the air inside your own home, even though studies have found the dry-cleaning solvent perchloroethylene at levels up to nine times those found outdoors.

Repeated studies, including some in the Bay Area, have found that indoor levels of many toxic chemicals—released by such everyday products as air fresheners and paint strippers—typically exceed outdoor levels, even near chemical plants and refineries. And their effects are magnified since most Americans, even activity-conscious Californians, spend about 90 percent of their time indoors.

All told, the studies found that health risks from indoor air pollution are greater than living next to virtually any hazardous-waste dump or drinking all but the most contaminated water. Yet nearly a decade after these dangers became recognized, there still is no comprehensive state or federal effort to monitor or reduce the indoor risks Americans face at home and work.

"If the outdoor air is bad for you, the indoor air certainly has to be worse," said Lance Wallace of the U.S. Environmental Protection Agency, one of the nation's leading researchers on indoor air pollution. "Concentrations of almost everything are three to four times higher (indoors)."

AN UNPUBLICIZED DANGER

Experts say the government pays little attention to indoor air pollution because people don't understand the risks; and there's little pressure from environmental groups active on other issues, and even less from elected officials. While the EPA ranks indoor air pollution among the nation's biggest environmental threats, it spends hundreds of millions, even billions more to control less dangerous risks.

"It's hard for a lot of people to imagine that things in their own personal environment are hazardous to them," said Adam Finkel of Resources for the Future, a Washington, D.C. think tank. "We're used to

(continued)

thinking of pollution as something outside and something that other people impose on us."

Indeed, some of the most dangerous sources of indoor air pollution are very familiar. Dry cleaning emits perchloroethylene; some room air fresheners and mothballs are made almost entirely of paradichlorobenzene. Both compounds cause cancer in laboratory animals.

One of the most pervasive indoor pollutants is methylene chloride, a potent animal carcinogen whose emissions by factories are regulated by state and federal governments. A 1987 EPA survey found about one-third of 1,000 common household products—paint strippers, spray paints, shoe polishes, lubricants and others—contained methylene chloride; some paint strippers and spray paints were pure methylene chloride. But only about half these products listed the chemical on their labels because the government doesn't require it.

HIDDEN INGREDIENTS

Wallace began uncovering these hazards a decade ago, in studies comparing chemical levels outdoors—particularly in neighborhoods near large oil refineries and chemical plants—with those inside nearby homes. One study included 71 homes near Pittsburg and Antioch in Contra Costa County.

The researchers looked at about two dozen known or suspected carcinogens found throughout the environment, many regulated by hazardous-waste, drinking-water or outdoor-air laws.

In nearly every case, researchers found the highest concentrations of these chemicals indoors. For the perchloroethylene found in dry cleaning, indoor levels averaged $1\frac{1}{2}$ to nine times outdoor levels. For benzene, a component of gasoline emitted by cars and refineries which is known to cause leukemia, indoor levels were three times higher. Major indoor sources of benzene include cigarette smoke and gasoline evaporating in attached garages.

Other studies found particularly high concentrations—as much as 100 times outdoor levels—of these chemicals in new homes and office buildings. Researchers suspect the high concentrations occur as chemicals used to make carpets, upholstery and other furnishings break down and evaporate. Three months after an office building opens, concentrations typically decline to about 10 times outdoor levels; after six months, they tend to settle at about three times outdoor levels.

(continued)

"SICK BUILDING SYNDROME"

Some blame these emissions for outbreaks of "sick building syndrome," where dozens of employees suffer from headaches, colds or respiratory ailments. The Solano County district attorney has sued California home builders to require them to warn prospective purchasers of new homes about these risks.

But testing lags so far behind that it is hard to link symptoms with specific emissions. Researchers admit they don't know the sources of many of the chemicals they find indoors, and don't even have standard tests to look for them.

"We're caught in a position where we don't know what to do because we haven't looked at the problem carefully," said Joan Daisey, head of the indoor air program at Lawrence Berkeley Laboratory.

The threat of indoor air pollution is not limited to these synthetic chemicals. The recent outbreak of Legionnaires' disease at a government office building in Richmond, which killed one worker and sickened eight others, shows the dangers of bacterial contaminants. Others include radioactive radon seeping up from the ground, asbestos once used as insulation, carbon monoxide from gas-burning appliances and tobacco smoke.

OTHER INDOOR POLLUTANTS

Like the toxic chemicals Wallace's group studied, these pollutants all tend to dissipate outdoors to harmless levels; confined and recirculated indoors, however, they can be deadly.

Federal officials have taken limited steps to deal with some of these risks, setting standards for radon and requiring the removal of damaged asbestos, for example. Likewise, lawsuits filed under California's Proposition 65—most of them by private environmental groups—have spurred manufacturers to remove cancer-causing chemicals from correction fluids, spot removers and waterproofing sprays. But the law relies on these product-by-product challenges, a slow process.

Meanwhile, government has largely ignored the indoor threats of synthetic chemicals—even though top EPA officials twice in the last four years ranked indoor air pollution among the nation's biggest environmental threats and acknowledged that the problem was not receiving as much money and attention as it deserved.

(continued)

Despite recent increases, EPA still spends only about $5 million a year trying to control and reduce indoor pollution, and another $5 million on research like Wallace's. That compares with about $390 million on outdoor air problems.

It's the same story in Sacramento. Two state agencies—the Department of Health Services and the Air Resources Board—have small contingents working on indoor air pollution. Together, the units employ 13 people, compared with about 800 who work on outdoor air issues for the air board.

Defending their priorities, regulators say they take instructions from elected officials. And lawmakers say they hear little from individuals or environmental groups.

"As far as I know, we have never been approached to carry a bill in that area," said a staffer for the state Senate Toxics and Public Safety Committee.

With limited ammunition, EPA officials are not even aiming to regulate emissions from specific products or individual chemicals. Robert Axelrad, head of the agency's indoor air pollution program, says it would take years of potentially inconclusive tests, followed by court battles, to ban specific carpets or glues.

Instead, he says, the EPA will seek voluntary agreements by manufacturers to reduce the use of toxic chemicals and perhaps new ventilation standards for commercial buildings.

"Rather than spend 20 years trying to understand every indoor air pollutant, the better approach is to take steps that will reduce people's exposures to all pollutants across the board as quickly as possible," Axelrad said.

Axelrad and Steve Hayward, head of the indoor air program for the state Department of Health Services, expect marketplace competition, where firms try to outbid each other for the "cleanest" products, to prod manufacturers to reduce use of toxic chemicals.

But skeptics say environmental virtues can be muddled by deceptive advertising; they believe business needs a governmental push to reform.

Take DAP Weldwood Plastic Resin Glue, for example. Touted as being part of an "Enviro-Line" of products because it does not contain chemicals that form smog, the glue does contain formaldehyde, a powerful animal carcinogen.

Dawn of the "Tiny Tech" Age

By Lowell Ponte

Paul swallows the pill prescribed for his sleeping disorder. It is not a medicine but a tiny machine, and as it travels inside Paul's body, the device transmits a continuous temperature record to a receiver on his belt to help doctors monitor his bodily rhythms.

At a nearby hospital, surgeons begin an operation by inserting into Harold's heart a blood-pressure sensor so small that three of the tiny machines could fit on a pinhead.

Blocks away, Ellen is telling friends of the high-speed head-on crash she survived unhurt. At the instant the automobiles collided, a sensor micromachine almost too small to see triggered her car's air bag fast enough to save her life.

These devices, all gaining wider use, are among the first fruits of a major scientific revolution. We are leaving the Industrial Age, when nations took pride in building the biggest machines possible, and entering the Tiny Tech Age, where power and prosperity will go to those who make machines smaller and smaller.

Microscopic Motor

On May 27, 1988, at the University of California at Berkeley, graduate students Long-Sheng Fan and Yu-Chong Tai took a giant step into the miniworld. They switched on the voltage to an experimental device roughly 3/1000 inch across and, via an electron microscope, watched the first microscopic electric motor start to spin.

The motor was fabricated like a computer chip. By this method a paper-thin silicon wafer four inches in diameter is encased in glass (silicon dioxide) and coated with a photoresist, a liquid plastic that breaks down into microscopic patterns when exposed to light. Etching chemicals then dissolve unprotected parts of the glass layer, carving circuits onto the silicon wafer. By repeating the process, layer upon layer of interconnected circuits can be built onto the chip. Next, by fashioning a silicon base, hub and rotor, and using acid etch to dissolve the glass layer between them, Fan and Tai created a motor.

"Think what this means!" says George Hazelrigg, research engineer at the National Science Foundation in Washington, D.C. "Each wafer can contain up to 200 chips, and each chip can carry hundreds of micromachines."

(continued)

"With this new technology," adds Richard Muller, a director of the Sensor & Actuator Center at the University of California at Berkeley, "we can create on the same chip a computer brain and micromachine sensors and actuators to be the eyes, ears and hands of the computer." Within a decade, Hazelrigg predicts, doctors could be using such devices with microscopic cutters and manipulators to perform what is now major surgery without cutting open a patient's body.

Micromachines inhabit a world where the unit of measure is the micrometer—roughly one-millionth of a yard. (The average human hair is 70 to 100 micrometers wide.) In this Alice in Microland realm, the rules of the "big" world do not always apply—insects walk on water, ants carry many times their own weight and invisible specks of airborne dust can become giant monkey wrenches jamming unprotected gears in micromotors. Those who work in micromachine labs must don surgical masks, lest they inadvertently inhale the tiny devices from work surfaces.

As explorers in this strange microworld, scientists and engineers are seeking new ways to build machines. Some metals that seem sturdy in our "big" world become soft and weak at microscale; but other materials too fragile for big machines—such as silicon—become powerful in microland. "At this tiny scale," says Muller, "silicon is as mechanically strong as some types of steel."

Heart-Rate Wristwatch

Until recently the use of micromachines had been limited to specialized fields such as medicine and aerospace. By the mid-1990s, however, you will be able to buy a "smart" car carrying micromachine sensors that automatically monitor air pressure in tires and tension in shock absorbers. Micromachines may even prompt your car's onboard microcomputer to adjust the engine continuously as temperature and humidity change—so you can drive from a sunny beach to a mountain ski resort and your car will respond to the changed conditions.

In the coming decades micromachines will be able to make your home "smart" too. They will tell your furnace and air conditioner how to adjust themselves for efficiency and minimal pollution. Sensors will sound an alarm if a child falls into the swimming pool. In earthquake-prone areas, microsensors could detect a tremor and automatically disconnect and shut off water, electricity and gas lines to prevent water damage, electrocution and fires.

In the recreation field, your home-exercise machine will use microsensors to automatically record effort and performance. And golf clubs with sensors to improve your game are in development. Also on

(continued)

or near the marketplace are the first microsensor wristwatches that monitor heart rate, pulse and blood pressure, and—with pressure sensors—altitude and underwater depth.

A Robot Named "Squirt"

Tiny technologies are a key to the future of robots. The old science-fiction notion of a robot pictured a large machine that mimicked human form and thinking. But a visitor to the Artificial Intelligence Laboratory at the Massachusetts Institute of Technology in Cambridge gains a truer glimpse of our robotic future.

Out of the corner of his eye, the visitor sees something zip across the floor and duck under a chair. It is Squirt, a robot cockroach scarcely a cubic inch in size, designed not to think but to react in preprogrammed ways to stimuli. Turn on the lights or make a noise and Squirt scurries for the nearest dark place it "sees." Turn off the lights and Squirt comes out a few minutes later in the direction from which its twin-microphone "ears" last picked up noise.

For M.I.T. scientists gathered by "Mobot lab" founder Rodney Brooks, the idea is to invent microrobots that are fast and cheap and that work without human supervision. Squirt developer Anita Flynn foresees "gnat robots," micromachine servants mass-produced so cheaply that when one finally wears out, you'll be able to throw it away and buy a new one.

In this world of nanotechnology (from the Greek *nanos*, or dwarf), "some ideas are far-fetched, others are realistic," says John Foster, a physicist at IBM's Almaden Research Center in San Jose, Calif. "I have a modern chemistry textbook that states: 'Of course, we will never actually be able to see atoms.'" Foster laughs. "In our lab I see atoms *every day*"—through IBM's scanning tunneling microscope (STM).

Tunnel of Electrons

This microscope is an achievement so great that the two scientists in IBM's Zurich Research Laboratory who invented it in 1981 were later awarded the Nobel Prize. The STM works by moving a tiny probe tip to within a nanometer (about a billionth of a yard) of a surface to be studied. When voltage is applied to the probe tip, electrons tunnel through the gap between the tip and the surface. As the tip moves above the surface, this tunneling current rises and falls as the size of the gap changes. A computer draws a map of the shape and contour of atoms on the surface as if they were mountains and valleys.

(continued)

Even more amazing, the STM's tunneling current can prod atoms to move across a surface. Last November, IBM-Almaden researchers Donald Eigler and Erhard Schweizer used an STM to build the first man-made structure, atom by atom: the shape spelled out the letters "IBM." From 35 xenon atoms, they created letters only 200-billionths of an inch tall. To understand what this means, calculates Eigler, the process would fit the text of 2000 issues of this magazine within the space of one of the periods on this page.

As an old saying goes: we shape our tools, and thereafter our tools shape us. It is already clear that our tiny new tools will shape the way we live in astonishing ways. By "thinking small," scientists and dreamers are enlarging and enriching the future for all of us.

Reader's Digest, Nov. 1990: 25–31.

the world, as the popular media—most notoriously children's cartoons—have tended to depict them. We can learn a great deal about how people "take" to science—why they find it exciting. Galileo, for example, had a penchant for teaching, for getting students excited about new ideas, no matter how unsettling they might be, no matter how they might contradict the revered authorities like Aristotle or the Church of Rome. (Ironically, Galileo was convinced that nothing he discovered with his telescope could possibly contradict Holy Writ; any apparent contradictions had to be the result of *human* misinterpretation.) Einstein was a dreamer, a mystic, a deeply religious man who, like Newton, imagined the universe to be the manifestation of a Divine Intelligence, as rich in mystery as in what is revealable to the probing eye. Marie Curie was as intuitive as she was rational, able to interact well with fellow physicists and chemists in her pioneering work with radioactive elements, work that helped lay the groundwork for nuclear physics.

But you need not restrict yourself to the greatest men and women of science to find fascinating personalities. Let's take a look at a contemporary biologist at a relatively early stage in her career (pages 464–465).

The author, Sabrina Brown, engages the attention of her readers (all the members of a university community, including alumni), with a clever allusion to fearsome, invading insects: "First it was fire ants, then killer bees." Such an opening not only kindles curiosity but establishes the proper context for introducing Dr. Janice Edgerly-Rooks, an entomologist, and her work with mosquitos—specifically the Asian Tiger mosquito, a known transmitter of serious tropical diseases. What is interesting about Brown's approach is her emphasis on the scientist as much as the scientist's object of study. Far from being some disembodied laboratory wizard, Dr. Edgerly-Rooks emerges as a real person: her views are quoted, not just summarized; the scene of her campus laboratory and her interaction with students are mentioned.

WRITING ARTICLES ON SCIENCE AND TECHNOLOGY: A CHECKLIST

- ☐ Do I have a particular audience in mind for my article?
- ☐ Is my approach to the topic equally informative and interesting?
- ☐ Have I carefully defined unfamiliar terms?
- ☐ Have I used such rhetorical devices as analogy and metaphor to help readers comprehend complex ideas?
- ☐ Does my opening paragraph arouse curiosity?

Process Articles

- ☐ Have I introduced the process engagingly?
- ☐ Have I described all facets of the process?
- ☐ Have I included sufficient background information?

Procedural Articles

- ☐ Does my opening clearly and interestingly introduce the task to be performed?
- ☐ Have I listed all necessary equipment and materials?
- ☐ Have I included all necessary steps in the procedure?
- ☐ Have I included appropriate visual aids?

Investigative Articles

- ☐ Have I effectively communicated the need for investigating the problem in question?
- ☐ Have I covered all relevant aspects of the problem?

Discovery Articles

- ☐ Do I make a clear connection between the new discovery and its possible importance for advancing knowledge or for human welfare?
- ☐ Do I allude to the scientists involved in the discovery and how they went about making the discovery?

Biographical Articles

- ☐ Have I brought the scientists or technologists into central focus by quoting them? By describing the importance of their research? the manner in which they are conducting their experiments?
- ☐ Have I clearly explained the technical aspects of their work so laypersons can understand it?

Biologist's Work Could Help Control
Invading Asian Tiger Mosquitoes

Sabrina Brown

First it was fire ants, then killer bees.

Next on the list of invading insects seems to be Asian Tiger mosquitoes; and SCU [Santa Clara University] biologist Janice Edgerly-Rooks, with funding from an $89,000 National Institutes of Health grant, is doing work that could help to understand—and eventually control—them.

The Asian Tiger, nicknamed for its black-and-white stripes, is more formally known as *Aedes albopictus*. After entering this country in 1985—probably in shipments of used tires from Japan—the mosquito has already spread to 18 states and is known to be a transmitter of diseases such as dengue, a tropical disease characterized by severe pains in the joints and back, fever, and rash; dengue hemorrhagic fever; and yellow fever.

A hardy variety of mosquito able to survive freezing temperatures, the Asian Tiger is a more aggressive biter of humans than mosquitoes already established in the United States, said Edgerly-Rooks, an assistant professor.

It is the Asian Tiger's interaction with resident U.S. mosquitoes in breeding habitats that intrigues the SCU biologist.

"We're looking at mechanisms that might regulate the [mosquito] population," said Edgerly-Rooks, whose work could be described as the first link in a long chain of research that could lead to controlling the dangerous pests.

Research conducted by Edgerly-Rooks and her colleagues Todd P. Livdahl and Michelle Willey at Clark University in Worcester, Mass., shows that the Asian Tiger exhibits characteristics that may enable it to dominate *Aedes triseriatus* (the Eastern treehole mosquito) and *Aedes aegypti* (a mosquito species found in the South).

Specifically, their research investigated how the presence of larvae affects the number of eggs hatched when the three species occupy the same breeding habitat. Different combinations of fertilized eggs and larvae were placed together, and egg hatch was monitored.

Asian Tiger eggs were the least inhibited from hatching by the presence of larvae, and Asian Tiger larvae caused the most inhibition on the hatching of other species' eggs, said Edgerly-Rooks.

(continued)

"There's a living larva in the egg responding to its environment," she said. "Given certain stimulation, it will hatch."

Other research by Edgerly-Rooks may give some clues as to what is just the right stimulation.

Working in her Santa Clara laboratory with student assistants (the studies with Livdahl and Willey were conducted at Clark), Edgerly-Rooks has focused on the Eastern treehole mosquito. One study, which has been accepted for publication in *Ecological Entomology*, combined eggs and varying numbers of larvae in environments where they were either in the same liquid but separated or in contact with one other.

In addition to the effects of the larvae on the eggs, she was interested in the part that the growth of microbial organisms (a food source) plays in triggering egg hatch. Edgerly-Rooks and Michelle Marvier '90 found that abundant microorganisms and a small number of larvae (four in the study) stimulated egg hatch in groups where larvae were in contact with the eggs. The combination of the two seems critical because the lowest egg-hatch rate occurred when there was abundant food, but no larvae present.

"This hatching response may have evolved because both abundant micro-organisms and a moderate number of larvae reflect a habitat of good quality," said Edgerly-Rooks.

However, it is a delicate balance, she said. When numerous larvae contacted eggs and grazed micro-organisms from their surfaces, egg hatch was inhibited.

When there was no contact between larvae and eggs, egg hatch increased with the number of larvae present, indicating that larvae may produce some chemical that triggers hatch when combined with abundant micro-organisms, Edgerly-Rooks added.

As with most basic research, a layperson might wonder what significance these findings have in the eventual goal of controlling the Asian Tiger.

"If larvae are crowded, we could end up with a weakened mosquito population, whereas the best mosquitoes are produced when there are less larvae in the habitat," Edgerly-Rooks said.

"We look at it from a basic ecological point of view, but the pest-control people will need that information if they're going to do something to control them in an educated manner."

■ CHAPTER SUMMARY

Popular articles on topics relating to science and technology play an important role in closing the gap between specialists and general readers, enabling the widespread dissemination of information that should be familiar to all persons, not just scientists and technicians. Articles that explain how a phenomenon, mechanism, or activity works; articles that give step-by-step instructions on how to perform a task; articles that disclose essential information about an urgent problem; articles that introduce readers to a fascinating new discovery or invention; and articles that offer a behind-the-scenes look at men and women in science and technology are always in demand and can contribute much to a deeper understanding and appreciation of what it means to be a scientist, engineer, or inventor.

■ FOR DISCUSSION

1. After reviewing the six articles presented in this chapter, summarize their similarities as well as differences. Are there elements present in one feature that ought to be present in another?

2. Locate an article in one of the following popular science or technology magazines. After summarizing the feature's contents and identifying its type, explain how the author helped you to understand the principle(s) discussed; look especially for attention-grabbing introductions, clear explanations of technical concepts, and an engaging style.

Air & Space	*Personal Computing*
Archaeology	*Popular Electronics*
Astronomy	*Popular Photography*
Discover	*Popular Psychology*
Earth	*Popular Science*
Homeowners	*Sky and Telescope*
Natural History	*Smithsonian*
Oceans	*Technology Review*
Omni	*Today's Health*

3. Read the draft on pages 472–474 of a how-to feature by student writer Paul Freitas. Using the criteria for good feature writing discussed in this chapter, discuss what the author might do to strengthen the draft.

4. Discuss the rhetorical techniques employed by the author of the excerpt from an article on research being conducted on the mechanics of cell division (see pages 475–479). How would you describe the intended audience? How can you tell? What assumptions does the author make about

his readers' scientific background, for example? How does the author maintain interest in the topic being discussed?

■ FOR YOUR NOTEBOOK

1. The diagram on page 468 indicates ways in which air pollution can affect the human body. Keep an ongoing record in your notebook of information you come across that goes into detail about any of the physiological effects indicated. Consider using your entries to develop an article on one of them.

2. Do some background reading on a facet of technology that interests you—a facet of space technology, perhaps, such as artificial satellites. Record facts that interest you; for example, How many types of satellites are there? How many are restricted to military use? How long do they function? How expensive are they? How have they influenced our way of life? Don't think about article possibilities until you have compiled several pages of notes.

3. (Continuation of No. 2) Now that you have accumulated several pages of notes on artificial satellites, brainstorm for possible topics on the subject. For example, "The Amazing Legacy of America's Oldest Functioning Satellite—Pioneer 6" (which was launched in 1965, and was still transmitting data about solar radiation on its 25th anniversary on Dec. 16, 1990).

4. Over the next week or so, begin listing problems you discover or suspect either at home, at work, or on campus (e.g., in the dorms, in the library, in a laboratory or storage facility), and which ought to become better known via an article in the campus newspaper.

5. Speculate on technological developments over the next ten to fifty years. Some possibilities: lunar or Martian colonization; improving public transportation systems; combatting deforestation or soil deterioration in developing countries.

■ WRITING PROJECTS

1. Write a descriptive article on a medical or health-related topic, such as "How the kidneys filter impurities from the blood"; make use of an illustration or two (see the illustration on page 469, which accompanied an article on the kidney-filtering topic). Before you begin, follow these steps:
   a. Prepare an audience profile in which you list the background characteristics of your target audience (see Chapter 2).

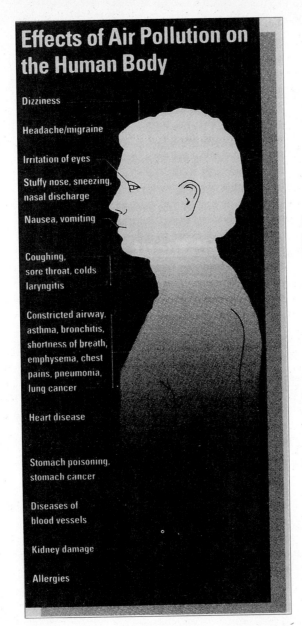

Effects of Air Pollution on the Human Body

Dizziness

Headache/migraine

Irritation of eyes

Stuffy nose, sneezing, nasal discharge

Nausea, vomiting

Coughing, sore throat, colds laryngitis

Constricted airway, asthma, bronchitis, shortness of breath, emphysema, chest pains, pneumonia, lung cancer

Heart disease

Stomach poisoning, stomach cancer

Diseases of blood vessels

Kidney damage

Allergies

Source: Jon Naar, *Design for a Livable Planet*. Harper & Row, 1990: 80.

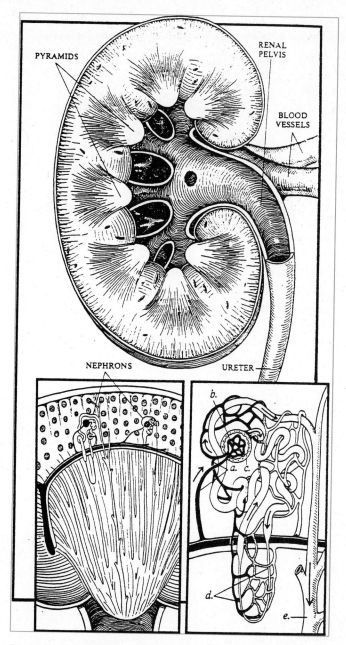

Pyramid-shaped units (left) in the kidney contain the tangled nephrons (right) that cleanse the blood. Vessels (a) carry blood to and from the glomerulus (b), where water and other molecules are drawn into Bowman's capsule (c) and through tubules linked by the loop of Henle (d). Most of the fluid reenters the blood; the remainder goes into a collecting duct (e) and is excreted.

Illustration by Michael Reingold, *Science Digest*, November, 1983: 93.

b. In the library, gather facts about kidneys and kidney function. Begin with an encyclopedia article or your introductory biology textbook; look for bibliography entries relating to kidneys in these works. The caption beneath the illustration on page 469 will give you some of the facts you need to know, but not all.

c. Create a flowchart (see Chapter 11) that depicts the filtering stages.

d. Create an outline for your intended article. Remember to involve the diagrams in the discussion, not just tag them on at the end.

2. Following the strategies described in topic 1, write a descriptive article that focuses on one of the following topics:
   a. How a cellular telephone works.
   b. How archaeologists excavate a site.
   c. Techniques of modern-day weather forecasting.
   d. Communicating with chimpanzees.
   e. How chocolate is made.

3. Write a procedural (how-to) article on one of the following topics:
   a. How best to sharpen household cutting tools (lawn-mower blades, kitchen knives, etc.).
   b. How to install or repair a household mechanism, such as a
      showerhead
      garage-door opener
      toilet
      dishwasher
      kitchen faucet
      smoke detector
      thermostat
      sprinkler system
   c. How to create your own
      bracelets or earrings
      bonsai
      terrarium
      picture frames
   d. How to get started in
      wind sailing
      Easter-egg decorating
      ballroom or disco dancing
      white-water river rafting
      fossil hunting

   Be sure to do an audience profile beforehand. Next, jot down notes on the following:
   • What will your readers need to know beforehand, if anything, in order to complete the project successfully?

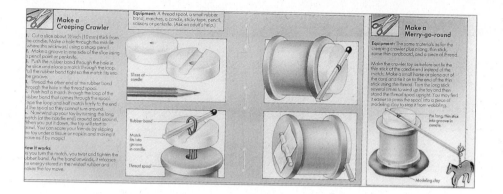

- What tools and materials will they need to have on hand?
- What tables, diagrams, charts might you wish to include?
- What cautionary reminders will you need to include?

4. Think up an amusing yet educational device to create for children (or for adults in playful moods, for that matter!)—a compass, a kite, a simple electrical game, or a mechanical toy—and develop a feature consisting mainly of drawings and captions. Create your own format! For example, you might divide the feature into "YOU'LL NEED . . ."; "HERE'S WHAT TO DO"; and "WHY DID IT HAPPEN?" segments—an arrangement similar to that of the feature reproduced above.

5. Investigate a household problem (potential fire hazards, inadequate insulation, leaks in the plumbing, etc.) and write an article that clearly identifies the problem and all its aspects—the importance of what, when, where, and the whys of the potential dangers—and how the problem might be solved.

6. Interview a faculty member in one of the science or engineering departments on your campus, focusing on that person's current research project. Use the information you gather to write a profile of that scientist or engineer, the details of the project, and its importance.

7. Consult one of the science or engineering professors at your college about a new discovery or invention. Write an article describing it, and speculate on its potential benefits—or dangers.

8. Write a speculative essay based on one of the "futuristic topics" you have been brainstorming about in your notebook—or select a topic from the following list:
   a. Videophones: Advantages and disadvantages.
   b. Books on tape and CD–ROM: An end to the conventional book?
   c. Cloning: Where should we draw the line?
   d. Virtual-reality technology: Assets and dangers.

Paul Freitas
English 179A/Technical Writing
Professor White
April 16, 1991

Hot Waxing Your Own Skis

Every experienced skier knows how important ski wax is.
Unwaxed skis are slow and hard to control. They make
turns more difficult, and can stop a skier dead on
shallow slopes. Unwaxed skis are also very tiring, because of all of the extra work a skier needs to do
while skiing (pushing, leaning, twisting, etc.).
Sometimes just having wax isn't good enough; most waxes are only effective in specific snow conditions, so
a new wax is needed when conditions change. To truly
enjoy skiing, whether you are an expert skier or just
a beginner, you need properly waxed skis.

There are two ways to wax skis. Cold waxing, or
rubbing, involves rubbing a bar of cold wax on the base
of the ski. Cold waxing is simple, but it is only effective as a stopgap measure. Hot waxing is, by far,
the most effective way to wax skis. Hot waxing involves
melting wax onto the ski base and spreading it with a
hot iron. Afterwards, the excess wax is scraped from
the base with a plastic scraper. Hot waxing leaves a
flat, smooth ski base with a thin layer of wax impregnated in the base material.

Every ski shop routinely waxes skis, but there are
a number of reasons why you should do it yourself.
First, there is cost. Once you have the necessary
equipment, a hot wax costs one-tenth of what a ski shop
would charge. The second reason is quality. The quality of a "professional" waxing job sometimes varies
widely; oftentimes, too much or too little wax is left
on the skis. By doing it yourself, you can be certain
of a quality finish every time. Lastly, the procedure
is so simple that literally anyone can do it.

Very little equipment is needed to hot wax skis.
First, you will need a workbench. Second, you will need
a good pair of ski vises, to hold the skis while you

wax them. Ski vises can be obtained from many ski shops, or by mail order. Third, you will need a clothes iron. I prefer to use a non-steam iron, but these are difficult to find these days. Any steam iron will work almost as well. Do not use a travel iron; it might fall apart while you are using it. Also, don't use the same iron for both waxing and ironing, because you can never get all of the wax out of an iron once you've used it for waxing. You will need a plastic scraper (not a metal one), and a bottle of wax remover or base cleaner. And, of course, you'll need wax. Wax is formulated for specific snow conditions; be sure to use the wax that is best suited for the conditions you will be skiing in. If you have doubts, use a universal wax.

The first step of waxing is mounting the vises on your workbench. They usually attach with a simple clamp mechanism. Next, mount one ski in the clamps. Now you are ready for base preparation. Using your base cleaner, remove the old wax from the bases of the skis. Follow the instructions on the package. Be careful that you don't use too much cleaner; it can dry out the base and make it difficult to wax. Let the ski dry for at least 15 minutes before going any further. You may want to clean the other ski while you're waiting. Next, inspect the base for damage and wear (like gouges, rusty edges, concave or convex bases, etc.). If you find any problems, you should tune the ski before waxing. If you can do the tune-up yourself, do so; if not, bring your skis to a ski shop for professional tuning.

After the bases have been prepared, you are ready to wax. Plug in your iron and turn it on. Set it for the "wool" setting. When it is warm enough to melt wax, hold the iron over the ski base (point down) and touch a bar of wax to it. Melted wax should start dripping from the iron. Move the iron so that at least one drop falls in each square inch of the base material. If you have a groove running the length of the ski base, you don't need to put wax in it, because the groove doesn't come in contact with the snow. After you have finished, remove the bar and iron the base so that the wax covers the entire base (except for the groove). Keep the iron moving at all times; letting the iron sit could melt the base material itself! When wax covers all of the base surface, you are finished with applying the wax. Wait 30 minutes (at least) before continuing. You may want to wax your other ski while you're waiting.

The final step of waxing is scraping. The base is now coated with a thick, uneven layer of wax, most of which is unnecessary. The excess wax must now be removed. If the plastic scraper isn't sharp (its edges should all be square), sharpen it with a file. Now, scrape the base of the ski with the scraper. Make sure that you remove all of the wax from the steel edges, and that the central groove is clean. Remove as much wax as you possibly can. The only wax you need is impregnated in the base material, and it can't be removed with a scraper.

You are almost finished. The last thing you must do is go skiing, but you already knew that, didn't you?

Dangerous Dance of the Dividing Cell

Most cells divide at just the right moment, and in just the right place. Discovering the molecules that control this timing may help us to understand cancer.

Stephen Young

Few phenomena in biology are quite as dramatic as the cell cycle, the ordered sequence of events by which cells grow and divide. Under the microscope the high point of the cycle is undoubtedly division itself, especially the stately dance of the chromosomes which ensures that each new cell is born with a full set of genetic instructions. This spectacle—mitosis—is followed by a much quieter, but no less important, period during which the new cells grow and copy their DNA, ready for the next division and another performance of the mitotic dance.

The cycle is controlled by a biochemical machine of exquisite design. In recent years researchers have taken much of this machinery to pieces, showing how it works in normal cells and hoping to shed light on its relation to cancer. They have been particularly interested in the main transitions of the cycle: the entry into mitosis, the exit from mitosis and the point at which the cell starts copying its DNA. The most remarkable discoveries have centred on the control of mitosis.

There are many components in the mitotic machine, but a protein called cdc2 (cell division cycle) is particularly important. When suitably activated, cdc2 is a powerful and versatile molecular tool, whose forte is to spur other molecules into action, an effect which it achieves chemically by "persuading" them to pick up and bind a phosphate group. The addition and removal of phosphate groups is a standard technique by which cells switch proteins from dormant to active states and vice versa. The process is common in many different areas of cell chemistry.

The cdc2 protein is behind many of the spectacular events of cell division. Although its exact role in the living cell is still being unravelled, the evidence from the test tube is persuasive. The protein adds phosphate groups to a network of linked proteins called nuclear lamins, which line the membranous envelope around the cell's nucleus. Lamins respond by severing links with their neighbours, an effect which dissolves the nuclear envelope.

Another of cdc2's roles is to add phosphate groups to histone H1, one of a pack of proteins that cling to DNA, sculpting its helices into various shapes. This chemical transaction appears to be necessary for DNA to collapse into compact chromosomes ready for the events of mi-

(continued)

tosis. cdc2 also has a hand in the construction of the transient structure—a spindle—that controls the movements of a cell's chromosomes during mitosis.

Although the details of mitosis vary from organism to organism—in yeasts, for example, the nuclear membrane does not disintegrate—there seems to be a broad consensus over the role of cdc2. Five years ago, Melanie Lee and Paul Nurse demonstrated the point in an extraordinary fashion during the course of research at the Imperial Cancer Research Fund's laboratory in London. Their work used a single-celled organism, the yeast *Schizosaccharomyces pombe*—also known as fission yeast to distinguish it from the budding yeast familiar to bakers and brewers.

Lee and Nurse worked with a mutant strain of yeast in which the gene responsible for producing cdc2 was defective, causing cell division to be blocked. Remarkably, they found that they could overcome the block by inserting a stretch of human DNA into the yeast—a stretch which they subsequently found to be the human version of the *cdc2* gene. The fact that this DNA worked normally in such an alien environment provided dramatic proof that evolution has been remarkably conservative in its dealings with the *cdc2* gene.

Hundreds of millions of years have elapsed since our ancestors parted company from those of the yeast—years that saw all the great radiations of life on earth. And yet our cdc2 protein still bears a striking resemblance to the yeast's. When the researchers worked out the structure of human cdc2 and compared it to cdc2 from yeast, they found that they were 63 per cent similar. This fraternal similarity, in both form and function, opened up the possibility that organisms of all kinds might control their cell cycles with the same basic machinery. That idea turned out to be correct. The cdc2 protein is now regarded as a common thread that joins all organisms whose cells have nuclei—protists, fungi, plants and animals.

Although the cdc2 protein is the centrepiece of the cell cycle, it is only a part of the overall design. The concentration of this influential protein does not vary during the cell cycle and yet its effects are unleashed only at certain times. This implies that it is switched on and off by companion molecules. Recent research has uncovered a retinue of such molecules, which hold cdc2 in check, spur it on, or work with it.

One approach focuses on a specific group of proteins whose concentration varies during the cell cycle. A decade ago Tim Hunt, then at the University of Cambridge, and his colleagues found the first such protein in developing embryos of a sea urchin. The protein accumulated throughout the cell cycle, right up until the moment of division, when it suddenly disappeared. The same behaviour was repeated in each successive cycle. Hunt and his colleagues christened the protein

(continued)

cyclin, an apt name for a molecule with such habits. Tests on other embryos revealed more cyclins with the same basic properties.

The team's belief that the cyclins would turn out to play a role in cell division was amply rewarded. Researchers have since found a vast array of cyclins of various types in all sorts of organisms, including yeasts, and a consensus of sorts has emerged about their role: they attach themselves to the cdc2 protein and help activate it. So the agent that sets mitosis in motion is not cdc2 pure and simple; it is a complex formed from the temporary marriage of a cdc2 molecule and a cyclin molecule. In this way the ebb and flow of the cyclins can influence the progress of the cell cycle. (Other influences also come into play, of which more later.)

A striking demonstration of the role of cyclins has come from an extraordinary "stripped down" version of the cell, whose behavior has been explored by researchers such as Andrew Murray and Marc Kirschner at the University of California, San Francisco. The technique involves preparing extracts of the eggs of *Xenopus laevis* (an African frog much used in laboratory work) and then adding nuclei from sperm. The result is a mixture in a test tube that will undergo three or more cell cycles at the behest of a researcher, while remaining amenable to all manner of chemical treatments. Such mixtures do many of the things that intact cells do, including the production of proteins.

SHOOTING THE MESSENGER

Murray and Kirschner were able to show that the extracts must be allowed to make cyclin if they are to undergo mitosis. Genes make proteins via an intermediary called messenger RNA, so the first step was to destroy all the mRNA in the extracts, by means of a selective poison, and see what happened. The extracts were unable to switch into mitosis. The researchers then added to these disabled extracts a form of mRNA carrying the instructions for making a cyclin molecule (from sea urchins, as it happened, but cells are not fussy about the origin of molecules involved in the cell cycle, as we have already seen). This treatment, which allowed the extracts to start making cyclin again, was enough to restart mitosis.

Cyclin clearly plays an important part in controlling the cell's entry into mitosis. Does it play a comparable role in the exit from mitosis? Recall that the exit from mitosis is accompanied by a precipitous decline in the concentration of cyclin. If this decline is prevented, by giving cells an artificial cyclin that is unusually resistant to destruction, mi-

(continued)

tosis cannot be completed. Destruction of cyclin, it seems, is what turns off cdc2, allowing cells to finish mitosis and get on with the next phase of the cycle. This vital demolition job is the work of a protein-destroying enzyme (a protease) which, some researchers believe, could have the additional task of splitting the chromosomes in two and permitting them to separate. So the cdc2 protein, with its cyclin, engineers the onset of mitosis; the protease destroys the engineer and completes mitosis.

Cyclins are necessary for the smooth operation of the mitotic machinery but they are, again, only a part of the overall picture. Work on a variety of organisms, notably fission yeast, has uncovered a network of additional regulatory molecules, whose function is to adjust the timing of mitosis.

How do these regulators work? Ultimately their role is to modify the cdc2 protein, according to Nurse and his colleagues at the ICRF's Cell Cycle Group at the University of Oxford. The cdc2 protein can be kept dormant if a phosphate group is installed at a certain site within the molecule, at an amino acid called tyrosine. Removal of the phosphate group has the opposite effect: it activates cdc2, allowing mitosis to take place. By balancing these two effects against one another, the cell can control its entry into mitosis.

A protein called cdc25 is the molecular "accelerator"; it removes phosphate from cdc2, either directly, or through an intermediary—probably the former, according to recent evidence. Mutant strains of fission yeast that make too much cdc25 are too keen to divide; those with a faulty version of the gene cannot enter mitosis. Human cells also make cdc25 and this substance can activate cdc2 from starfish eggs—another example of the universal nature of the cell cycle.

Fission yeast also has two molecular "brakes," wee1 and mik1, whose roles have been explored by Karen Lundgren and her colleagues at the Howard Hughes Medical Institute in Cold Spring Harbor, New York state. The wee1 and mik1 proteins work against cdc25, making sure that cdc2 retains its phosphate and so keeping it in check. There are also molecules that regulate the regulators.

While researchers have been building up this picture of events around mitosis, there has been a parallel effort to understand the entry into S phase, the part of the cycle during which the cell copies its DNA ready for division. In an extraordinary development, it now seems that cdc2 itself controls entry into this phase of the cycle. This conclusion has emerged from a number of studies, including work on yeasts which carry a mutated form of the *cdc2* gene and which consequently cannot duplicate their DNA. The situation in higher organisms, such as frogs and humans, is a little more complicated. In such creatures a close relative of the cdc2 protein operates at this point, not cdc2 itself, but the

(continued)

biochemical properties of this molecule are similar to those of its more celebrated relative.

As this picture has emerged, researchers have wondered whether cyclins might also have a role to play in the entry into S phase. Their suspicions have proved well founded, for several different types of cyclin have now come to light in addition to those involved in mitosis (mitotic cyclins). Some of these cyclins accumulate in what biologists call the G1 phase of the cycle and disappear once the cell has begun to copy its DNA. The behaviour of these G1 cyclins echoes that of the mitotic cyclins around mitosis, so they may perform a similar job: joining forces with the cdc2 protein, activating it and perhaps helping to direct it at an appropriate set of targets. Those targets are still a matter for debate, but they could include proteins that react to the addition of phosphate by switching on genes whose products are essential for the copying of DNA.

Much of the early evidence for G1 cyclins came from budding yeast, but recent developments have again revealed a common thread that runs right through the living world. A case in point is the work of Daniel Lew, Vjekoslav Dulic and Steven Reed at the Scripps Research Institute in La Jolla, California. The team worked with a strain of budding yeast whose genes for G1 cyclins had been disabled so as to block cell division. They added pieces of human DNA to these cells, looking for genes that would correct the deficiency. They found several. Of these, one in particular, cyclin E, is a promising candidate for a G1 cyclin in human cells.

If these ideas prove correct the result will be a symmetrical picture, in which cdc2 (or a close relative, depending on the cell in question) together with various different cyclins, controls all the main transitions of the cell cycle. It is an elegant, persuasive idea and one which has wide support. However, many researchers are cautious about drawing too close a parallel between the two transitions until more evidence is to hand on the role of cdc2 and its allies in the S phase of the cycle.

That evidence may come all the sooner thanks to recent work on budding yeast by Stephen Bell and Bruce Stillman of the Cold Spring Harbor Laboratory in New York state. Last month the researchers reported in *Nature* that they had isolated a complex of proteins that bind to chromosomes at the special sites where the copying of DNA begins. The implication is that these proteins may set the copying process in motion and hence play a key part in touching off the S phase of the cell cycle.

20

Writing and Designing Brochures and Fact Sheets

Brochures may well be the most popular formats of all for communicating technical information to lay audiences. Thousands of them are published regularly by government and nonprofit agencies, as well as by public and private enterprises. Health organizations such as the American Heart Association and the U.S. Department of Health and Human Services; environmental groups such as the Union of Concerned Scientists and the U.S. Fish and Wildlife Service; private businesses such as computer and automobile manufacturers, printing and photoduplication services—all publish brochures to disseminate vital information or attract clientele. What is it about brochures that makes them so popular? How does one go about writing and designing a brochure?

THE NATURE AND PURPOSE OF BROCHURES

Brochures are important because they provide key information about a subject for lay readers—information that can help them choose a course of action. For example, a brochure on hypertension, by suggesting a low-sodium diet, can help readers reduce their high blood pressure. Brochures designed to convey essential information about a topic in a clear, visually attractive format are usually distributed free of charge, *pro bono publico,* or as a promotion for or introduction to a product or service. Because they are almost always intended for lay readers, any specialized terms or concepts must be defined in context.

Typically consisting of a single sheet ($8\frac{1}{2} \times 11$ or $8\frac{1}{2} \times 14$), brochures are folded into thirds or fourths to produce six or eight narrow pages of text and graphics. Documents that are stapled instead of folded are usually referred to as booklets; these typically range between ten and fifty pages.

Brochure writing and design often are regarded as practical skills in the workplace. Many organizations use several kinds of brochures as an inexpensive and colorful way to publicize their products and services. And of course the federal government, through its many agencies, publishes them by the thousands.

THE RHETORICAL ELEMENTS OF A BROCHURE

Although brochures vary widely in both form and content, they do share common rhetorical and design features:

- A distinctive front page, which often displays a company or insitution logo or slogan.
- An introductory segment (defining the concept).
- The body of text, often divided into segments, sometimes taking a question-answer or steps-to-follow format.
- A concluding segment (summary, look ahead, and/or where to obtain further information).

Consider, for example, how these structural features work together in a six-page brochure on secondhand smoke, published by the American Lung Association (Figure 20.1).

PLANNING THE LAYOUT OF A BROCHURE

In order to combine concise and informative writing with visual attractiveness—the twin goals of any kind of brochure writing—careful planning is important and ought to be perceived as a three-part process:

1. Outline the text, noting inclusion of visuals.
2. Prepare the visuals.
3. Create the layout design.

1. Outline the Text

Because the space in a brochure is so limited, you need to work out an outline with those limitations in mind. The text of a brochure is itself a kind of outline, which is why subheads are essential. Notice how prominently the subheads are presented in the following brochure on secondhand smoke. Also notice how short most of the paragraphs are—usually no more than two sentences, themselves quite short. As in newspapers, normal-length paragraphs in a brochure would appear disproportionately long because of the narrow columns.

2. Prepare the Visuals

Use visuals sparingly, if at all, because of space limitations. Skillfully designed textual elements can constitute a visual element in their own right. The visual

This page
Conspicuously formatted title, the producer's logo, and an eye-catching illustration that trigger reader interest.

Lead segment
Defines secondhand smoke; also includes facts about the toxic properties of tobacco, necessary for understanding smoke is a problem.

Principal text:
First segment
Distinction is made between "mainstream" smoke, both of which inhale; explanation of why side-stream smoke is especially dangerous.

What is Second-hand Smoke?

Tobacco smoke contains about 4,000 chemicals, including 200 known poisons. Every time someone smokes, poisons such as benzene, formaldehyde and carbon monoxide are released into the air, which means that not only is the smoker inhaling them but so is everyone else around him. Many studies now show that this secondhand smoke can have harmful effects on nonsmokers and even cause them to develop diseases such as lung cancer.

Americans are beginning to recognize how hazardous smoking can be to everybody's health. National surveys show that most nonsmokers and even the majority of smokers themselves — believe that people should not smoke when they are around nonsmokers. Clearly, in our society, causing other people to be exposed to secondhand smoke is becoming less and less acceptable.

Side-stream Smoke

Every time anyone lights up a cigarette, cigar, or pipe, tobacco smoke enters the air from two sources. The first is mainstream smoke, which the smoker pulls through the mouthpiece when he inhales or puffs. Nonsmokers are also exposed to mainstream smoke after the smoker exhales it. The second, and even more dangerous source, is sidestream smoke, which goes directly into the air from the burning tobacco.

Sidestream smoke — which a nonsmoker inhales whenever he's around someone who's smoking — actually has higher concentrations of some harmful compounds than the mainstream smoke inhaled by the smoker. Studies show that there are several cancer-causing substances, as well as more tar and nicotine, in sidestream smoke as compared to mainstream smoke. In addition, carbon monoxide, which robs the blood of oxygen, can be two to fifteen times higher in sidestream smoke.

Most of the smoke in a room results from sidestream smoke. When nonsmokers breathe in this type of smoke from other people's cigarettes, cigars, and pipes, it is often called involuntary or passive smoking.

(continued)

AMERICAN LUNG ASSOCIATION®
The Christmas Seal People®

facts about...
SECOND-HAND SMOKE

FIGURE 20.1   Brochure: "Facts About Secondhand Smoke."

Principal text:
Major areas of concern
Each of the remaining segments in the principal text link secondhand smoke to the medical and social problems it generates:
(1) link to lung cancer

(2) respiratory problems of children exposed to secondhand smoke

(3) changing perceptions and policies re smoking in work environments

(4) effects of tobacco odors on nonsmokers; why smokers are usually unaware of the bad effects of tobacco odor

Second-hand Smoke and Lung Cancer

The fact that cigarette smoking is the main cause of lung cancer in smokers is well-known. In 1986 the Surgeon General of the United States reported that involuntary smoking can cause lung cancer in healthy nonsmokers. Recent studies also indicate that secondhand smoke causes death from heart disease.

What this could mean is that tobacco smoke and radiation may have this in common: there are just no safe levels of exposure.

Effects on Children

Secondhand smoke has an especially bad effect on infants and children whose parents smoke. A number of studies show that in their first two years of life, babies of parents who smoke at home have a much higher rate of lung diseases such as bronchitis and pneumonia than babies with nonsmoking parents.

A study involving children ages five-to-nine showed impaired lung function in youngsters who had smoking parents as compared with those whose parents were nonsmokers. And smoking by pregnant women seems to predispose premature babies to respiratory distress syndrome.

Parents who smoke at home can aggravate symptoms in some children with asthma and even trigger asthma episodes. Parents should only smoke outside the home or, better yet, quit smoking altogether.

Even among children without asthma, a team of researchers found that acute respiratory illnesses happen twice as often to young children whose parents smoke around them as compared to those with nonsmoking parents.

The American Lung Association is encouraging smoke-free families so that children can have the best possible chance to grow up healthy.

Smoke at the Workplace

The Surgeon General's report of 1986 established that the simple separation of smokers and nonsmokers within the same air space may reduce but *not* eliminate the risk of exposure to environmental tobacco smoke. As a result, an increasing number of state and local laws now restrict smoking at the workplace. The idea behind these laws is that the preferences of both nonsmokers and smokers should be considered, whenever possible. However, when these preferences conflict, the health and preferences of nonsmokers should come first.

More and more private companies are also adopting policies that restrict smoking and protect nonsmokers at work.

Tobacco Odors

Burning tobacco smoke creates bad odors which also cling to people's clothes, hair, and even their skin. This contamination is so intense that when someone smokes in an air-conditioned room, the air-conditioning demands can jump as much as 600 percent in order to control the odors.

The bad odors created by tobacco smoke also linger on. Long after a person has left a smoke-filled room, they may still have the odor of cigarettes on their bodies and in the fabric of their clothes. This is because while certain chemicals created by burning tobacco cause bad odors, other chemicals actually help the odors to hold onto the surface that they penetrate.

Smokers themselves usually are not sensitive to these odors because of the destructive effects that the smoke from their own cigarettes has on the inner linings of the smoker's nose.

FIGURE 20.1 (continued)

Principal text:
Importance of working toward the Surgeon General's goal placed after the facts about second hand smoke for greater impact.

Principal text:
How to reach that goal. What can be done to decrease exposure to secondhand smoke.

Concluding segment:
Call to action
Most people are aware of the problem.
What the ALA can do to help.

"Smoke-Free Family" logo with ALA slogan to reinforce memory. Conspicuous acknowledgments note also has persuasive appeal.

A Smoke-Free Society

More than 41 million Americans have kicked the cigarette habit. Millions more are trying. Overall, less than one out of three people in this country still smokes.

Clearly, people who don't smoke are the majority, and they are concerned about being able to breathe clean air, free from harmful and irritating tobacco smoke. Even most smokers agree that smoking is hazardous to the health of nonsmokers as well as to their own health.

These are among the facts that have led the Surgeon General to propose that America become a Smoke-Free Society by the year 2000. If we were a smoke-free nation, we would be helping to protect everybody's health.

Clean Air for Everyone

Being able to breathe clean air, free from harmful, irritating tobacco smoke is a serious issue for everyone. At home, at work, and in other public and private places it is important to speak up about how dangerous smoking can be to smokers and nonsmokers alike.

Here's what you can do to help:

☐ Let family, friends, co-workers, and others know that you mind if they smoke.

☐ Put stickers, buttons, and signs in your home, car, and office. Ask to be seated in nonsmoking sections when you travel or dine.

☐ Support legislation to restrict smoking or to set up smoke-free areas in public places and at the workplace.

☐ Ask your doctor and dentist to restrict smoking in their waiting rooms and to help establish no-smoking regulations in all health-care facilities, including hospitals.

☐ Propose no-smoking resolutions at organization meetings. Encourage hotels and restaurants to establish no-smoking areas.

☐ Encourage management and unions where you work to establish a policy to protect nonsmokers on the job.

☐ Help to promote the concept of smoke-free families in your community.

☐ Contact your nearest American Lung Association office. They have the facts about smoking, and a network for action.

A Challenge for the Future

A recent Gallup survey conducted for the American Lung Association revealed that the majority of both smokers and nonsmokers believe that smoking can damage the health of people who don't smoke.

Translating this belief into social action is the challenge we all face as we head into the 1990's. Your local American Lung Association can help: they have a wide variety of programs to help people quit smoking. They can also assist you in finding out about ways to protect nonsmokers at work.

JOIN THE · SMOKE-FREE FAMILY®

It's a Matter of Life and Breath®

This publication was made possible in part by your support of Christmas Seals® and other contributions to the American Lung Association.

AMERICAN ✚ LUNG ASSOCIATION®
The Christmas Seal People®

Published by American Lung Association

0006

4/92

elements on the first page of the brochure on secondhand smoke include a stylized silhouette of a couple seated at a table, together with the no-smoking icon and the logo of the American Lung Association, the producers of the brochure. The only other visual elements are on the back page: the logo of the smoke-free family and a smaller reproduction of the A.L.A. logo. There are no visuals in the body of the brochure.

3. Create the Layout Design

If you work with desktop publishing software, you can create page layouts—column sizes, presentation of text, interweaving of text and visual elements—and easily readjust them as many times as you wish; that is, you can make your existing columns wider or narrower, or change two-column format into single-column format, simply by selecting options from a pull-down menu.

Without a desktop publishing program, the work might be more time-consuming but just as much fun. After writing out the text of the brochure and selecting the visual aids you want to include with it, you create what are called page dummies—mock-ups of each brochure page, consisting of printouts of the text and the pasted-in visuals. If the page dummies are "camera-ready," they go directly to the printers; if not, they go to a professional graphic designer, who prepares the camera-ready copy.

THE ELEMENTS OF A FACT SHEET

Fact sheets, which are very similar to brochures in elements and purpose, have the distinction that everything must appear on the front and back of a single sheet. Whereas a brochure is folded so that a complete, visually attractive title page is all that potential readers will notice when it is displayed, a fact sheet must devote a comparable space at the *top* of the front side of the sheet for its title. Also, a pictorial graphic, conspicuously positioned, will enhance the visual impact of the fact sheet.

As with a mainstream brochure, the text should be divided into easily digestible subheads, and each segmented portion of the text should itself be visually appealing, using a bulleted breakdown of points that substantiate opening assertions.

The following fact sheet, "Transportation," prepared by Bay Area Action, a Palo Alto–based environmental action group, demonstrates one way in which these structural elements can be presented (Figure 20.2).

Transportation

Cars and the Environment

America's love affair with the automobile has a heavy impact on the environment. Burning gasoline emits pollutants into the air we breathe; chlorofluorocarbons (CFCs) from leaky car air conditioners deplete the ozone layer; oil and other automotive fluids contaminate water and soil; and large tracts of land are lost as they are covered with asphalt to make roads and parking lots. Despite the magnitude of these problems, more people are driving greater distances. The problem will only get worse unless changes are made in our transportation priorities. We desperately need better public transportation, improved carpooling programs, increased interest in biking and walking, and higher gas mileage standards for automobiles.

Automobiles use approximately half of all the oil consumed in the United States. To keep up with this demand, oil companies are drilling in sensitive natural areas, such as in offshore waters and in the pristine wilderness of Alaska. With the Earth's known usable oil reserves expected to be depleted by the year 2040, oil is becoming harder to extract and the process is inflicting greater damage to the environment. As the U.S. supply dwindles (it is expected to run out by 2020), more oil will be imported from foreign sources across great expanses of water, leading to increased oil spills. The recent Exxon Valdez oil spill in Alaskan waters clearly illustrated the scale of the risks involved.

When oil is burned, large amounts of carbon dioxide, the major "greenhouse gas," are released into the air. Greenhouse gases trap heat from the sun in the Earth's lower atmosphere, causing temperatures to rise, a process known as global warming. Automobiles are responsible for about 20 percent of carbon dioxide emissions in the United States, with the average car releasing about five tons every year. Automobiles

also emit about 40 percent of the nitrogen oxides that contribute to acid rain, as well as poisonous carbon monoxide and hydrocarbons that cause smog. 158 million Americans live in areas that violate the Clean Air Act standards.

Automobile air conditioners use CFCs (Freon) that, when discharged into the atmosphere, destroy the Earth's protective ozone layer that shields us from cancer-causing ultraviolet light. In 1985, a hole the size of the continental United States was discovered in the ozone layer over Antarctica. CFCs also are responsible for as much as 25 percent of the global warming trend. Although there are far fewer CFCs than carbon dioxide molecules in the atmosphere, each CFC molecule is up to 15,000 times more efficient at trapping heat. One charge of CFCs from an automobile air conditioner contributes as much to global warming as the carbon dioxide emitted from an average new car driven 20,000 miles.

More than 60,000 square miles of land have been paved in the lower 48 states to accommodate America's 135 million cars. This amounts to 2 percent of the total land surface—an area the size of Georgia. Close to half of the land area in most cities goes to providing roads, highways and parking lots for automobiles, and two-thirds of Los Angeles is paved.

The Solution

Raising the fuel efficiency standard for automobiles will cut down on air pollution by requiring less gasoline to be burned per mile driven. There are already cars on the market that get 50 miles per gallon or more. Converting segments of our transportation system to cleaner burning fuels, such as compressed natural gas, methanol and ethanol, may also improve air quality. However, methanol is a questionable fuel because when it is derived from coal it releases twice as much carbon

dioxide as oil. Furthermore, alternative fuels do not address the problem of traffic congestion and highway expansion. The true solution to our transportation problem lies with improvements in public transportation rather than with improvements in public transportation and carpooling programs, and increased interest in biking and walking.

According to the American Public Transit Association, commuting on mass transit in place of driving cuts hydrocarbon emissions that produce smog by 90 percent, carbon monoxide emissions by more than 75 percent, and nitrogen oxides emissions by up to 75 percent. Despite these impressive figures, only one penny of the nine cents per gallon federal gasoline tax is used to improve mass transit.

Railway Age magazine points out that a single highway lane can accommodate 2,250 people per hour in automobiles, 9,000 in buses, 15,000 on a light rail line and 34,000 people per hour on a heavy rail line. The newest French train is capable of traveling at a speed of more than 180 miles per hour while saving energy and providing a safe, comfortable ride.

What You Can Do

- Walk or bike for close errands.
- Arrange to carpool with your co-workers.
- Use public transportation whenever possible.
- If it's reasonable, ask your employer to allow you to work at home one or two days a week.
- Encourage your employer to offer a financial incentive in place of a parking permit.

- Take a job close to your home or move closer to your place of work.
- Enjoy local recreational activities rather than traveling long distances for entertainment.
- Urge your local officials to improve and promote public transportation, carpooling programs and bicycle lanes. Write your elected officials and urge them to support legislation to raise the fuel efficiency standard for automobiles and to put funding towards public transportation rather than highway expansion.
- If you are buying an automobile, consider a model that:
  - Gets good gas mileage (at least 35 miles per gallon).
  - Doesn't have an air conditioner.
  - Has radial tires with a high tread rating for longer use.
- For proper driving and maintenance:
  - If your car has an air conditioner, make sure the CFCs are recycled anytime it is serviced and before the car is scrapped.
  - Have your car smog checked and install pollution-control equipment if necessary.
  - Keep your car tuned up and the tires properly inflated.
  - Call ahead before shopping and consolidate errands.
  - Avoid quick acceleration and deceleration and keep your speed under 60.
  - Avoid "drive-through" lines where your car engine must idle for long periods.
  - Recycle used motor oil, transmission fluid, brake fluid and antifreeze.
  - Turn in your old battery when you buy a new one.

For More Information

- Bay Area Action
  504 Emerson Street
  Palo Alto, CA 94301
  (415) 321-1994
- BART
  P.O. Box 12688
  Oakland, CA 94604
  (415) 464-6000
- California Department of
  Transportation
  P.O. Box 7310
  San Francisco, CA 94120
  (415) 923-4444
- Greenpeace
  139 Townsend St.
  San Francisco, CA 94107
  (415) 512-9025

- Metropolitan Transportation
  Commission
  MetroCenter
  8th and Oak Streets
  Oakland, CA 94607
  (415) 464-7700

- Peninsula Rail 2000
  P.O. Box 193552 Rincon Annex
  San Francisco, CA 94119
  (415) 737-1144
- Planning and Conservation League
  909 12th St. #203
  Sacramento, CA 95814
  (916) 444-8726
- Regional Bicycle Advisory Council
  3313 Grand Ave.
  Oakland, CA 94610
  (415) 452-1221
- Rides for Bay Area Commuters
  60 Spear Street, #650
  San Francisco, CA 94105
  (415) 861-POOL

BAY AREA ACTION
504-A EMERSON, PALO ALTO, CA 94301 (415) 321-1994

FIGURE 20.2   Fact sheet: "Transportation."

Analysis

Content: A fact sheet is tight organization in a nutshell. The plan of "Transportation" is simple: Problem-solution method: Why passenger cars are an environmental problem; what needs to be done to make them less of a problem; and what we as individuals can do to help.

Design: The designer effected a pleasing balance between text (divided into conspicuous subheads) and visuals—the Bay Area Action logo alongside the title, as well as the artwork in the center of the page, which serves to break up text that would otherwise have seemed too monotonous to catch the eye.

PROMOTIONAL FACT SHEETS

At first sounding like a contradiction in terms, a promotional fact sheet, while certainly trying to persuade the reader to purchase a product, generally includes more factual information about the product than does a run-of-the-mill magazine ad. It truly is factual in the sense that the technical "specs" of the product are listed as they would be in the owner's booklet. A promo sheet measures $8\frac{1}{2} \times 11$ (so it can fit into a marketing rep's notebooks) and has four pages; in other words, the sheet actually measures 17×11 but is folded in half.

Study the following promo sheet to see how advertising rhetoric mingles with factual rhetoric (Figure 20.3).

Analysis

"Microsoft Project" opens not with an introduction but with a pitch—and in boldface 18-point type to make sure that you stay alert while reading it. The Overview, too, is not free of pitch rhetoric—"extraordinary project management tool"; "no matter how large your projects are"; "exceptionally versatile"; etc.—but it does introduce the product's key features clearly. For example: "It's easy to customize fields and filters and to redefine menus to accommodate your industry standards. Split the screen to add a new perspective." Also important to the reader is how the product can fit in with other applications; this is answered in paragraph 5.

The Highlights section provides a fairly detailed rundown of Project's capabilities. Note the attractively formatted text and the sample screens. Finally, on the back page, the Performance Features section gives the technical specifications.

Microsoft Project

Business Project Planning System

With Microsoft Project, you can manage projects in more ways than you ever imagined. You get an unprecedented number of project perspectives. All the powerful tools you need to successfully juggle schedules and resources. And an array of charting and formatting options to ensure your reports communicate effectively.

Overview

Microsoft Project is an extraordinary project management tool for the Macintosh that gives you the flexibility to enter, view, schedule, summarize, and manage your projects exactly as you want.

Enter project information through a Gantt Chart, a PERT Chart, or spreadsheet-like views. It's easy to customize fields and filters and to redefine menus to accommodate your industry standards. Split the screen to add a new perspective and resolve any potential resource or scheduling problems. Or collapse and expand the level of detail within project tasks by just clicking a button.

No matter how large your projects are, Microsoft Project can handle them. Microsoft Project easily schedules and manages resources across multiple projects.

You'll also create polished, high-quality output, with incredible style options to choose from. And you can easily customize your report styles to use over and over again.

Microsoft Project is currently available for the Microsoft Windows™ graphical environment and the OS/2® operating system, so you can easily share files among coworkers—Microsoft Project even reads and writes Microsoft Excel files.

Whether you're new to project management or an experienced pro, you'll find Microsoft Project an exceptionally versatile, easy-to-use product that will become an indispensable part of your business.

Apple Macintosh Series

Version 1.1

Highlights

View your project from every possible perspective
- Gantt Chart: show project schedules graphically on a timescale (minutes to years scaling)
- PERT Chart: show task relationships for better decision making
- Outlining: easily group and arrange project tasks in a hierarchical order; collapse and expand levels of detail by clicking one button
- Filters: use and define filters so that you can view selected information exactly the way you want to see it
- Resource Usage view: view resource workloads to quickly identify resource availability and costs across multiple projects
- Resource Histogram: graphically display resource workloads and over-allocations across multiple projects
- Split view: view any two screens simultaneously and use the combinations to help solve allocation or scheduling problems interactively
- Task and Resource Sheets: enter and view data in a familiar row-and-column spreadsheet format
- Task and Resource Forms: view and enter information about a specific task or resource
- Task PERT Chart: view the immediate predecessors and successors to a specific task

Unprecedented flexibility lets you customize any way you want
- Custom fields: enter up to six custom task fields and two custom resource fields per project
- PERT Palette: display up to five data fields in each node (choose from more than 60 optional field selections) then choose different text styles, border styles, and colors to build your chart
- Gantt Palette: choose up to 8 colors, 7 bar sizes, 13 patterns, and a variety of symbols and annotations
- Custom tables: choose from over 60 fields to build tables for reporting and viewing data
- Custom menus: define menu selections for commonly used views, tables and filters
- Save your custom views, tables, filters and formatting to create specialized reports that you can access regularly or easily adapt

Delivers boardroom-quality reports and presentations
- Choose from a wide variety of font styles, point sizes, shading, and colors to clearly communicate your project information
- Enhance charts with special symbols, patterns, colors, and selected annotations
- Use print preview to view your chart or report

It's easy and intuitive to use
- Group link: save keystrokes by just clicking the on-screen button to automatically link two or more selected tasks
- Group edit: speed up data entry by just clicking the button to automatically make changes to a selected group of tasks or resources
- Build a PERT Chart interactively on-screen with just a few mouse clicks
- Easily adjust task durations, dates, and percentages completed by dragging the bar on the Gantt Chart
- Enter and edit project information in the way that's easiest for you: Gantt Chart, PERT Chart, Edit Form, Task and Resource Forms, Task Entry view, Task and Resource Sheets, and Resource Usage view
- Context-sensitive Help puts answers just a point and click away
- Classroom training materials are also available

Works across platforms and applications
- Microsoft Project is available for the Macintosh, Windows and OS/2 platforms
  - Use the same command sequences and files
  - Minimize training costs in mixed-platform workgroups
- Microsoft Project reads and writes Microsoft Excel files

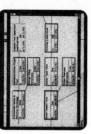

Microsoft Project Gantt Chart graphically displays project tasks on a timescale. Create custom bars to highlight specific conditions in your schedule.

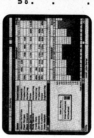

The PERT Chart lets you view and quickly define task relationships. Just point and click to easily create and link tasks. And select specific data fields that you want to display or edit.

Split the screen to get two perspectives at once. Add a Resource Histogram to graphically highlight workloads, so you can easily view resource allocation across projects.

FIGURE 20.3 Promo fact sheet: computer software.

Performance Features (cont.)

Organization and retrieval of data
- Outlining (five levels)
- Subprojects
- WBS (work breakdown structure) codes
- Filters (and/or logic)
- Sorting (three levels)
- Notes can be stored with each task or resource
- Go To: locate with task ID or date
- Find: search for project data that meets defined criteria (next or previous)

Data exchange with other popular applications
- Microsoft Excel
- Text (ASCII)
- CSV (comma-separated value)
- MPX (for data exchange with minicomputers and mainframe systems)

Functional capabilities
- 2000 tasks per project (random-access-memory-dependent)
- 20 resources per task
- 2000 resources per project (RAM-dependent)
- Calendar dates from 1984 through 2050

Specifications

System requirements
- Macintosh Plus, Classic, LC, SE, Portable, IIsi or other II-family personal computers
- Minimum of 1 MB of memory, 2 MB if using MultiFinder or System 7.0
- System 6.04 or later
- One 800K floppy disk drive and one hard disk drive

Options
- All Macintosh-compatible printers

Microsoft Corporation • One Microsoft Way • Redmond, WA 98052-6399
Version 1.1
0591 Part No. 098-22319

Microsoft®
Making it all make sense™

Performance Features

Scheduling
- Critical path method (CPM) scheduling
- Finish-to-start, start-to-start, start-to-finish, finish-to-finish relationships
- Lead/lag (negative and positive)
- Free, total, and negative slack calculated
- Supports the following constraints: ASAP (As Soon As Possible), ALAP (As Late As Possible), Must Start On, Must Finish On, Start No Earlier Than, Start No Later Than, Finish No Earlier Than, Finish No Later Than
- Resource-driven scheduling
- Task-driven scheduling
- Deadline (backward) scheduling
- Minutes, hours, days, weeks, and elapsed-duration scheduling

Resource management
- Partial resource assignments
- Standard work and rates
- Per-use rates
- Overtime work and rates
- Cost accrual at start, end, or prorated
- Automatic resource leveling resolves conflicts across multiple projects
- Selective resource leveling for a group of resources according to priority, date, slack, and ID
- Supports both resource-constrained and task-constrained leveling
- Individual resource calendars
- Group (department) resource calendars
- Resource identification codes

Multiple projects
- Selectively link projects together to share a common resource pool
- Open, arrange, and view multiple projects on-screen simultaneously
- Subdivide large projects into manageable subprojects
- Save and retrieve multiple projects simultaneously using a workspace

Progress tracking
- Earned value reporting
- Track actual, plan, scheduled, and remaining cost, work, and durations
- Variance reporting on cost, work, and schedule
- Track percent of task complete
- Set a baseline plan and save up to three "snapshots" of your project as it evolves over time

Click the group edit button to make changes to a selected group of tasks or resources automatically.

Click the group link button to link selected tasks automatically.

Click the outline buttons to group and arrange project tasks easily and logically—both top-down or bottom-up.

Produce boardroom-quality reports using fonts, colors, borders, shading, and special symbols to clearly communicate your project information.

Create custom menus to easily access your most commonly used filters and tables. Create interactive filters that prompt you to supply your own selective criteria.

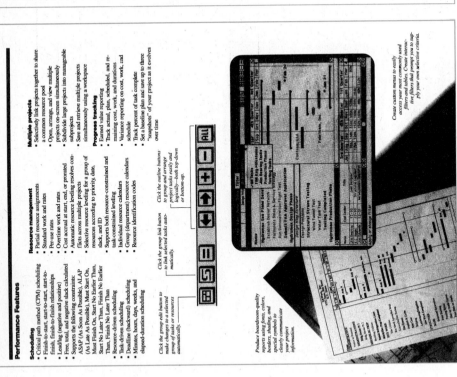

FIGURE 20.3 (continued)

WRITING THE TEXT OF A BROCHURE

Before actually getting down to writing a draft of a brochure, it is important to have a good idea of the slant you want to give to the subject. Let's say the subject for which you have been asked to prepare a brochure is a common health problem—influenza. Before you can begin writing, you need to decide who will be your primary audience: those who already have come down with the flu? Or those who are concerned about coming down with it during the flu season and are wondering what precautions, if any, they ought to take? Will your readers be primarily adults or young people or both (i.e., a "general" audience)? How much detailed information should the brochure contain? That is, do you see your readers as being intrinsically interested in influenza? Or do you see them as purely pragmatic in their desire for information, as people who couldn't care less about all the medical details? How you answer these questions will have a significant influence on the writing of your brochure.

Let's assume that you define your audience as follows: readers of all ages who wish to take precautions for what threatens to be a high-risk flu season, and who are interested only in useful information.

Your next task is to structure the text in light of your readers' needs: general definition first, followed by an elaboration of the definition (i.e., symptoms of the flu, seriousness of the flu); followed by a discussion of the flu's effects (Who is most likely to catch it? Can one catch it more than once? Is it preventable?); and concluding with ways of treating the flu (Figure 20.4).

WRITING AND DESIGNING A BROCHURE OR FACT SHEET: A CHECKLIST

☐ Is my topic and the approach I am taking on the topic suitable for a brochure or fact sheet?

☐ Do I have a clear sense of my intended readers?

☐ Have I worked out a suitable design for the brochure or fact sheet?

☐ Have I organized my topic into distinct subsections?

☐ Did I choose appropriate visuals?

☐ Do text and visuals interact well?

facts about...

INFLUENZA (FLU)

What Is Flu And How Is It Caused?

Influenza is a contagious disease caused by a virus. A virus is a germ that is very small. Influenza viruses infect many parts of the body, including the lungs.

When someone who has the flu sneezes, coughs, or even talks, the flu virus is expelled into the air and may be inhaled by anyone close by. Flu may be transmitted by direct hand contact.

What Happens When You Get The Flu?

When flu strikes the lungs, the lining of the respiratory tract is damaged. The tissues become swollen and inflamed. Fortunately, the damage is rarely permanent. The tissues usually heal within two weeks.

Influenza is often called a respiratory disease, but it affects the whole body. The victim usually becomes acutely ill with fever, chills, weakness, loss of appetite, and aching of the head, back, arms, and legs. The flu sufferer may also have a sore throat and a dry cough, nausea, and burning eyes.

The fever mounts quickly — temperature may rise to 104° — but after two or three days, it usually subsides. The patient is often left exhausted for days afterwards.

Is Flu Considered Serious?

For healthy children and adults, influenza is typically a moderately severe illness. Most people are back on their feet within a week.

For people who are not healthy or well to begin with, influenza can be very severe and even fatal. The symptoms described above have a greater impact on these persons. In addition, complications can occur.

Most of these complications are bacterial infections, since the body can be so weakened by influenza that its defenses against bacteria are low. Bacterial pneumonia is the most common complication. But the sinuses and inner ears may become inflamed and painful.

Who Gets The Flu?

Anyone can get the flu — especially when it is widespread in the community. In a flu epidemic year, from 20 to 50 percent of those not immunized may contract influenza.

People who are not healthy or well to begin with are particularly susceptible to the complications that can follow. These people are known as "high risk." For anyone at high risk, influenza is a very serious illness. You may be at high risk if you:

- have chronic lung disease such as asthma, emphysema, chronic bronchitis, bronchiectasis, tuberculosis, or cystic fibrosis
- have heart disease
- have chronic kidney disease
- have diabetes or other chronic metabolic disorder
- have severe anemia
- have diseases or are having treatments that depress immunity
- are residing in a nursing home or other chronic care facility
- are over 65 years of age

A physician, nurse, or other provider of care to high-risk persons should be immunized to protect high-risk patients.

How Are Flu And Complications Prevented?

Influenza can be prevented when a person receives the current influenza vaccine. This vaccine is made each year so that the vaccine can contain influenza viruses that are expected to cause illness that year.

The viruses in the vaccine are killed or inactivated so that someone vaccinated cannot get influenza from the vaccine. Instead the person vaccinated develops protection in his or her body in the form of substances called antibodies.

The amount of antibodies in the body is greatest 1 or 2 months after vaccination and then gradually declines. For that reason and because the influenza viruses usually change each year, a high-risk person should be vaccinated each fall with the new vaccine.

(continued)

FIGURE 20.4 Fact sheet: "Flu."

November is the best time to get your flu shot. Such a yearly vaccination has been found to be about 75 percent effective in preventing flu. It also may very well reduce the severity of flu and be lifesaving in vaccinated persons.

A drug called amantadine also can be used to help prevent flu. It is discussed later in the section on treatment.

What About Reactions To The Vaccine?

Most people have little or no reaction to the vaccine.

One in four might have a swollen, red, tender area where the vaccination was given.

A much smaller number, probably more children than grownups, might also develop a slight fever within 24 hours. They may have chills or a headache, or feel a little sick. People who already have a respiratory disease may find their symptoms worsened. Usually none of these reactions lasts for more than a couple of days.

In addition, adverse reactions to the vaccine, perhaps allergic in nature, have been observed in some people. These could be due to an egg protein allergy, since the egg in which the virus is grown cannot be completely extracted. These people should be vaccinated only if their own physician believes it necessary and if the vaccine is given under close observation by a physician.

Who Should Be Vaccinated?

People at high risk should be vaccinated yearly against flu. In addition, those who provide care to high-risk patients should be vaccinated.

If you are not in a high-risk group, ask your doctor if you need the vaccine.

Can You Have A Recurrence Of Flu?

A person can have influenza more than once. Here's why:

The virus that causes influenza may belong to 1 of 3 different flu virus families, A, B, or C. Influenza A and influenza B are the major families.

Within each flu virus family are many viral strains, like so many brothers and sisters. Both A and B have strains that cause illnesses of varying severity. But the influenza A family has more virulent strains than the B family.

If you have the flu, your body responds by developing antibodies. The following year, a new family member or a member of another family may appear. Your antibodies are less effective or ineffective against this unfamiliar strain. If you are exposed to it, you may come down with flu again.

How Are Flu And Complications Treated?

For uncomplicated flu, your doctor will probably tell you to stay in bed at home as long as the sickness is severe — and perhaps for about two days after the fever is gone.

The drug called amantadine is useful for treating someone who develops influenza A, particularly if it is given as soon as possible after the onset of flu. Amantadine also can be used as a preventive, but for prevention it must be taken daily as long as flu cases continue to occur in a community. Your doctor would have to decide whether to use amantadine either for prevention or treatment. If it is used for treating an early case of flu, it may shorten this illness and reduce the severity. Amantadine works only against influenza A viruses and should be used only if influenza A is suspected.

Amantadine sometimes causes side effects such as difficulty in sleeping, tremulousness, or depression; these are usually mild and often go away even when the medicine is continued.

The treatment of non-bacterial complications varies with the illness. If you should develop a bacterial complication, however, your doctor can give you an antibiotic.

Why In Some Years Do More People Get Flu Than In Others?

Every 10 years or so, a flu virus strain appears that is dramatically different from the other members of its family. When this major change occurs a worldwide epidemic — called a pandemic — almost inevitably follows. Few people have antibodies that are effective against the new virus.

One such virus caused the 1918 flu epidemic that swept the world and left in its wake more than 20 million dead. Fear of a similar outbreak in the fall of 1976 inspired a mass vaccination effort. Fortunately, no epidemic developed.

Key Points To Remember

• If you are a high risk person, you should get your yearly flu shot.
• You should know about treating flu in case it occurs despite your vaccination.
• Discuss this with your doctor.

It's a Matter of Life and Breath®

This publication was made possible in part by your support of Christmas Seals® and other contributions to the American Lung Association.

■ CHAPTER SUMMARY

Brochures and fact sheets are effective ways of introducing concepts and issues, products and services to the public. Because they are folded into thirds or fourths and because both sides of the sheet are used, brochures can pack a lot of information onto a single sheet of paper. The texts of brochures and fact sheets are tightly organized, with distinct subheads or question-and-answer format. Creating a brochure or fact sheet is as much designing as writing. One needs to work out a general layout of each page: how much text to use with which visual aids, if any; how best to format the text.

■ FOR DISCUSSION

1. Critique the flu brochure in terms of clarity, organization, and use of document design techniques.
2. Study the brochure on microwave oven radiation (pages 496–498), published by the Public Health Service. Be prepared to comment on the overall design, the organization of text, the use of subheads, and the appropriateness of the visuals. Also suggest alternative design possibilities.
3. Locate a brochure on your own, make transparency copies of each page, and using the overhead projector, discuss the layout with the class.
4. Why are brochures such a popular form of publication?
5. Are promotional fact sheets more like advertisements or like conventional fact sheets? Compare the promo fact sheet for Microsoft® Project with the conventional fact sheet on Transportation. What are the similarities and differences? Do you see any persuasive rhetoric in the fact sheet on Transportation? In what ways is its rhetoric similar or dissimilar to that of the software promo fact sheet?

■ FOR YOUR NOTEBOOK

1. Collect several types of brochures and take notes on their similarities and dissimilarities.
2. Sketch out an alternative design for a brochure you have located.
3. Jot down several ideas for brochures that you would enjoy creating. Choose the most interesting one, outline the text, and sketch out a design.

■ BROCHURE AND FACT SHEET WRITING PROJECTS

1. Create a finished brochure for the plan you worked out in For Your Notebook, topic 3.

2. Prepare a brochure about a health problem, such as AIDS, alcoholism, Alzheimer's disease, diabetes, emphysema, asthma, or allergies.

3. Prepare a brochure, aimed at pet owners, on how to care for a new dog or cat (one or the other).

4. Prepare a fact sheet on an evironmental or campus-related issue.

5. Using a computer program you are thoroughly familiar with, introduce that program in a promotional fact sheet. Perhaps you can envision certain upgrades for the program; if so, introduce the product in a new version.

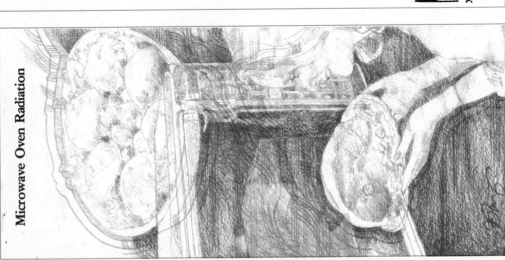

Microwave Oven Radiation

Microwaves are used to detect speeding cars, to send telephone and television communications, and to treat muscle soreness, to dry and cure plywood, to cure rubber and resins, to raise bread and doughnuts, and to cook potato chips. But the most common consumer use of microwave energy is in microwave ovens. That use has soared in the past decade.

The Food and Drug Administration has regulated the manufacture of microwave ovens since 1971. On the basis of current knowledge about microwave radiation, the Agency believes that ovens that meet the FDA standard and are used according to the manufacturer's instructions are safe for use.

What is Microwave Radiation?

Microwaves are a form of "electromagnetic" radiation; that is, they are waves of electrical and magnetic energy moving together through space. Electromagnetic radiation ranges from the energetic x rays to the less energetic radio frequency waves used in broadcasting. Microwaves fall into the radio frequency band of electromagnetic radiation. Microwaves should *not* be confused with x rays, which are more powerful.

Microwaves have three characteristics that allow them to be used in cooking: they are reflected by metal; they pass through glass, paper, plastic, and similar materials; and they are absorbed by foods.

Cooking with Microwaves

Microwaves are produced inside the oven by an electron tube called a magnetron. The microwaves bounce back and forth within the metal interior

Electromagnetic Radiation

X Rays   Light   ← Microwaves   Radio Frequency →

until they are absorbed by food. Microwaves cause the water molecules in food to vibrate, producing heat that cooks the food. That's why foods high in water content, like fresh vegetables, can be cooked more quickly than other foods. The microwave energy is changed to heat as soon as it is absorbed by food. Thus it can not make food radioactive or "contaminated."

Although heat is produced directly in the food, microwave ovens do not cook food from the "inside out." When thick foods like roasts are cooked, the outer layers are heated and cooked primarily by microwaves while the inside is cooked mainly by the slower conduction of heat from the hot outer layers.

Microwave cooking can be more energy efficient than conventional cooking because foods cook faster and the energy heats only the food, not the oven compartment. Microwave cooking does not reduce the nutritional value of foods any more than conventional cooking. In fact, foods cooked in a microwave oven may keep more of their vitamins and minerals, because microwave ovens can cook more quickly and without adding water.

Glass, paper, ceramic, or plastic containers are used in microwave cooking because the microwaves pass through them. Although such containers cannot be heated by microwaves, they can become hot from the heat of the food cooking inside. Some plastic containers should not be used in a microwave oven, as they can be melted by the heat of the food inside. Generally, metal pans or aluminum foil should also not be used in a microwave oven, as the microwaves are reflected off these materials causing the food to cook unevenly and possibly damaging the oven. The instructions that come with each microwave oven indicate the kinds of containers to use. They also cover how to test containers to see whether or not they can be used in microwave ovens. FDA recommends that microwave ovens not be used in home canning. It is believed that neither microwave ovens nor conventional ovens produce or maintain temperatures high enough to kill the harmful bacteria that occur in some foods while canning.

(continued)

(continued)

are being produced. There is no residual radiation remaining after microwave production has stopped. In this regard a microwave oven is much like an electric light that stops glowing when it is turned off.

All ovens made since October 1971 must have a label stating that they meet the safety standard. In addition, FDA requires that all ovens made after October 1975 have a label explaining precautions for use. This requirement may be dropped if the manufacturer has proven that the oven will not exceed the allowable leakage limit even if used under the conditions cautioned against on the label.

To make sure the standard is met, FDA tests microwave ovens in commercial establishments, dealer and distributor premises, manufacturing plants, and its own laboratories. FDA also evaluates manufacturers' radiation testing and quality control programs. When FDA finds a radiation safety problem in a certain model or make of oven, it requires the manufacturer to correct all defective ovens at no cost to the consumer.

Although FDA believes the standard assures that microwave ovens do not present any radiation hazard, the Agency continues to reassess its adequacy as new information becomes available.

Microwave Oven Safety Standard

All microwave ovens made after October 1971 are covered by a radiation safety standard enforced by the Food and Drug Administration. The standard limits the amount of microwaves that can leak from an oven throughout its lifetime. This limit is 5 milliwatts of microwave radiation per square centimeter at approximately 2 inches from the oven surface. This is far below the level known to harm people. Furthermore, as you move away from an oven, the level of any leaking microwave radiation that might be reaching you decreases dramatically. For example, someone standing 20 inches from an oven would receive approximately one one-hundredth of the amount of microwaves received at 2 inches.

The standard also requires all ovens to have two independent interlock systems that stop the production of microwaves the moment the latch is released or the door opened. In addition, a monitoring system stops oven operation in case one or both of the interlock systems fail. The noise that many ovens continue to make after the door is opened is usually the fan. The noise does not mean that microwaves

Microwave Ovens and Health

Much research is under way on microwaves and how they might affect the human body. It is known that microwave radiation can heat body tissue the same way it heats food. Exposure to high levels of microwaves can cause a painful burn. The lens of the eye is particularly sensitive to intense heat, and exposure to high levels of microwaves can cause cataracts. Likewise, the testes are very sensitive to changes in temperature. Accidental exposure to high levels of microwave energy can alter or kill sperm, producing temporary sterility. But these types of injuries—burns, cataracts, temporary sterility—can only be caused by exposure to large amounts of microwave radiation, much more than can leak from a microwave oven.

Less is known about what happens to people exposed to low levels of microwaves. To find out, large numbers of people who had been exposed to microwaves would have to be studied for many years. This information is not available. Much research has been done with experimental animals, but it is difficult to translate the effects of microwaves on animals to possible effects on humans. For one thing, there are differences in the way animals and humans absorb microwaves. For another, experimental conditions can't exactly simulate the conditions under which people use microwave ovens. However, these studies do help to better understand the possible effects of microwaves.

One experiment, for example, showed that repeated exposure to low-level microwave radiation (less than 10 milliwatts per square centimeter) does not cause cataracts in rabbits. On the other hand, some animals display an avoidance reaction when exposed to low levels of microwaves—that is, they try to get away from the microwaves. Other effects noted in experimental animals include a decreased ability to perform certain tasks, genetic changes and an "immune response" (the body acts as if it were responding to protect itself from a disease). While these and similar effects have been observed in animals, their significance for human health remains unclear.

These kinds of findings, together with the fact that many scientific questions about exposure to low-levels of microwaves are not yet answered, point to the need for FDA to continue to enforce strict radiation controls. They also underscore the need for consumers to take certain common sense precautions.

Have Radiation Injuries Resulted From Microwave Ovens?

There have been allegations of radiation injury from microwave ovens. The injuries known to FDA, however, have been injuries that could have happened with any oven or cooking surface. For example, people have been burned by the hot food, splattering grease, or steam from food cooked in a microwave oven.

Ovens and Pacemakers

At one time there was concern that leakage from microwave ovens could interfere with certain electronic cardiac pacemakers. There was similar concern about pacemaker interference from electric shavers, auto ignition systems, and other electronic products. Because there are so many other products that also could cause this problem, FDA does not require microwave ovens to carry warnings for people with pacemakers. The problem has been largely resolved since pacemakers are now designed so they are shielded against such electrical interference. However, patients with pacemakers may wish to consult their physicians about this.

Checking Ovens For Leakage

There is little cause for concern about excess microwaves leaking from ovens unless the door hinges, latch, or seals are damaged, or if the oven was made before 1971. In FDA's experience, most ovens tested show little or no detectable microwave leakage. If there is some problem and you believe your oven might be leaking excessive microwaves, contact the oven manufacturer, a microwave oven service organization, your State health department, or the nearest FDA office. Some oven manufacturers will arrange for your oven to be checked. Many States have programs for inspecting ovens or they may be able to refer you to microwave oven servicing organizations that are equipped to test ovens for excessive emission. A limited number of ovens are also tested in homes by FDA as part of its overall program to make sure that ovens meet the safety standard.

A word of caution about the microwave testing devices being sold to consumers: FDA has tested a number of these devices and found them generally inaccurate and unreliable. If used, they should be relied on only for a very approximate reading. The sophisticated testing devices used by public health authorities to measure oven leakage are far more accurate and are periodically tested.

Tips on Safe Microwave Oven Operation

• Follow the manufacturer's instruction manual for recommended operating procedures and safety precautions for your oven model.

• Don't operate an oven if the door does not close firmly or is bent, warped, or otherwise damaged.

• Never operate an oven if you have reason to believe it will continue to operate with the door open.

• To add to the margin of safety already built into the oven, don't stand directly against an oven (and don't allow children to do this) for long periods of time while it is operating.

Other Tips for Microwave Oven Use

• Some ovens should not be operated when empty. Refer to the instruction manual for your oven.

• Clean the oven cavity, the outer edge of the cavity, and the door with water and a mild detergent. A special microwave oven cleaner is not necessary. Do not use scouring pads, steel wool, or other abrasives.

For more consumer information on microwave oven radiation, write to Microwave Ovens, HFX-28, Bureau of Radiological Health, Food and Drug Administration, Rockville, Maryland 20857, or to your State health department or your local FDA office.

U.S. Department of Health and Human Services
Public Health Service
Food and Drug Administration
Rockville, Maryland 20857
HHS Publication No. (FDA) 80-8120

21

Newsletter Writing and Editing

L ike brochures and fact sheets, newsletters abound. Any amateur society from baseball-card enthusiasts to wine connoisseurs; any business or professional group from a neighborhood chiropractor to an international association of archaeologists quite likely produces a newsletter or is planning to.

THE NATURE AND PURPOSE OF NEWSLETTERS

A newsletter, to put it in simplest terms, is an "insider" publication—a "private newspaper" as Marvin Arth and Helen Ashmore have termed it in their *Newsletter Editor's Deskbook;* it typically consists of information of practical use to its intended readers. Members of an organization, especially a small or newly launched one, are often encouraged to contribute ideas; a newsletter can thus be a significant means by which the group's identity evolves.

Although it is difficult to generalize about what a "typical" newsletter contains, articles can be categorized as straight news, editorials, announcements, personal testimony from members (sometimes in the form of letters to the editor), and general features. Whatever is included, the newsletter should convey an element of the personal. As graphic designer Jan V. White puts it, "A true newsletter is simple, informal, relaxed. Its effectiveness lies in the illusion it creates of being a person-to-person letter."[1]

More specifically, newsletters include material like the following:

News

- information about new products or opportunities
- company policy changes, newly appointed staff, reports of product sales
- disclosures, such as reorganization, mergers

Announcements

- society-sponsored outings
- meetings, lectures, visitors
- special rates for insurance, loans, etc.

---

[1]Jan V. White, *Mastering Graphics: Design and Production Made Easy.* New York: R. R. Bowker Co., 1983: 2.

Features

- profiles of personnel
- "in-group" humor, cartoons

Small newsletters are usually staff written, although short items may be solicited from the readership. These days, newsletters are designed on a computer, using a desktop publishing program. These versatile programs enable the individual user to create full-fledged layouts of each page, from column widths to contouring ("scalloping") text around an illustration, to creating and inserting visuals.

While it is true that newsletters contribute to the coherence and efficient operation of a company by keeping its reader-members informed about institutional matters, newsletters are not merely fact sheets; they possess a certain entertainment value as well. Their recipients enjoy reading about the significant activities of their company and about the individual accomplishments of their fellow employees. Such news items go a long way toward boosting and maintaining morale. Even in large organizations, newsletter reports and features can stimulate increased participation and even increased financial support (alumni newsletters being an obvious case in point).

Most newsletters are intended for nontechnically trained readers, even though they may be greatly interested in the specialized activities of the institution in question. Consider, for example, the following two-page newsletter, *TopHealth* (Figure 21.1), issued ten times a year, which slants health and medical information for people in organizations and corporations. This particular issue contains two feature articles—one on heath-care fraud, the other on the dangers of hearing loss. The shorter pieces are tied together by their seasonal slant as well as by their practicality. Most newsletters maintain a balance between short and long pieces; between text and visuals.

DESIGNING A NEWSLETTER

Before reading on, take another look at the issue of *TopHealth*. What are your initial impressions of its overall design? Now look more closely at its design features: How many different kinds of design features do you see? How do they relate to each other? Which features dominate? Which items catch your eye?

Here are some of the characteristics you may have noticed in the *TopHealth* newsletter:

TopHealth

We Care About Your Health

August 1991

The Health Promotion and Wellness Letter

Healthcare Consumerism
How to Duck Quackery

Perhaps you've had a chronic disease that does not appear to resolve with conventional medical treatment. Or maybe you know a friend, relative or co-worker with such a problem. If so, you've probably felt the lure of quackery—or unproven and sometimes dangerous remedies—even if you've never given in to it.

Quack products and treatments prey on people who feel frustrated about having to endure their incurable conditions. Especially after years of pain and suffering, they may feel desperate for any glimmer of hope. Pushers of unproven remedies often exploit these feelings by bashing the "medical establishment" for "overlooking" or "holding back" their products.

Healthcare quackery has become much more sophisticated than in the heyday of snake oil salesmen. Some hucksters still give themselves away with references such as "miracle cure," "amazing breakthrough," "exclusive discovery" and "secret formula." But some now also use scientific-sounding terms. They may band together in false groups with names that have been chosen to appear misleadingly impressive: "associations," "colleges," even "academies." And, sad to say, some may even have scientific training.

Fraud Markets
Some of the biggest areas attracting health quackery:
1. aging
2. arthritis
3. cancer
4. Allergies
5. weight loss
6. men's dysfunction
7. sexual dysfunction

IF IT SOUNDS TOO GOOD TO BE TRUE, IT PROBABLY IS.

Red Flags

How to avoid getting duped? Realize that not all advertising is closely regulated. The FDA (Food and Drug Administration) and Council of Better Business Bureaus say beware if a product's label or ad:

- pledges instant, effortless results
- promises a "cure"—or complete relief from pain
- offers a "money-back" guarantee
- includes testimonials from anonymous "satisfied users"
- says it is available by mail from only one source
- claims it can treat a wide range of ailments
- boasts of FDA approval (as law forbids makers of drugs and medical devices from mentioning the FDA to suggest marketing approval)

Continued on next page

SEASONAL HEALTH SNIPPETS

Sweat-Proof Sunscreen Now's the smart time to wear sunscreen every day. But if you wash off as you sweat, you may find most sunscreens wash off too, right into your eyes, causing stinging and worse. The solution: Look for new "sports" sunscreens, made to stay put and take a little more waterproof.

Traveler's Alert Blood clots in the legs, especially for people with excess weight, are a little-known but real risk of flying. But this fear of flying is easy to prevent: periodically stretch your legs and relax the muscles.

Poison Ivy Leaves of three: Let them be. If you ever touch any part of the plant, oily poison can contact an oily poison that causes itchy skin blisters and can be soothed by calamine lotion; severe ones may respond to hydrocortisone cream, available over the counter.

Swimmer's Ear Avoid swimmers risk developing a condition called swimmer's ear, as microbes thrive in wet, dark places like the inside of the ear. The solution: Dry your ears out with vinegar or rubbing alcohol whenever you fear water's trapped inside them.

RINGING IN THE EARS > > >

Hearing Protection
Ringing in the Ears

Have you ever noticed a ringing in your ears after you used loud equipment or listened to music with the volume pumped? About 37 million Americans are estimated to suffer from some chronic tinnitus, or head noise—that is, a ringing, buzzing, hissing, clicking or whistling noise in one or both ears.

For most of these people, tinnitus is a minor, whispering nuisance that stops soon after it starts. But over seven million are more seriously disturbed by the noise they can't escape from. It can become severe enough to threaten their well-being. Many find that persistent tinnitus disrupts their concentration and their sleep, making them feel depressed and sometimes even paranoid or suicidal.

Tinnitus often accompanies other ear problems, such as becoming hard of hearing. But it can also occur on its own. Sometimes doctors can find the cause of the tinnitus and get rid of it. Here are examples of conditions that can provoke tinnitus:

- exposure to loud noise
- high blood pressure, or hypertension
- an allergic reaction
- impacted wax in the canal of the outer ear — that's the part of the ear you can see
- a hole in the eardrum
- infection of the middle ear—the eardrum and trio of tiny bones just beyond it
- damage to the hearing nerve in the inner ear
- a tumor or growth in or near the ear

In most cases, the underlying reason for the tinnitus remains a mystery. Still, several approaches can provide relief. Sometimes certain medications can help mute the noise. Some people with impaired hearing and tinnitus can reduce head noise by wearing their hearing aids. And treatment called TENS—transcutaneous (through-the-skin) electrical nerve stimulation—has recently been reported to help silence the tinnitus.

Even in quiet surroundings, where tinnitus is most bothersome, people can learn to ignore their head noises, with the help of biofeedback (concentration and relaxation exercises) and "masking." The principle behind masking is that familiar exterior sounds are easier to ignore than tinnitus is.

Masking uses soothing background sounds called "white noise." These sounds can be as simple as static from a radio or the ticking of a clock. Or they can come from a tinnitus masker, a small electronic device built into a hearing aid. It emits a constant, low-level sound that drowns out tinnitus and makes people less aware of it.

The good news: Most people with tinnitus say their condition gets better with time. And you can do a lot to prevent it from starting. □

Healthcare Consumerism *(continued)*

Remember, if it sounds too good to be true, it probably is. If you're tempted to try something unorthodox, it pays to ask your doctor about it first. Ideally your personal physician has received training from a reputable institution and is a board-certified or board-eligible healthcare practitioner. He or she can guide you toward high-quality care—and away from fraud and quackery, and having your hopes cynically raised and dashed. □

INFOLINE If you think you've been bilked by a mislabeled or misrepresented product, complain to the Food and Drug Administration, Consumer Affairs and Information, 5600 Fishers Lane, HFC-110, Rockville, MD 20857.

TopHealth™ is published 10 times a year to promote the good health of members and employees of organizations and corporations. It is written under the supervision of physicians from leading medical institutions to provide the information and motivation necessary for healthy and realistic lifestyle.

TopHealth is not intended to provide medical advice on individual health problems. Such should be obtained directly from a physician. The subscribing company/organization is not responsible for the content herein and does not endorse any product or service advertised or mentioned. Reproduction in any form of any part of this material without prior written permission is strictly prohibited. ©Copyright The Health Source Corporation, 1991. TopHealth is also available in Spanish.

For subscription information contact TopHealth, 74 Clinton Place, P.O. Box 203, Newton, MA 02159, (617) 244-8865. Editor: Amori I. Dreyfuss, M.D.

FIGURE 21.1 A health-care newsletter.

- *An eye-catching banner* (designer jargon for presentation of the newsletter's name). Take a lot of time in choosing a name and deciding how to render it graphically as a banner. "Your name is your trademark. It is the major symbol by which you are recognized."[2]
- *Lively visuals of various sizes and types* (but not so varied as to seem gaudy).
- *An attractive layout* (a design that captures and holds the reader's attention). Its elements typically include **balance** (of text, visuals, and white space) and **variety** (of visual aids, type fonts and typestyles, and text design) to prevent the page from becoming monotonous.
- *Attractively formatted text* The text of each news item is formatted to stimulate reader attention: bulleted text, subheads, columns scalloped to accommodate visuals.

Clearly, *TopHealth* is a model of creative design and economy. On just a single double-sided sheet are two substantive articles, four health-news bulletins under the heading Seasonal Health Snippets, and an "Infoline" reminder following the lead article on quackery. Even with all this material, the layout makes effective use of white space, incorporates several types of visual aids, and uses a variety of text-design strategies, all without the slightest trace of clutter. In this regard, note how the white space serves a dual purpose: (1) to highlight the distinctive design elements, and (2) to shield the reader from distraction, or *feeling* of clutter, while he or she is trying to focus on a particular section.

Steps in the Newsletter Design Process

Newsletter layout is no haphazard process. It is important, first of all, to begin with a dummy sketch of each page. The first page, for example, will contain these elements:

- Banner (newsletter title, usually between 48 and 72 pt. bold with a customized font (see samples, Figure 21.2); accompanied by the logo or motto and date of issue.
- Headline of lead feature.
- Text of the feature or a portion thereof (with continuation on another page).
- Headline and text of secondary feature.
- At least one large visual.

---

[2]Jan V. White, 14.

FIGURE 21.2 Sample newsletter banners.

Figures 21.3 and 21.4 show typical dummy sketches of the first page of typical newsletters. Compare the sketches with the printed newsletter (Figure 21.1).

WRITING AND EDITING A NEWSLETTER: A CHECKLIST

☐ Is the name I have given my newsletter conspicuous and memorable?

☐ Does the newsletter convey a personal character in both content and style?

☐ Have I worked out a balance of news, announcements, and features for my newsletter?

☐ Do the textual and visual elements interact effectively?

☐ Have I divided lengthy pieces with appropriate subheads? Are article titles and subheads distinctive?

FIGURE 21.3 Dummy for first page of UNICEF newsletter.

■ CHAPTER SUMMARY

Although it has no fixed definition, a newsletter tends to be more informal, more personal than any single item in the whole. That is, all the elements in the newsletter work together to create a distinctive character. Typical newsletters will contain a blend (if not an equal balance) of news, announcements, and features. They are ideal publications for professionals in a given field, for hobbyists, members of clubs, political groups, professional societies, and the like. Thanks to desktop publishing technology, anyone can learn to produce professionally designed newsletters.

■ FOR DISCUSSION

1. Compare the character of a newsletter to that of a brochure or fact sheet (see Chapter 20).
2. Critique the organization and design of the newsletter from an environmental studies program (page 508).

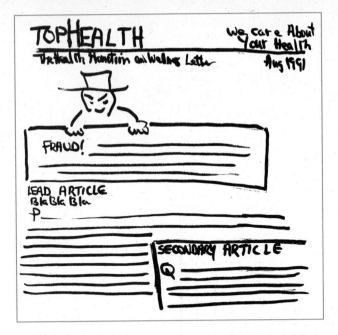

FIGURE 21.4 Trial dummy sketches for Top Health newsletter.

3. Study the newsletter on page 509, published by a chiropractor. Imagine that Dr. Irvin has asked you to prepare the next issue for her and to redesign it, expand it, or reformat it any way you wish. Working in groups of three, come up with some possible layout designs, additional features or visuals; different type fonts and sizes, alternative banners.

4. Discuss the usefulness of newsletters in science and technology.

■ FOR YOUR NOTEBOOK

1. Make a list of names you would like to give your newsletter. You might also sketch possible logos to use.

2. Work out a table of contents and table of figures for a newsletter you would like to publish.

3. Sketch some dummy layouts for the pages of the newletter for which you have already tentatively determined the contents.

■ NEWSLETTER WRITING AND DESIGNING PROJECTS

1. Using your preliminary work in For Your Notebook, write and design a complete four-page newsletter.

2. Redesign an existing newsletter according to your taste. Edit, rearrange, or replace the existing contents.

3. Prepare three alternative first pages for a newsletter addressed to a technological or scientific society of some sort.

4. Working in collaboration with one or two other classmates, prepare a newsletter appropriate for your major field of study. Appoint a "design manager" to allocate specific duties to each member of the group.

EnviroNews

The Environmental Studies Newsletter

Winter 1993 · Santa Clara University · Volume 1

THE CONCEPT

Designed for students interested in participating in an environmental studies minor, this quarterly newsletter will contain messages from the co-directors, brief listings of internship and research opportunities, information regarding the quarterly colloquia, summaries of future courses offered within the minor, and articles submitted by students who wish to share their environmental research and internship experiences.

AN ADVISORY NOTE

All those who have declared, or are interested in declaring this minor, please see Dr. Janice Edgerly-Rooks (x2709), in room 100B of the Alumni Science building. She will assign advisors to those who wish to declare the minor. The advisors will help students coordinate the Environmental Studies curriculum with the student's major course of study. In addition, they will offer valuable insights regarding career planning and graduate study within the diverse fields of environmental studies. Career Services will inform students of rewarding opportunities within the field. Environmental Studies advisors are available in many departments throughout the campus, from Anthropology to Political Science. For more information please contact Dr. Janice Edgerly-Rooks or call x2709 for an appointment.

OPPORTUNITIES IN THE FIELD

Internship Workshop: February 23

Many internships, careers and graduate opportunities are available to students in environmental studies. Patti Wilson of the SCU Career Services will present an overview of these opportunities at a workshop on February 23 between 5-6pm in Alumni Science 102. Ms. Wilson will discuss programs, such as the Environmental Career Center, which help place juniors, seniors and graduates in environmental internships throughout the nation. Students can gain invaluable experience as well as credit in the minor by participating in an internship program. (Please remember that all scholarly activities off campus must be approved beforehand for credit to be received).

VISITING SCHOLAR

Every quarter, Environmental Studies invites an active scholar within an environmental field, to share their knowledge with the students, faculty, and friends of Santa Clara University. Students participating in the Environmental Studies minor are required to attend six of these presentations. The event will occur in the Williman Room of Benson Center, and will be preceded by a reception open only to the students and faculty members involved with the minor. The reception serves as an informal gathering where students, faculty, and the invited guest can openly converse and become aquainted with one another.

ANNE EHRLICH BIOLOGICAL EFFECTS OF GLOBAL CHANGE	February 22, 1993
	Desert Reception
	Benson Parlors
	6:30 pm
	RSVP by Feb. 10, call x2709
	Lecture
	Williman Room, Benson
	7:30 pm

This quarter's visiting scholar is Anne H. Ehrlich. A senior research associate in biology and policy coordinator of the Center for Conservation Biology at Stanford University, she will deliver a presentation entitled Biological Effects of Global Change. Unnatural mobilization of elements, the toxic effects of novel materials, such as synthetic pesticides and CFCs, and the harmful disruption of terrestrial vegetative patterns, degradation of land and soils, decimation of biodiversity and changes in atmospheric composition will be discussed. She will then address the possibility of prevention and mitigation, and discuss some reform programs presently in action. In addition, she will suggest new methods of reform, and warn listeners of the future consequences for humanity if such problems fail to be addressed. The Environmental Studies program at Santa Clara University is proud to present such an environmentally active and aware member of this society and hopes that all will attend.

Printed on 100% chlorine-free recycled paper

FIELD NOTES

Last summer, I spent eight weeks in a field study summer school session through the U.C. Santa Cruz extension program. It was the most rigorous educational experience I have had in college. By their nature, field studies almost always educate students in a more comprehensive way. The outdoors as a classroom provided me the opportunity to study the environment in which I would live for the summer.

The program, titled Mountain Ecology, dealt solely with relationships in our world. Through the use of textbooks, photocopied readers, two knowledgeable teachers, fourteen other students, and the Sierra Nevada as classroom, I discovered some of the unbelievable intricacies of existing on a life-sustaining planet.

For two months, the fifteen of us (five men and ten women) studied and lived together. A typical day began at sunrise and ended after sunset. Academic studies, the main focus of the program, were nearly constant. There was individual work to do, such as bird watching, plant keying, reading, and paper writing, as well as group presentations and research projects. Each day there were two or three lectures given by either teacher or other students. One lecture given on Fire Ecology took place about ten yards away from a burning forest fire, while another on glaciation was given at 10,000 feet overlooking a panorama of glacial valleys.

All the learning processes combined textual and personal sources with 24 hour learning experiences. Many of the experiences were not academic, so to speak, but were completely relevant to a field course. Dealing with a mother black bear and her two cubs at a three in the morning is not something a student can really be tested on. However, when such situations arose, the proper actions mean the difference between the bear family eating a fine breakfast, and our group not having enough food for the next nine days.

Overall, the Sierra Institute field study provided an incredible experience which broadened my education academically and personally. It offered insights into many different aspects of today's environmental issues, and required that I become knowledgable about multifaceted current problems. If you might be interested in a field study program and how to receive academic credit for such programs, talk to your academic advisor, Dr. Janice Edgerly-Rooks, or Dr. Amy Shacher in the Environmental Studies Program.

Written by Tony Facchino, Class of 1995

If you are interested in writing a short article for next quarter's newsletter, please call x2709.

SPRING QUARTER NEWS

Capstone Seminar (Envs 101): an opportunity for seniors, from diverse backgrounds, to collaborate their efforts and utilize their critical thinking skills to tackle current environmental problems. Discussions will be led by Drs. Amy Shachter (Chemistry) and Fred White (English). This course is required for seniors in the minor.

Resources, Food & the Environment (Agribus. 101): this course fulfills the "Economic Dimensions" requirement of the minor. It is now available in the Spring Quarter, in addition to Winter. Hopefully, this will alleviate scheduling problems some students may be experiencing.

New course: Animals in Antiquity: Classical Origins of Western Attitudes Towards Animals (Classics 178). Dr. John Heath of the Classics Department will offer this course as an elective for ES minors. Employing lecture, discussions, slides, films, and guest speakers, the course will examine the literal and metaphorical manipulation of animals in ancient Greek and Roman cultures. Spring 1993.

New course: Introduction to Environmental Law and Regulation. Ken Manaster, School of Law, will offer an undergraduate environmental law course next winter (1994). This course will fulfill the "Political/Legal Dimensions" component of the requirements for the minor.

Spring Quarter E.S. Core Courses:
** = new course
- Statistics
- Populations & Resources (Sociology)
- Environmental Ethics
- Populations (Biology)
- Resources, Food & the Environment (Agribusiness)
- **Seminar in International Relations (Ecopolitics)**

Spring Quarter E.S. Electives:
- **Classics 178**
- Economic Botany
- Economics I or II
- Field Botany
- General Ecology
- Technical Writing

Special thanks to Alexandra Mendrinos, Class of 1993, for preparing this newsletter.

Dr. Kristina Irvin, D.C.
CHIROPRACTOR

■ 1786 Technology Drive

■ San Jose, California 95110

■ (408) 441-6551

June 1993

Special discount coupons for referrals of new patients.

Herbal Healing

Before the rise of pharmaceutical companies in the early 1900's, herbs were an integral part of medicine. Herbal medicine and folklore wisdom on how to use plants for healing is just about as old as mankind. The knowledge was passed down from generation to generation. When drugs came into vogue, the public lost both its knowledge and trust in herbs—a tribute to the power of advertising and industrialized society's obsession with instant results for less effort.

During the 1960's, there was a reappearance of the "back to nature" era that brought herbal and alternative medicine slowly back into favor. Natural healing arts such as herbal medicine have since been rising in popularity. Today, more consumers are becoming fed up with rising medical costs and an unresponsive medical system. Pharmaceutical companies, with their big bucks and government connections, unfortunately highlight isolated incidents of adverse reactions from herbs. Yet how many people do you know that are "allergic" to penicillin? There are herbs that are poisonous or have serious side effects. They have been identified and banned by the U.S. Food and Drug Administration (FDA) long ago and also through trial and error of folklore history.

Recently, there has been a push by the U.S. FDA to regulate herbs and vitamin/mineral supplements. The Generally Regarded As Safe (GRAS) List developed by the FDA contains about half of the 500 herbs on the market. The other half is not necessarily dangerous—just unclassified. Three years ago, Congress passed the Nutrition, Labeling & Education Act which intends to regulate herbs like drugs. This Act means that in order to make a health claim about herbs, scientific research would be required. The financial outlay would be substantial, and herbal companies do not have the financial backing that drug companies have developed. Since herbs are plants and not really patentable, there is little incentive to do research.

Scientific research of herbs is difficult and often flawed. Research is usually done on only the most active component and not the entire plant. Plants are complex living organisms with many active chemical components. These chemicals often nullify each others qualities through synergistic relationship with one another. This means that the sum of these chemicals determines the plants action.

As herbs is true alternative to allopatic drugs? Yes, there are natural remedies as effective as drugs. Along with symptomatic relief, many also promote systemic correction and support to overcome the problem. Drugs and herbs are different. Drugs are

continued on page 2

Tantalizingly Tart Cranberry Muffins
by Kirra

When you split these treats open, they look like jewels. You may make them vegan (non-dairy) or vegetarian (dairy)—the choices all yours. Enjoy!

1 1/4 cups unbleached white flour
1/2 tsp. ground cinnamon
1/2 tsp. ground cardamom
1/2 tsp. ground nutmeg
1 tsp. baking powder
1/2 tsp. salt
1/2 tsp. baking soda
1/2 cup finely chopped pitted prunes
1/2 cup rolled oats
3/4 cup soymilk (or buttermilk)
3 Tbsp. canola oil
6 Tbsp. maple syrup
1/2 cup fresh or frozen cranberries
3 egg whites, lightly beaten (or Egg Replacer equivalent to 2 eggs)

Preheat oven to 400 degrees. Line a 12 cup muffin tin with muffin papers (or lightly oil muffin tin or spray with vegetable spray).

In a large bowl, sift together flour, cinnamon, cardamom, nutmeg, baking powder, baking soda, and salt. Add prunes and oats, and toss to coat. In another bowl, whisk together milk (soymilk or buttermilk), oil, maple syrup and cranberries. In a third bowl beat Egg Replacer or egg whites to soft peaks; set aside.

Lightly combine dry and wet ingredients, then fold in egg whites until just incorporated. Spoon batter into muffin tins, filling 3/4 full. Bake for 12 to 15 minutes, or until muffins are light and springy to the touch.

Hint: Because cranberries are seasonal ingredients, make sure to stock up at winter holiday time so you have a year-round supply. They keep for several months in the freezer. ◆

Herbal Healing...

continued from page 1

isolated, synthetic chemicals to which the body reacts, whereas herbs are recognized by the body because they are a complete food spectrum containing micronutrients. Herbs are processed like food that allows the system to select the nutrients as needed and eliminate unneeded portions through normal elimination. Rarely do herbs accumulate to toxic levels or cause side effects. Drugs are not assimilated through digestion or normal assimilation routes. Drugs have effects on the body whether they are needed or not and are difficult for the body to control. This situation lends easily to toxic build up and frequent side effects.

Many acute problems are dealt with quite well through drugs. For example, bacterial infections or pain from injuries. But the allopathic drugs have been a near complete failure with chronic or degenerative type diseases and problems. Western medicine has increased in awareness of that situation, and we are beginning to realize the power behind diet and plant remedies for such problems as diabetes, heart disease, and arthritis.

Another common issue against herbs is the quality. Since they are produced by nature, the potency, and thus results, could vary depending on sun, moisture, soil conditions, maturity, air pollution, and processing. Herbs are best when grown organically or wild crafted. When bought packaged at the store, the quality is controlled. This results in consistent product dosage and effects. ◆

Herbology has returned as an answer to disease. Check into herbs for yourself and you may find the answers you've been seeking. ◆

Dr. Kristina Irvin, D.C.
CHIROPRACTOR

1786 Technology Drive
San Jose, California 95110

Bulk Rate
U.S. Postage
PAID
San Jose, CA
Permit No. 904

Hiking Club

The first adventure for the hiking club was a great success. A few hardy souls showed up for the hike in Quicksilver County Park in South San Jose. We hope that all who have expressed an interest will have free time on Sunday morning, June 20th at about 9 a.m. for a hike in Rancho San Antonio Park outside Cupertino. There is a flat 2 mile round trip to a farm. All those with children should seriously consider this hike. It really is a nice area with many miles of trails. Remember, there will be bonus dollars towards treatments for those who attend. Spouses and friends are welcome. Call for directions and details. ☆

Personal

The orthopedics program is finally over in June. After 3 long years it feels like I've been in school all over again. Now I'm just studying for the board exam.

Running has had its ups and downs. The uphill was that I was 45 minutes faster on a 30 mile training run over last year's time, but I fell and injured my foot. I am finally getting over the strain and trying to stay mentally and physically strong for the 100 miler on June 26th. This foot strain has forced me to turn to bicycling for training rides and into the pool for longer workouts. The race in May was in Quicksilver Park. This race was quite hilly and turned out to be a rather rough day for me—but I still ran for 10 hours and 20 minutes. My time still placed me within the top 4 women for my age group. ◆

APPENDIX

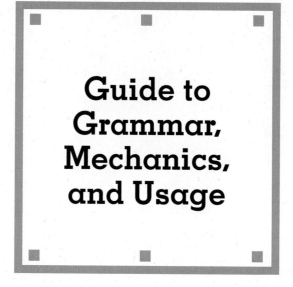

Guide to Grammar, Mechanics, and Usage

A(n)

Indefinite article. See Article.

Abbreviations, Common Technical

NOTE: Abbreviations which could be confused with actual words use periods; others do not.

Å, angstrom unit

ac/dc, alternating current/direct current

amp, ampere

at. wt., atomic weight

au, astronomical unit [93,000,000 miles]

bl; bbl, barrel; barrels

bp, blood pressure

BTU, British Thermal Unit

bu, bushel

c or cal, calorie

C, Celsius; Centigrade

c, speed of light [186,000 miles per second]

cc, cubic centimeter

cm, centimeter

db, decibel

DNA, deoxyribonucleic acid

dpi, dots per inch [measure of printer resolution]

eta, estimated time of arrival

etd, estimated time of departure

F, Fahrenheit

gb, gigabytes [1 billion bytes]

g, gram

g, unit of gravitational pull [Earth normal = 1g]

gal., gallons

GDP, Gross Domestic Product

GNP, Gross National Product

HD, hard disk

hp, horsepower

Hz, Hertz

K, Kelvin

kb, kilobytes [1 thousand bytes]

kg, kilogram

km, kilometer

kph, kilometers per hour

kw, kilowatt

l, liter

m, meter

ml, milliliter

mm, millimeter

oz, ounce(s)

pH, degree of acidity of a solution

r, roentgen(s)

UHF, ultrahigh frequency

v, volt(s)

VHF, very high frequency

yd, yard(s)

w, watt(s)

Abbreviations (Common Editing)

abr, use/incorrect use of abbreviation

abs, revise level of abstraction

adj/adv, revise use of adjective/adverb

agr, error in agreement: subject-verb; noun-pronoun

amb, revise to remove ambiguity

bf, use/incorrect use of boldface

br, use/incorrect use of brackets

cap, revise capitalization

chop, revise to remove choppiness from sentences

coh, revise to improve coherence

col, use/ problem with use of colon

com, revise comma use

cs, revise to eliminate comma splice

dm, revise to eliminate dangling modifier

doc, problem with documentation format; documentation needed

frag, revise to eliminate sentence fragment

fus, revise fused or run-on sentence

incl, use inclusive language

ital, use/problem with use of italics

jgn, revise to eliminate inappropriate jargon

lc, use lowercase letters

mod, problem with modification

num, use/problem with use of numbers

par, use/problem with use of parentheses

prg, [¶] problem with paragraphing

prl, [||] problem with parallelism

q, quotation marks needed/used incorrectly

r, revise to eliminate redundancy

rep, revise to eliminate repetition

semi, use/problem with use of semicolon

sp, spelling error

t, revise tense or shift in tense

trns, use/problem with use of transition

w, revise to improve word choice

wdy, revise to improve conciseness

x, typographical error

Abstract, Summary

An **informative abstract** is a condensed version of the purpose, methods, findings, conclusions, and recommendations of a report, and is placed at the beginning of the report. See "Writing the Abstract" in Chapter 15. A **descriptive abstract** is a brief statement describing the nature of the report itself rather than summarizing its specific contents. Here is a descriptive abstract for the report,

"Restoring Diversity in Salt Marshes," which appears in Chapter 16, pp. 369–379:

> This report examines the problems with existing coastal wetlands restoration projects, and gives reasons why biodiversity continues to decline. The report also discusses the most feasible methods of restoring habitats based on current understanding of species transplantation, minimum wetland area required, long-term assessments, and protection of endangered species.

Informative abstracts of research reports are sometimes published independently of the report, by abstracting services. See, for example, *Biological Abstracts* and *Chemical Abstracts* in the reference room of your library. Descriptive abstracts are often used for annotated bibliographies.

A **summary** resembles an abstract in that it includes only the key information in a document. It typically appears in the concluding section of the document, along with recommendations and conclusions. However, one type of summary, known as **executive summary,** appears at the beginning of reports in which new policy has been argued for or against. Unlike an abstract, it often calls attention to the rationale behind the new policy. An executive summary for a report on ways to improve worker productivity on the job might look like this:

> This report investigates the problem of decreasing productivity at Technocraft Corp. It first isolates four causes:
>
> 1. Inadequate interaction with project supervisors.
> 2. Workplans are not sufficiently detailed.
> 3. Timetables are not given.
> 4. Progress reports are not required.
>
> The report then strongly advocates the adoption of the following measures to counteract the problem:
>
> 1. Retrain supervisors so that they can interact more efficiently with their technicians.
> 2. Require each supervisor to distribute weekly workplans to each technician and to make sure that the plans are understood.
> 3. Workplans should also include timetables.
> 4. Each technician should be required to fill out a daily progress report.
>
> All four measures need to be adopted for any chance of productivity to improve. With no measures taken, productivity will surely continue to decline.

Accept, Except

To **accept** (verb) is to receive voluntarily: Will you **accept** the package? **Except** (preposition) means "with the omission of": Every package was delivered **except** mine.

Acronyms

An acronym is an abbreviation spoken as a word:

CAD	Computer Aided Design
NAFTA	North American Free Trade Agreement
NASA	National Aeronautics and Space Administration.
NATO	North Atlantic Treaty Organization
NOW	National Organization for Women
OPEC	Organization of Petroleum Exporting Countries
PIN	Personal Identification Number
RAM	Random Access Memory
ROM	Read-Only Memory
SAT	Scholastic Aptitude Test
WYSIWYG	What You See Is What You Get

Active Voice, Passive Voice

See Chapter 8, page 181–182.

Adapt, Adopt

To **adapt** is to change or adjust:

Jill **adapted** this report so outsiders can understand it.

To **adopt** is to acquire to one's own use:

My colleague **adopted** one of my lesson plans for his class.

Advice, Advise

Advice (noun) is synonymous with suggestions or recommendations:

Frank offered us useful **advice** on how to prepare our annual report.

Advise (verb) is the act of giving advice:

> Did Frank **advise** you on how to prepare your annual report?

Affect, Effect

See Chapter 9, page 194.

Agreement

The influence of one grammatical element over another, such as the influence of a singular or plural noun or pronoun over the verb:

> The total **is** correct.
>
> The numbers **are** correct.

The influence of first, second, or third person upon the verb:

> First person singular: I **am** fluent in three languages.
>
> First person plural: We **are** fluent . . .
>
> Second person: You **are** fluent . . .
>
> Third person singular: He (She) **is** fluent . . .
>
> Third person plural: They **are** fluent . . .

A compound subject receives a plural verb:

> The last train and the last bus **arrive** at the same time.

When subject and verb are separated by a phrase or clause, the verb must agree with the subject and not with any noun within the phrase or clause:

> The last train, which happens to be the most punctual of all the trains, **arrives** at midnight.

All Ready, Already

All ready (adverb + adjective) is an informal phrase meaning prepared, or completely ready:

> We were **all ready** to begin work at seven A.M.

Already is an adverb meaning prior to expectation:

> We were **already** in the office by six-thirty.

All Right, Alright

Always two words in standard usage: The photographs came out **all right.** "**Alright**" is nonstandard, sometimes used in dialogue to indicate harshness or laxity of speech: "**Alright,**" he grumbled, "let's get going."

All Together, Altogether

In **all together** (adverb + adverb) the first adverb intensifies the condition of togetherness:

> Protestors young and old marched **all together.**

Altogether (adverb) means "entirely," "completely," and in most cases immediately precedes the adjective or verbal adjective it modifies:

> The protestors were **altogether** determined to complete their long march to the munitions factory.

Allusion, Illusion

An **allusion** is a reference, either direct or indirect:

> Many politicians make **allusions** to Abraham Lincoln in their campaign speeches.

An **illusion** is a false impression or image of something:

> The magician created the **illusion** of sawing a person in half.

Alot

Misspelling of **a lot** (always two words).

Among, Between

Use **among** when referring to three or more objects or persons:

> The workload was divided evenly **among** the six employees.

Use **between** when referring to two persons:

> The workload was divided evenly **between** Natasha and Joe.

Amongst

Antiquated. Use **among.**

Amount, Number

Amount is used with nouns that are not counted: **amount** of sugar.

　　Number is used with plural nouns that are counted: **number** of cups of sugar.

> NOTE: Cups (or spoons, etc.) are counted, but the sugar itself is not.

Apostrophe

The apostrophe has three uses:

> to form contractions: can**'t** for cannot

> to pluralize letters and numerals: A**'s,** 7**'s;** IRS**'s**

>> NOTE: The apostrophe here is optional, except where confusion could result: YES: 2**s,;** NO: As

> to indicate possession: Joe**'s** account; the bank**'s** address; the IRS**'s** tax forms.

>> NOTE 1: For plural possessives ending in **s,** place the apostrophe after the **s:** the boys**'** gym (optional for names: Mars**'** atmosphere or Mars**'s** atmosphere). For plural possessives not ending in **s,** place the apostrophe between the last letter and the **s:** the childre**n's** toys.

>> NOTE 2: Possessive pronouns do not take apostrophes: yours, hers, its, theirs.

Appositive

A noun or noun phrase placed alongside the noun or noun phrase it amplifies or identifies, separated by commas: Reggie, **our assistant technician,** worked overtime. The elements, of course, can be reversed, so that the original noun becomes the appositive: Our assistant technician, **Reggie,** worked overtime.

Article

Articles designate nouns or noun phrases and are either definite or indefinite.

Indefinite article: **a** or **an.** If the word following the indefinite article begins with a vowel, use **an:**

> We made **an** error in our calculations.

> If the word following the indefinite article begins with a consonant, use **a:**

> We found **a** bug in our program.

Definite article: **the.** We found **the** bug we were looking for.

As, Like

See Chapter 9, page 194.

Beside, Besides

Beside means "alongside" or "next to":

> We placed the plotter **beside** the printer.

Besides is an informal synonym for "in addition":

> **Besides** excelling in math, Chris is skillful with languages.

Brackets

Brackets have two major uses: (1) to set off a word, phrase, or number within a parenthetical expression:

> Our flights (no. 224 out of LAX [6:20 A.M.]; no. 350 out of O'Hare [12:45 P.M.]) were all on time.

(2) to add a word to a direct quotation that was not part of the original quotation:

> According to Albert Einstein, "The most beautiful and deepest experience [we can have] is a sense of the mysterious."

See Ellipses.

Can, May

Can means [be] able:

> **Can** I walk to my friend's house? (i.e., Am I able to walk that far?)

May means [be] permitted:

> **May** I walk to my friend's house? (i.e., Will you permit me to walk there?)

Capitalization

Names, Including Brand Names: Our dog, Eliot; the constellation Orion; Jell-O; Walden Pond

Titles of Books and Articles: First and last words are always capitalized; all other words in the title are capitalized except articles, the infinitive **to,** prepositions less than 5 letters, and coordinating conjunctions:

> Jackson's "The Lottery"
> Asimov's *The Stars in Their Courses*
> Shakespeare's *Romeo and Juliet*

Titles in Formal Address: Professor Jones; Lieutenant Andrews

> NOTE 1: When the titles are used in ways other than direct address, they are not capitalized: R. Jones is our professor.
> NOTE 2: Collegiate years or rankings are not capitalized: freshman, sophomore, junior, senior, graduate

Days, Months of the Year: Monday; October

Holidays or Special Occasion Days: Fourth of July, Veteran's Day, Thanksgiving, Earth Day.

> NOTE: Seasons are not capitalized: autumn, winter, spring, summer.

Geographical Regions: the South, the Midwest, Portugal, Northern Ireland, Crater Lake, the Badlands of South Dakota

Name of the genus, but not the species, in scientific nomenclature: Felis concolor; Homo sapiens neandertalis

Languages and Dialects: Hindi, Gaelic, French, Pidgin English

Institutions, Departments, Offices: Xerox Corporation, Department of Defense Sociology Department, Society for the Prevention of Cruelty to Animals

> NOTE: If referring generically to an institution, department, or office, use low-

ercase (unless another capitalization rule applies): today's modern colleges; foreign defense, departments of sociology, modern languages, and English; animal protection societies.

Religious, Ethnic, and Racial Groups: Mexican-American, Pacific Islander, Vietnamese, Muslim, Buddhist, Jew, Caucasian

NOTE: Commonplace words used to refer to race are not capitalized (black, white, yellow, red), though these terms are generally avoided in current usage in the United States.

Clause

A clause is a sequence of words containing a subject and predicate, and functioning as a member of a complex or compound sentence. There are two kinds: main clauses, which can stand alone grammatically as complete sentences, and subordinate (or dependent) clauses, which cannot. Consider the following sentence:

We had no idea our computer would quickly become obsolete after we purchased it.

This sentence consists of a main clause, "We had no idea our computer would quickly become obsolete," and a subordinate clause, "after we purchased it." The subordinate clause (which always begins with a subordinating conjunction) establishes, in this case, a temporal connection with the main clause. Subordinate clauses can also establish causal, spatial, conditional, and relational connections:

Causal: I majored in communication *because I am interested in broadcast journalism.*

Spatial: Did you see *where she was going?*

Temporal: Nobody knew *when they were departing.*

Conditional: *Unless we hire additional help,* we will not meet our deadline.

Relational: We wrote all of the proposals *that got accepted.*

Inexperienced writers sometimes neglect to establish clear relationships in their paragraphs; they will string simple or compound sentences together (that is, sentences consisting of just one main clause, or two main clauses connected with "and" or "but") and assume that readers can infer the connections.

Consider: "It was snowing and I stayed home." Does this mean the speaker stayed home *because* it was snowing (i.e., because of the bad weather), or *even though* it was snowing (the person usually goes skiing when it snows)?

Clause (Relative)

A relative clause modifies the subject of the sentence and begins with the word *who* or *that*. There are two types of relative clauses: restrictive and nonrestrictive:

> *Restrictive:*
>
> The children **who designed the best science project** received a commendation from the principal.

The clause's reference is "restricted" to those children who designed the best science project.

> *Nonrestrictive:*
>
> The children, **who designed the science project,** received a commendation from the principal.

In this case *all* the children were involved in the science project. The clause functions much like an appositive.

> NOTE: Nonrestrictive clauses are separated by commas; restrictive clauses are not.

Colon

Colons are used in the following situations:

1. *To introduce a series:*

   Do not operate the machinery until you are wearing these items: safety goggles, hardhat, gloves.

2. *To call special attention to a fact or observation:*

   The supervisor repeatedly stressed one point above all others: Double-check your computations.

3. *To separate hours, minutes, and seconds:*

   Liftoff occurred at 7:15:01 E.S.T.

Colons are *not* used where a period would not be used; that is, use of a colon must not result in a sentence fragment:

> NO: Each camper brought along: a sleeping bag, canteen, and three days' rations.

YES: Each camper brought along a sleeping bag, canteen, and three days' rations.

YES: Each camper brought along the following gear: a sleeping bag, canteen, and three days' rations.

Comma

Commas *are* used in the following situations:

a. Before and after an appositive:

Mr. Parks, the company's founder, retired last week.

b. After an opening subordinate clause:

When the shuttle lifted off, everyone cheered.

c. After each item except the last in a string of three or more items:

Basil, nutmeg, and sage are my favorite spices.

d. After a conjunctive adverb:

I own a cheap camera; nevertheless, it takes excellent pictures.

e. After nonrestrictive relative clauses [see Clause (Relative)]

Commas are *not* used in the following situations:

a. Between a subject and its verb:

NO: Mary, Sam, and John, are enrolled in advanced chemistry.

YES: Mary, Sam, and John are enrolled in advanced chemistry.

b. Between fewer than three items in a series:

NO: Broccoli, and Brussels sprouts are rich in nutrients.

YES: Broccoli and Brussels sprouts are rich in nutrients.

c. After a conjunction that begins a subordinate clause:

NO: They arrived an hour late because, their flight was delayed.

YES: They arrived an hour late because their flight was delayed.

Complement, Compliment

To **complement** (verb) means *to augment or accompany:*

Good conversation **complements** a good dinner.

A **complement** (noun) means *an augmentation:*

> We planted crysanthemums to **complement** the poppies.

To **compliment** (verb) means *to praise:*

> Our supervisor **complimented** us for producing such a successful proposal.

A **compliment** (noun) is a statement of praise:

> The more **compliments** I receive, the harder I work.

Concave, Convex

A **concave** lens is inwardly curved, like the apparent inner curve of a crescent moon. A **convex** lens is outwardly curved, like the surface of a sphere.

Conscience, Conscious

Conscience is a noun referring to moral judgment:

> Obey your **conscience.**

Conscious is an adjective referring to a state of wakefulness:

> Children don't look fully **conscious** in front of the TV.

Continual, Continuous

Continual refers to something that occurs regularly but not unstoppingly:

> Young people need to read **continually.**

Continuous refers to that which occurs without interruption:

> The faucet dripped **continuously** for a week.

Criteria, Criterion

There may be several **criteria** (rules, standards) one must follow in preparing a formal report; but there may be one **criterion** that seems the least necessary.

Dash

See Hyphen, Dash

Data, Datum

Datum is singular but rarely used as such; one would say "one bit of data" rather than "one datum." Nonetheless, **data** requires a plural verb in formal usage: The **data** from the space probe are now being received.

Definitely

Avoid using this intensifier superfluously, as in "I have definitely applied for the job" or "I am definitely opposed to spot quizzes."

Different From, Different Than

Standard usage favors the former: Venus and Earth are the same size but radically **different from** each other.

Dominant, Dominate

Dominant is an adjective describing a state of being most noticeable or influential:

   Yuen's is the most **dominant** voice in the chorus.

Dominate is a verb, meaning to command most attention or influence:

   The children **dominated** everyone's attention at the party.

Due To

Use as a synonym for "caused by." Avoid using it as a synonym for "because."

   YES: The excessive pressure buildup was **due to** overheating.

   NO: **Due to** the fact that I overslept, I missed the exam.

Each Other, One Another

When referring to two persons or objects, use **each other:**

The two crustal plates are rubbing against **each other.**

When referring to more than two persons or objects, use **one another**:

People were pushing against **one another** in the hallways.

e.g.

Abbreviation for "for example":

The severe pollution in some lakes (**e.g.,** Lake Erie) has been substantially reduced.

See also **i.e.**

NOTE: Except where space is at a premium, it is generally best to limit use of these abbreviations to accompany items in parentheses.

Ellipsis

An ellipsis is an omission. When quoting a passage, omitted words must be indicated by ellipsis dots [. . .]. Note that one uses only three dots, each separated by a single space. If the words omitted occur after the end of a sentence, the three ellipsis dots are placed after the period ending the sentence, with one space between that period and the first ellipsis dot. Ellipses representing text preceding or following the quotation are not necessary. See Brackets.

et al.

Abbreviation for "and others."

etc.

Abbreviation for "and so on." Be sure not to misuse this abbreviation. See Chapter 9, page 194–195.

Every Day, Everyday

When the intention is to use **every** (adjective) to modify **day** (noun), as in the expression "each day," two words must be used:

We can park here every day.

"We can park here **everyday**" is incorrect. **Everyday** is an adjective that modifies another noun and means "daily":

> Exercise is part of my **everyday** routine.

Evidence, Proof

Evidence refers to material or arguments used to support a claim; it may or may not be conclusive. **Proof** refers to indisputably conclusive evidence.

Farther, Further

Use **farther** to indicate distance:

> Jupiter is **farther** away from the sun than Mars.

In formal usage **further** is a synonym for *additional:*

> These prices are subject to change without **further** notice.

As a verb, it can be used as a synonym for *enhance* or *advance:*

> I took a course in computer science to **further** my education.

Fewer, Less

See Chapter 9, page 195.

Fragment, Sentence

A sentence fragment is part of a sentence (e.g., a phrase or a clause without the verb) that is used as if it were a grammatically whole sentence. In the following passage,

> The screws' diameters were too wide. For the holes.

the second sentence is a fragment because "for the holes" contains neither subject or verb; it is only an isolated phrase. It should be part of the preceding sentence:

> The screws' diameters were too wide for the holes.

Sometimes an infinitive phrase can be mistaken for a whole sentence:

The astronauts repaired the space telescope. To replace its faulty mirror.

The second sentence lacks a finite verb (a verb that takes a subject and conveys one tense (e.g., she **goes;** she **replaced**). Because an infinitive (*to* + verb: e.g., **to go, to replace**) or an infinitive phrase (infinitive + modifiers: e.g., **to replace its faulty mirror**) cannot take a subject, it cannot stand alone as a complete sentence:

The astronauts rendezvoused with the space telescope to replace its faulty mirror.

NOTE: Occasionally, fragments are used for dramatic emphasis in informal writing:

We couldn't find an available seat. Not one.

Gerund, Participle

A **gerund** is a verbal (i.e., a nonfinite verb) that functions as a noun; it is characterized by the *-ing* suffix:

Welding together pieces of junk metal satisfied her deeply.

In this sentence, "welding together pieces of junk metal" is a gerund phrase, the subject of the verb, *satisfied.*

A **participle** is that "part" of a progressive verb tense, past or present, ending in *-ing* (for present participles) and *-d* or *-ed* (regular verbs); *-t, -en,* or *-n* (irregular verbs) for past participles The other part of the verb is the auxiliary.

Maureen's new invention is (was; has been; had been) **taking** the country by storm.

Emergency food supplies were **airlifted** to Somalia.

The semimolten metal was **beaten** into odd shapes.

NOTE: When you begin or end a sentence with a participial phrase, make sure that it clearly and logically modified the noun it is intended for. Consider the following example:

Driving to work, my hubcap fell off.

Such an error is called a **dangling modifier.** Even though the author's intention is clear, the sentence is ungrammatical because "driving to work" is not the intended modifier of "my hubcap." Revised, the sentence should read,

Driving to work, I lost my hubcap.

or

As I was driving to work, my hubcap fell off.

Hardly

Because this word indicates negation, do not use it with a negative such as *couldn't* or *could not.*

> NO: The diver **couldn't hardly** see the sunken ship.

> YES: The diver **could hardly** see the sunken ship.

His, His/Her, His or Her

See discussion of inclusive language, Chapter 9, pages 196–197.

Hopefully

Its proper usage is adverbial:

> The peddler gazed **hopefully** at the potential customers.

In informal usage, it substitutes for the subject and verb:

> **Hopefully** it will rain soon. (instead of **We hope** it will rain soon)

Hyphen, Dash

These punctuation marks are easily confused with each another. The hyphen is the shorter of the two and is used to connect words together that are too linked to stand separately, but would look awkward as a single word:

> ex-president

> two-by-fours

Some adjective + noun or adverb + adjective combinations are hyphenated when the combination is used adjectivally before the noun it modifies:

> a castle from the twelfth century → a twelfth-century castle

> this report was well written → a well-written report

> NOTE: Some forms are hyphenated according to established usage. When in doubt, consult a dictionary.

A dash, by contrast, indicates a sudden dramatic emphasis, is twice the length of a hyphen, and appears in typescript as a double hyphen (--):

> The team lost again—but then, that didn't surprise us.

Dashes can enliven writing when used sparingly.

i.e.

Abbreviation for "in other words" or "that is"

> The surplus produce was judged to be overripe; **i.e.,** it was starting to rot.

Note that a comma follows the abbreviation.
> See **e.g.**

Imply, Infer

To **imply** is to hint at:

> My manager **implied** that I may be promoted soon.

To **infer** is to interpret what may not be explicit:

> I **inferred** from my manager's remarks that I would be promoted soon.

Inter-, Intra-

See Chapter 9, page 195.

Irregardless, Regardless

Only **regardless** is correct. The **ir-** prefix (meaning "not") is redundant because it does not alter the meaning of the word.

Its, It's

See Chapter 9, page 195.

Lay, Lie

See Chapter 9, page 195.

Lead, Led

Sometimes confused with each other because **lead,** as in "pencil lead," is pronounced the same way as **led,** which is the past tense of lead [pronounced *leed*], meaning "to direct."

> The chorus was **led** [not "lead"] by a famous conductor.

> The famous conductor will **lead** the chorus tonight.

Lightening, Lightning

The former (three syllables) is the opposite of darkening; the latter (two syllables) refers to bolts of electrical discharge during a thunderstorm.

Loose, Lose

The first word means "not tight"; the second, "misplace." The words can be confusing because **lose** is pronounced the same way as "choose."

Media, Medium

See Chapter 9, page 195.

Modifier

A word or phrase (adjective, adverb, adjective or adverb phrase) used to narrow the meaning of another word (a noun or another modifier). In the sentence, "Some fossils are microscopically small," the words "some" and "small" are modifiers of "fossils"; "microscopically" is a modifier of "small."

Mood

The three moods (or speaker/writer attitudes) in English are indicative (statement or question of fact), imperative (command), and subjunctive (expression of doubt, desire, or statement contrary to fact):

Indicative Mood:

> I am **taking** the 10 P.M. flight.
> Which flight are you planning to **take?**

(In each case the speaker is "indicating" an actual situation.)

Imperative Mood:

> **Take** the next flight to Paris.

(The subject is always understood to be "you.")

Subjunctive Mood:

> I wish I **were** on that flight to Paris.
> If I **were** going to Paris, I would have planned a party.

(In each case the utterance differs from the reality, which is that the speaker is not going to Paris.)

Nominalization

Nominalization refers to the use of an abstract noun that would come across more concisely and effectively as a verb.

Nominalization	Verb Equivalent
Their **assumption** is that . . .	They **assume** that . . .
I came up with a **definition** of . . .	I **defined** . . .
The board has come to the **conclusion** . . .	The board has **concluded** . . .
The substance had the **appearance** of being . . .	The substance **appeared** . . .

Nonrestrictive vs. Restrictive Clauses

See Clause (Relative).

Numbers

For nontechnical documents, the usual practice is to spell out whole numbers between one and ninety-nine, simple fractions (one-half, three-fourths, one-tenth, etc.), and large rounded numbers (one thousand, one billion, etc.). For technical documents, the practice is to use numerals for all numbers except ordinal numbers (first, twenty-second, etc.):

> The outer tip of the wing of a Boeing 707 is raised 39 inches when the plane is flying level in still air, but can flex upward more than 10 feet or down 3 feet if necessary.
>
> —Elizabeth A. Wood, *Science from Your Airplane Window.*
> New York: Dover, 1975: 6.

Other rules for numbers:
Spell out centuries, using lowercase letters:

> Imagine what computers will be like in the twenty-second century.

Spell out numbers that begin a sentence:

> Two hundred and ten years ago today our city was founded.

Use numerals if the sentence is reworded:

> Our city was founded 210 years ago today.

Often, Oftentimes

The latter word is antiquated and should be avoided.

Parallelism

See Chapter 8, pages 182–183.

Parentheses

Enclose words or passages in parentheses when they convey information important to the main text but could cause confusion if not so enclosed:

> For general cleaning, use one tablespoon Drain-Kleer (but see important exceptions below), wait five minutes, then flush.

Parentheses may also used to refer readers to other parts of the text:

> As discussed earlier (pp. 30–33; 78–81), children can absorb mathematical concepts at younger ages than most people realize.

Simple definitions can be presented parenthetically:

Mars's retrograde motion (seeming to move in reverse across the sky from night to night) baffled medieval astronomers.

Participle, Gerund

See Gerund, Participle.

Passive vs. Active Voice

See Chapter 8, pages 181–182.

Passed, Past

Passed is a verb meaning "completed successfully" or "moved through or beyond":

> Sheila **passed** her MCAT test with flying colors.

> The sailboat **passed** beneath the bridge.

Past is a preposition describing the act of moving beyond or an adjective describing an elapsed event:

> The ship moved **past** the bridge. [prep.]

> **Past** victories are still vivid in memory. [adj.]

Phenomena, Phenomenon

The former is plural; the latter singular:

> Many **phenomena** can be observed in the night sky.

> The aurora borealis (northern lights) is a stunning **phenomenon.**

Phrase

A group of words that lacks, by itself, a subject with a predicate.

> Noun phrase (noun accompanied by pronoun and/or modifiers): **my oldest sister**

Verb phrase (verb + noun phrase or prepositional phrase: **bought a Mercedes**

Prepositional phrase (preposition + noun phrase): **from our rich uncle**

Plurals

Most nouns in English become plural by adding an **s** or **es:**

> One dog; two dog**s**

> One peach; two peach**es**

A few nouns are exactly the same in plural form:

> One fish; two fish (fish**es** refers to different species of fish)

Some nouns are conceptually plural but grammatically singular:

> All of our equipment work**s** perfectly.

> Janet's team **is** most likely to take first place.

> The United States **is** highly diverse.

NOTE: Non-native speakers of English sometimes confuse the s/es plural formation of nouns and the third-person singular formation of verbs: The dog walk**s** (but: the dog**s** walk).

> See also Agreement.

Prefixes, Suffixes

Prefixes are syllables attached to the beginning of certain words and used to change meaning. The prefix *pre-* ("before"), attached to the word *view,* creates the word *preview* (a viewing of a work not yet released to the public). Suffixes are syllables attached to the end of certain words, and used to give grammatical function to (and thereby change the meaning of) that word. The suffix *-al* (state or condition of), attached to the noun, *nation,* for example, creates the adjective, *national.* See also Chapter 9, pages 193–194.

Principal, Principle

See Chapter 9, page 195.

Quotation Marks

Double quotation marks (". . .") are used as follows:

Around direct quotations:

> The President said, "I will continue to encourage the dismantling of nuclear missiles."

NOTE: Quotation marks are not used with indirect quotations:

> The President said that he would continue to encourage the dismantling of nuclear missiles.

Around titles of articles or chapters from books (except from the Bible, or when it is your own title on the first page of your article):

> The article I wrote for Professor Burke is titled "Dismantling Nuclear Missiles: A Top Priority."

> Dismantling Nuclear Missiles: A Top Priority
> By
> Joan Smith

> Last Sunday we read from the Book of Job and the Gospel of John.

Around dialogue:

> "Here is where we keep our spray paint," said the clerk, unlocking the metal cabinet.

NOTE: The closing quotation mark follows a comma, period, exclamation mark, or question mark. It precedes a colon or semicolon:

> He groaned, "I wish I'd never heard of the stock market": those were the exact words I'd been thinking.
> Jane said, "Please give me a hand"; but before I could do so, she lifted the box by herself.

Around words or phrases referred to as such:

> I was intrigued by the etymology of the word "plagiarism."

NOTE: If the last example were rephrased as a question, the closing quotation marks would precede the question mark:

> Are you intrigued by the etymology of the word "plagiarism"?

Around words or phrases used in an unconventional or nonliteral manner:

> The telepath "spoke" to her friends from a great distance.

As a ditto symbol:

> March 10, 1994: Staff Meeting
> April 22, 1994:   "   "

As the symbol for inches:

Legal size paper measures 8½″ × 14″.

Single quotation marks ('. . . ') are used in the following ways:

Around a quotation within a quotation:

My boss said, "When I asked Rosemary how long it would take her to debug the program, she replied, 'Maybe two days.'"

(Note that both the closing single and double quotation marks go outside the period.)

As the symbol for feet:

The kitchen measures 10' × 12'.

Raise, Rise

Raise is a transitive verb; that is, it must take an object:

The technician managed to **raise** the amplitude of the signal.

He **raised** the hundred-pound sack of flour over his head with one arm.

Rise is an intransitive verb; it does not take an object:

We'll need to **rise** at dawn if we decide to go fishing.

Before prices fall, they'll probably **rise** again.

Run-on Sentence, Comma Splice

A **run-on sentence** consists of two independent clauses jammed together with no intervening punctuation.

Run-on: Our tape recorder broke it was twenty years old.

Revised: Our tape recorder broke. It was twenty years old.

A **comma splice** is similar, except that the independent clauses are separated (spliced) only by a comma. They should instead be separated by a semicolon or become two separate sentences.

Comma splice: We worked overtime for a month, we saved enough money to go to Jamaica.

Revised: We worked overtime for a month; we saved enough money to go to Jamaica.

Semicolon

Use a semicolon in place of a coordinating conjunction to connect closely related independent clauses:

I tried to duplicate my experiment; it didn't work.

The sentence could be rewritten with a conjunction and comma:

I tried to duplicate my experiment, but it didn't work.

Use a semicolon to set off one series from another:

First we considered the impact of aerial spraying on rivers, lakes, and aquifers; then we considered the impact on soil and crops.

NOTE: Use commas, not semicolons, to separate items in a series (first, . . . second, . . .). Use a colon, not a semicolon, to introduce a series (The following items are required: . . .).

Sexist Language

See "Using Inclusive Language," Chapter 9, pages 196–197.

Slash

The slash mark [/] is used for the following:

1. To separate optional items:

The questionnaire consisted only of YES/NO questions.

Each worker received his/her bonus yesterday.

NOTE: The latter example should be used sparingly: his or her; or recast in the plural (The workers . . . their).

2. To indicate a fraction:

This bolt requires a 9/16 socket.

Some Day, Someday

If you are referring to a day within a given time frame, use two words:

Let's get together for dinner some day next week.

If you are referring to an indefinite time, use one word:

Someday I'd like to take you to dinner.

Subheads

Titles given to subdivisions of a text are called subheads. See Chapter 10, page 207.

Suppose, Supposed

Suppose (meaning to guess) is present tense; **supposed** is past tense:

> I **suppose** the contract should be drawn up right away.

> Yesterday, I **supposed** we were all in agreement.

As a synonym for "require," the word is **supposed** (to), not **suppose** (to):

> I am **supposed** to draw up this contract immediately.

Tense

Verb tenses indicate the time frame of the action or condition:

> *Present:* She drives to work.

> *Present perfect:* She has driven to work. (completed in the present)

> *Present progressive:* She is driving to work. (currently in progress)

> *Past:* She drove to work.

> *Past perfect:* She had driven to work. (completed in the past)

> *Past progressive:* She was driving to work (ongoing in the past but not in the present)

> *Past perfect progressive:* She had been driving to work. (ongoing in the past before another event in the past)

> *Future:* She will drive to work.

> *Future perfect:* She will have driven to work. (from the perspective of a particular future moment, this action is to have already taken place)

> *Future perfect progressive:* She will have been driving. (ongoing from past to present from the perspective of a future time)

That, Who

In formal usage, "that" refers to nonpersons:

> The dogs **that** had been abandoned by their owners were rescued by the SPCA.

Use who(m) when referring to persons:

The passengers **who** were on the train when it derailed escaped serious injury.

"The passengers that were on the train. . . " is only acceptable as informal usage.

Their, There, They're

Their is a plural pronoun: "The pilots were proud of **their** aircraft."

There is an adverb: "The tools are **there**" (on the table); "**There** are trillions of cells in the human body."

They're is a contraction for *they are* (pronoun + auxiliary verb): "**They're** (they are) the first to climb that mountain."

Transition

A **transition** is a word or phrase that helps readers make a logical connection between one passage and another; in so doing, it adds to the coherence and readability of the piece of writing. Common transitions include the following:

accordingly

as a result

at the same time

consequently

contrastingly

during this time

finally

for example

however

in addition

in conclusion

in other words

nevertheless

next

on the one hand . . . on the other hand

similarly

then

while this was taking place

NOTE: Be careful not to overuse transitions. Using transitions mechanically, such as at the beginning of every other sentence, can *interfere* with coherence.

Unique

See Chapter 9, page 196.

Use, Utilize

Use is the general word for most situations: We **used** (not "utilized") the finest ingredients for our recipe.

Utilize should be avoided except when making a general reference to large-scale operations: "Some day, fusion energy will be efficiently **utilized.**"

Very

Avoid using this intensifier superfluously, as in "very amazed" or "very unique."

Who, Whom

Who = subjective case (refers to the subject of a sentence or independent clause): "**Who** is on the phone?" "Ned is the customer **who** is on the phone.

Whom = objective case (refers to the direct or indirect object of the sentence): "**Whom** are you seeing today?" "Ned is the most recent customer for **whom** we opened an account."

Your, You're

Your is a possive pronoun: "Watch **your** hands when using the buzz saw."

You're is a contraction for *you are:* "**You're** not being careful enough."

APPENDIX

B

Documentation Formats (APA, CBE, MLA)

Documentation format guidelines vary not only among disciplines but also among professional journals. Generally speaking, the social sciences (education, psychology, sociology) follow guidelines prepared by the American Psychological Association (APA); the biological sciences follow guidelines of the Council of Biology Editors (CBE); and certain humanities (literature, linguistics, philosophy) follow the guidelines of the Modern Language Association (MLA). Other guidelines, not represented here, include those of the *Chicago Manual of Style* (CMS), commonly used in journalism, the fine arts, and history); the American Institute of Physics (AIP); the American Mathematical Society (AMS); the American Chemical Society (ACS); the United States Geological Survey (USGS); the American Medical Association (AMA), and the American Society of Civil Engineers (ASCE). Students should also be aware that guidelines are often revised.

AMERICAN PSYCHOLOGICAL ASSOCIATION (APA) STYLE

I. Citing Sources Within the Text

Option A: Following the quotation, paraphrase, or allusion: Surname(s) of author(s), comma, year of publication, page number(s) (including p. or pp. abbreviation).

> According to one expert, "Meteorite impacts occur more frequently on Mars than on Earth because of Mars's proximity to the asteroid belt" (Larkin, 1991, p. 73).

Option B: Surname(s) of author(s), year of publication in parentheses, followed by the quotation or paraphrase, followed by the page number (including p. or pp. abbreviation).

> According to Larkin (1991), Meteorites strike Mars's surface more frequently than Earth's because Mars is much closer to the asteroid belt (p. 73).

NOTE: If an author has published more than one work in a single year, follow the date with a, b, c, etc., in chronological order of publication:

> Larkin (1991c)

II. Formatting the Bibliography

General guidelines:

- Title the bibliography page "References."
- Double space; indent second and subsequent lines of each citation by three spaces.
- Alphabetize according to author's last name or first word in title (if no author is given), not counting "A" or "The."
- If there are two or more authors, alphabetize according to the first-listed author (last name first), followed by the names of the other authors (last names first as well).
- Use only the initials of the authors' first and middle names.
- Place year of publication in parentheses immediately after the name(s) of the author(s).
- Do not underline article titles; underline book titles and journal titles; underline volume numbers.
- Do not capitalize article and book titles (except for the first word, proper

nouns, and acronyms). Capitalize the titles of journals, newspapers, encyclopedias, and professional organizations.

A. ARTICLE

[From a Journal; Continuous Pagination]
Naiman, R. J., Meillo, J. M., and Hobie, J. E. (1986). Ecosystem alteration of boreal forest streams by beaver. *Ecology 67* (4), 1127–28.

[From a Journal; Noncontinuous (Issue-by-Issue) Pagination]
Lu, M-Z. (1990). Writing as repositioning. *Journal of Education 172,* 18–21.

[From a Weekly Magazine]
Wright, R. (1992, April 6). Report from Turkestan. *The New Yorker,* pp. 53–75.

[From a Monthly Magazine]
Soviero, M. (1993, December). Get ready for digital VCRs. *Popular Science,* p. 38.

[From a Newspaper]
Chui, G. (1993, December 7). Repairs humble Hubble helpers. *San Jose Mercury News,* p. 10A.

[From a General Encyclopedia]
Locomotive. (1991). In *Funk & Wagnalls New Encyclopedia.* Chicago: Funk & Wagnalls, Inc.

[From a Subject-Specific Encyclopedia]
Edwards, M. J. (1979). Matter Transmission. *The Science Fiction Encyclopedia.* Garden City, New York: Doubleday.

[From Published Proceedings of a Conference or Symposium]
Hawking, S. W., and Penrose, R. (1970). The singularities of gravitational collapse and cosmology. *Proceedings of the Royal Society* (London). (A314), pp. 529–48.

B. INDEPENDENTLY AUTHORED CHAPTER FROM AN EDITED BOOK

[One Author]
French, H. F. (1992). Strengthening global environmental governance. In L. R. Brown (Ed.), *State of the world, 1992.* New York: W. W. Norton & Co., pp. 25–55.

[More Than One Author]
Tipler, F. J., Clarke, C. J., and Ellis, G. F. (1980). Singularities and horizons: a review article. In A. Held (Ed.), *General relativity and gravitation.* New York: Plenum Press, pp. 97–206.

C. BOOK

[Original Language]
Pagels, H. R. (1988). *The dreams of reason: the computer and the rise of the sciences of complexity.* New York: Simon and Schuster.

[Translation]
Jung, C. G. (1958). *The undiscovered self* (R. F. C. Hull, Trans.). New York: New American Library.

[Edited Anthology]
Digby, J., and Brier, B. (Eds.) (1985). *Permutations: readings in science and literature.* New York: Morrow/Quill.

[Multivolume Work]
Copleston, F. J. (1964). *A History of Philosophy* (Vol, 6, Part II). Garden City: Image Books.

D. GOVERNMENT DOCUMENT

U. S. Department of Agriculture. (1983). *Using Our Natural Resources: Yearbook of Agriculture.* Washington, DC: U. S. Government Printing Office.

E. REPORT

Matrix Management Group. (1989). *Waste Stream Composition Study,* 1988–89. (Final Report). Prepared for the City of Seattle, Dept. of Engineering, Solid Waste Utility.

F. COMPUTER SOFTWARE

Grolier Electronic Publishing. (1988). *The electronic encyclopedia on CD-ROM.* (Computer software). Danbury, CT: Grolier Electronic Publishing.

G. UNPUBLISHED PRESENTATION

Hitchcock, R. K. (1991). Human rights, local institutions, and sustainable development among Kalahari San. American Anthropological Association Annual Meeting, Chicago, 23 November.

H. INTERVIEW

[Published]
Talking shop with Detroit's big three. (1993, December 13). [Interview with Robert J. Eaton, John F. Smith Jr., and Alex J. Trotman, automakers.] *Time,* pp. 66–67.

[Recorded]
Clarke, A. C. (1976, March 22). [Interview with Arthur C. Clarke, writer.] *The Tomorrow Show,* with Tom Snyder. NBC TV, New York.

[Personal]
Wu, J. (1993, Jan. 10). [Personal interview with Janet Wu.]

[Telephone]
Jones, L. (1991, Feb. 19). [Telephone interview with Lawrence Jones.]

I. MAP

Africa before partition by the European powers, 1800–1880. (1984). Map. *The Times atlas of world history* (Rev. Ed.). Maplewood, NJ: Hammond, p. 239.

COUNCIL OF BIOLOGY EDITORS (CBE) STYLE

I. Citing Sources Within the Text

Option A: Following the quotation, paraphrase, or allusion: Surname(s) of author(s), no comma, year of publication, page numbers (if appropriate), including p. or pp. abbreviation:

> According to one expert, "Meteorite impacts occur more frequently on Mars than on Earth because of Mars's proximity to the asteroid belt" (Larkin 1991, p. 73).

Option B: Surname(s) of author(s), year of publication in parentheses, followed by the quotation or paraphrase, followed by the page number(s), if appropriate, including the p. or pp. abbreviation:

> According to Larkin (1991), Meteorites strike Mars's surface more frequently than Earth's because Mars is much closer to the asteroid belt (p. 73).

NOTE: If an author has published more than one work in a single year, follow the date with a, b, c, etc.

> Larkin (1991c)

II. Formatting the Bibliography

General Guidelines:

- Title the bibliography page "Literature Cited" or "References Cited." The latter title must be used if unpublished citations (such as interviews), are included.

- Double space. Indent second and subsequent lines of each citation by three spaces.

- Alphabetize according to author's last name or first word in title, not counting "A" or "The."

- If there are two or more authors, alphabetize according to the first-listed author (last name first), followed by the names of the other authors, last names first as well.

- Use only the initials of authors' first and middle names.

- Do not enclose article titles in quotation marks; do not underline titles of articles, journals or books. Do not capitalize after first word in titles of articles or books, but do capitalize titles of journals, magazines, newspapers, and professional organizations.

- Cite volume number, issue number in parentheses, followed by a colon, followed by the page numbers. Do not use p. or pp. abbreviations. No spaces separate these elements.

- Place year of publication at the end. A semicolon and a space separates page number(s) from year.

A. ARTICLE

[From a Journal; Continuous Pagination]

Naiman, R. J., Meillo J. M., and Hobbie, J. E. Ecosystem alteration of boreal forest streams by beaver. Ecology 67(4):1127–1128; 1986.

[From a Journal; Noncontinuous (Issue-by-Issue) Pagination]

Lu, M-Z. Writing as repositioning. Journal of Education 172:18–21; 1990.

[From a Weekly Magazine]

Wright, R. Report from Turkestan. The New Yorker, 6 April:53–75; 1992.

[From a Monthly Magazine]

Eliot, J. L. Kodiak, Alaska's island refuge. National Geographic November:34–58; 1993.

[From a Newspaper]

Earth's bowels harbor hidden biosphere. San Jose Mercury News 28 Dec.:3E; 1993.

[From a General Encyclopedia]

Locomotive. Funk & Wagnalls New Encyclopedia; 1991 ed.

[From a Subject-Specific Encyclopedia]

Edwards, M. J. Matter transmission. The Science Fiction Encyclopedia. Garden City:Doubleday; 1979.

[From Published Proceedings of a Conference or Symposium]

Hawking, S. W., and Penrose, R. The singularities of gravitational collapse and cosmology. Proceedings of the Royal Society (London). A314:529–48; 1970.

B. INDEPENDENTLY AUTHORED CHAPTER FROM AN EDITED BOOK

[One Author]

French, H. F. Strengthening global environmental governance. Pages 25–55 in State of the World, 1992, edited by L. Brown. New York: W. W. Norton & Co.; 1992.

[More Than One Author]
Tipler, F. J., Clarke, C. J., and Ellis, G. F. Singularities and horizons: a review article. Pages 97–206 in General relativity and gravitation. Edited by A. Held. New York: Plenum Press; 1980.

C. BOOK

[Original Language]
Pagels, H. R. The dreams of reason: the computer and the rise of the sciences of complexity. New York:Simon and Schuster; 1988.

[Translation]
Jung, C. G. The undiscovered self (Transl. from German by R. F. C. Hull). New York:New American Library; 1958.

[Edited Anthology]
Digby, Joan, and Brier, Bob, eds. Permutations: readings in science and literature. New York:Morrow/Quill; 1985.

[Multivolume Work]
Copleston, F. J. Modern Philosophy, Part II: Kant. Vol. 6 of A History of Philosophy. Garden City: Image Books; 1964.

D. GOVERNMENT DOCUMENT

U. S. Department of Agriculture. Using our natural resources: yearbook of agriculture. Washington, DC:Government Printing Office; 1983.

E. REPORT

Matrix Management Group. Waste stream composition study, 1988–89. Final report. Prepared for the City of Seattle, Dept. of Engineering, Solid Waste Utility; 1989.

F. COMPUTER SOFTWARE

Grolier Electronic Publishing. The electronic encyclopedia on CD-ROM. Computer software. Danbury, CT:Grolier Electronic Publishing; 1988.

G. UNPUBLISHED PRESENTATION

Hitchcock, R. K. Human rights, local institutions, and sustainable development among Kalahari San. American Anthropological Association Annual Meeting, Chicago, 23 Nov.; 1991.

H. INTERVIEW

[Published]
Talking shop with Detroit's big three. Interview with automakers Robert J. Eaton, John F. Smith, Jr., and Alex J. Trotman. Pages 66–67 in Time 142:25; 1993.

[Recorded]
Clarke, A. C. Interview with Arthur C. Clarke. The Tomorrow Show, with Tom
Snyder. NBC TV, 22 March; 1976.

[Personal]
Wu, J. Interview with Janet Wu. 10 Jan.; 1993.

[Telephone]
Jones, L. Telephone interview with Lawrence Jones. 19 Feb.; 1991.

I. MAP

Africa before partition by the European powers, 1800–1900. Map. Page 239 of The
Times Atlas of World History, rev. ed. Maplewood, NJ: Hammond; 1984.

MODERN LANGUAGE ASSOCIATION (MLA) STYLE

I. Citing Sources Within the Text:

Option A: Following the quotation, paraphrase, or allusion: Surname(s) of
author(s); no comma; page number(s), without p. or pp. abbreviation. Year of
publication not stated in text unless reference is made to more than one document published in the same year by the same author.

> According to one expert, "Meteorite impacts occur more frequently on Mars
> than on Earth because of Mars's proximity to the asteroid belt" (Larkin 73).

Option B: Surname(s) of author(s), followed by quotation or paraphrase, followed by page number(s) in parentheses:

> According to Larkin, meteorites strike Mars's surface more frequently than
> Earth's because Mars is much closer to the asteroid belt (73).

II. Formatting the Bibliography

General Guidelines:
- Title the bibliography page "Works Cited."
- Double space. Indent second and subsequent lines of each citation by
  five spaces.
- Alphabetize according to author's last name or first word in title, not
  counting "A" or "The."
- If there are two or three authors, alphabetize according to the first listed

author (last name first), followed by the names of the other authors, first names first. If there are four or more authors, use only the name of the first-listed author (last name first), followed by et al. (Latin, *et alia,* "and all the rest").

- Use the full names (if known) of each author.
- Enclose article titles in quotation marks; underline titles of journals and books. Capitalize titles of articles, journals, books.

A. ARTICLE

[From a Journal; Continuous Pagination]
Naiman, R. J., J. M. Meillo, and J. E. Hobbie. "Ecosystem Alteration of Boreal Forest Streams by Beaver." *Ecology* 67 (1986): 1127–28.

[From a Journal; Noncontinuous (Issue-by-Issue) Pagination]
Lu, Min-Zhan. "Writing as Repositioning." *Journal of Education* 172.1 (1990): 18–21.

[From a Weekly Magazine]
Morrow, Lance. "To Conquer the Past." *Time* 3 Jan. 1994: 34–37.

[From a Monthly or Bi-monthly Magazine]
Robert Devine, "Botanical Barbarians." *Sierra* 79 Jan./Feb. 1994: 50–57; 71.

[From a Newspaper]
"Earth's Bowels Harbor Hidden Biosphere." *San Jose Mercury News* 28 Dec. 1993: 3E.

[From a General Encyclopedia; No Author Cited]
"Locomotive." *Funk & Wagnalls New Encyclopedia.* 1991 ed.

[From a Subject-Specific Encyclopedia; Author Cited]
Edwards, Malcolm J. "Matter Transmission." *The Science Fiction Encyclopedia.* Garden City: Doubleday, 1979.

[From Published Proceedings of a Conference or Symposium]
Hawking, Stephen W., and Roger Penrose. "The Singularities of Gravitational Collapse and Cosmology." *Proceedings of the Royal Society* (London). A314 (1970): 529–48.

B. INDEPENDENTLY AUTHORED CHAPTER FROM AN EDITED BOOK

[One Author]
French, Hilary F. "Strengthening Global Environmental Governance." *State of the World, 1992.* Ed. Lester R. Brown. New York: W. W. Norton & Co., 1992: 25–55.

[More Than One Author]
Tipler, F. J., C. J. Clarke, and G. F. Ellis. "Singularities and Horizons: A Review Article." *General Relativity and Gravitation.* Ed. A. Held. New York: Plenum Press, 1980: 97–206.

C. BOOK

[Original Language]

Pagels, Heinz R. *The Dreams of Reason: The Computer and the Rise of the Sciences of Complexity.* New York: Simon and Schuster, 1988.

[Translation]

Jung, C. G. *The Undiscovered Self.* Trans. R. F. C. Hull. New York: New American Library, 1958.

[Edited Anthology]

Digby, Joan, and Bob Brier, eds. *Permutations: Readings in Science and Literature.* New York: Morrow/Quill, 1985.

[Multivolume Work]

Copleston, Frederick J. *Modern Philosophy, Part II: Kant.* Volume 6 of *A History of Philosophy.* 9 vols. Garden City: Image Books, 1964.

D. GOVERNMENT DOCUMENT

U.S. Department of Agriculture. *Using Our Natural Resources: Yearbook of Agriculture.* Washington, DC: Government Printing Office, 1983.

E. REPORT

Matrix Management Group. *Waste Stream Composition Study, 1988–89.* Final Report. Prepared for the City of Seattle, Dept. of Engineering, Solid Waste Utility, 1989.

E. COMPUTER SOFTWARE

Grolier Electronic Publishing. *The Electronic Encyclopedia on CD-ROM.* Computer software. Danbury, CT: Grolier Electronic Publishing, 1988.

F. UNPUBLISHED PRESENTATION

Hitchcock, Robert K. "Human Rights, Local Institutions, and Sustainable Development Among Kalahari San." American Anthropological Association Annual Meeting, Chicago, 23 November, 1991.

G. INTERVIEW

[Published]

"Talking Ship with Detroit's Big Three." Interview with Robert J. Eaton, John F. Smith, Jr., and Alex J. Trotman. *Time,* 13 December 1993: 66–67.

[Recorded]

Clarke, Arthur C. Interview. *The Tomorrow Show,* with Tom Snyder. ABC TV, New York. 22 March 1976.

[Personal]

Wu, Janet. Personal interview. 10 January 1993.

[Telephone]
Jones, Lawrence. Telephone interview. 19 February 1991.

H. MAP

Africa Before Partition by the European Powers, 1800–1880. Map. *The Times Atlas of World History,* rev. ed. Maplewood, NJ: Hammond, 1984: 239.

APPENDIX

Formats for Report Elements

The following formats reflect a general standard, not an absolute one. Every organization tends to develop its own preferences. Your instructor may ask you to vary any or all of the formats presented here. When in doubt, always check with your instructor.

PRELIMINARY CONSIDERATIONS

PAPER: Use quality bond (20-lb or higher) white typing or computer paper.

REPRODUCTION: Print your final copy on a letter-quality, inkjet or laser printer; the printout should be dark black.

MARGINS: Margins should be 1" to 1½" on all sides.

REPORT COVER: Many corporations require that reports, especially long ones (exceeding 25 pages), be securely bound; e.g., in a vinyl cover with metal fasteners. Covers are usually optional in college writing courses.

TITLE PAGE

- Give the complete title of the report, single-spaced and centered, 5–6 spaces from the top of the page. Use all capitals; boldface is optional.
- Place author's name 12 spaces from the bottom of the page, centered or on the right. "By" is optional.
- Write the name or number of the assignment (for which the report has been prepared) one space below the author's name.
- Write the course number or title and the name of the instructor one space below the assignment name.
- Write the date on which the report is submitted one space below the course number.

BIOCONTROL AS A SUBSTITUTE FOR PESTICIDES:
AN ECONOMIC ANALYSIS

By Lisa M. Bonnell
Final Project
English 179A, Prof. White
5 June 1991

LETTER OF TRANSMITTAL

- Place author's address at the top of the page, flush right or centered, single-spaced.
- Place the date one space directly underneath author's address.
- Place the inside address (name and address of report recipient), flush left, two spaces below author's address.
- Place the Salutation with colon two spaces directly underneath the inside address.

- Begin the text of the letter two spaces below the salutation; single-space (double-space between paragraphs).
- Place the complimentary close two spaces below last line of text. The close may be indented or aligned with the salutation.
- Add the signature block (4–6 spaces for signature, followed by the typed name of the sender).

```
                              Lisa M. Bonnell
                               P.O. Box 130
                              Santa Clara, CA
                               5 June 1991


Ms. Victoria Kroll
Environmental Protection Agency
Washington, DC 20460


Dear Ms. Kroll:

Here, as promised, is my analysis of the economic
viability of biocontrol methods as substitutes for
chemical pesticide use. I hope my findings will
assist you in addressing the serious pesticide issue.

Biocontrol methods are becoming the focus of alterna-
tives to pesticide use. Because the direct costs
equal those of pesticides (and there are few, if any,
external costs to this method) biocontrol may easily
become the solution to the pesticide problem.

Because biocontrol is not the commonly used method of
pest management, there are few actual numerical
analyses of indirect costs. For this reason, the sec-
tion comparing the costs and benefits of biocontrol
lacks specific numerical data. However, it was easy
to assess that the problems imposed by pesticides far
outweigh those imposed by biocontrol methods.

Biocontrol is much more cost-efficient than pesti-
cides in that no human or environmental damage is
imposed. This method must be seriously considered as
the alternative to chemical use.
```

Thank you for your advice and encouragement in helping me to complete this report.

Sincerely,

L. Bonnell

Lisa M. Bonnell

TABLE OF CONTENTS

- Center title; use all caps; boldface is optional.
- Indicate all sections included in the report except the Title Page, the Letter of Transmittal, and the Table of Contents itself. This includes the major subdivisions of the Collected Data and Conclusion, the References section, and any Appendixes.
- Use leader dots to connect name of each section to the first page on which it appears in the report.
- Paginate the "front matter" (Table of Contents, List of Tables and Figures, and Abstract) with lowercase Roman numerals. Paginate the body of the report, including Appendixes, with Arabic numerals, begin-ning with page 1.

TABLE OF CONTENTS

LIST OF TABLES AND FIGURES

- Subdivide this page into "Tables" and "Figures." If you have only Figures in the report, then title the page "List of Figures"; if you have only Tables in the report, then title the page "List of Tables."
- Do not prepare this list unless you have at least two Tables or two Figures in your report.
- Identify each Table and Figure by number and title.
- Use leader dots to connect each item to the page on which it appears.

ABSTRACT

- Use an informative abstract rather than a descriptive abstract. An informative abstract condenses the essential points of the report into a single paragraph of 150–200 words; a descriptive abstract states the purpose and conclusion of the report in fewer than fifty words.
- Do not begin with the phrase, "In this report I . . . " Open instead with the first main point, such as the problem being addressed.
- Single-space the abstract.

NOTE: Sometimes managers will request an Executive Summary to be added to the report. These are similar to informative abstracts, except that they are longer, usually about three paragraphs—one full page—and are placed immediately before the Introduction. An executive summary includes more information about the collected data and conclusions but avoids technical specifics.

ABSTRACT

Because of the high costs resulting from the harmful environmental effects of pesticide use, the EPA has encouraged researchers to investigate the uses of biocontrol as an alternative to pesticides. Evidence suggests that biocontrol methods are not only more environmentally suitable than pesticides, but more economically sound as well. However, under current legislation, biocontrol cannot be implemented. Factors such as price-support programs, federal grading standards, and farmer dependence on pesticides make it difficult for biocontrol to be used as the sole method of controlling pests. For biocontrol to be implemented the following management system and legislation must change by (1) moving the biocontrol market from the public to the private sector; (2) imposing external costs of pesticide use on the farmers who use them; (3) changing price-support programs to encourage crop-rotation; and (4) reducing federal grading standards.

INTRODUCTION

- The introduction includes an explicit statement of the purpose of the report, background information necessary for fully understanding the problem or phenomenon, and a clear statement of the scope of the report (all that the report will discuss, in order of presentation).
- Do not include tables, figures, or any computations in the Introduction.

ECONOMIC ANALYSIS FOR BIOCONTROL AS A SUBSTITUTE FOR PESTICIDES

INTRODUCTION

The purpose of this report is to analyze the economic suitability of biocontrol as a substitute for pesticides. After presenting a brief overview of the problem, the report examines the costs and benefits of each method, then examines government policies that affect biocontrol implementation. Finally, the situation is evaluated and recommendations presented.

The method of biological pest control, or biocontrol, employs nature's own predators, parasites, and pathogens to do the work of pesticides, but without the harmful effects to both consumers and the environment from chemical use. Currently, farmers employ four different biocontrol methods: classical, rotational, microbial, and pheromonal.

Classical biocontrol. With this form of biocontrol, an imported organism is released into the environment where it establishes itself and begins controlling pests.

Crop rotation. This involves varying the type of crop planted in a given plot of land every two or three years. This method allows soil organisms to work against one or more of the crop's enemies. For example, by not growing wheat in the same field more than every second or third year, root parasites are forced into dormancy, giving other organisms a chance to sanitize the soil.

Microbial biocontrol. This is the method employed by using disease organisms as natural pest combatants. Disease organisms for a given pest, such as a single-celled protozoan, are developed in commercial labs, and from there distributed in the field. The parasitic protozoan then depletes the pests' fat stores, leaving them weakened, without energy to find food. With this type of biocontrol, only the harmful pest is eliminated.

Pheromonal biocontrol. Pheromonal biocontrol involves the use of insect sex attractants. Scientists analyze the pattern of sexual activity in specific insects and disrupt their mating patterns by filling the air around an infested tree with female insects' pheromones. As the male futilely attempts to mate with the nonexistent female, he becomes exhausted. These synthesized pheromones can prevent mating and reduce pest populations to practically zero.

COLLECTED DATA

- Use this section of the report to introduce and discuss particular findings. Use tables and figures to illustrate those findings.
- Be sure to annotate properly all information obtained from outside sources.
- Organize this section into discrete subtopics, each with its own subhead.

COLLECTED DATA

BIOCONTROL AND PESTICIDES COMPARED

In comparing the direct costs associated with both methods of pest control, the cost per unit of output is roughly equal; the cost of a pheromone strip or the importation of pest combatants equals the cost of chemical pesticide application. In order to be able to compare the two methods, the indirect costs associated with each must be examined.

Pesticide Cost/Benefit Analysis. Indirect costs associated with pesticide use are defined as any losses due to effects of pesticides on non-target organisms. The total estimated environmental and social costs in the United States for 1990 adds up to $839 million. For a numerical breakdown of costs, see Appendix A.

As Figure 1 indicates, the largest loss figure stems from the problem of reduced natural enemies of undesirable insects from pesticide use. It also includes the cost of target insects' pesticide resistance, which in turn requires heavier and repeated chemical applications.

Also to be included in the indirect costs of pesticide use is the incidence of cancer as a result of using pesticides. According to the Assembly

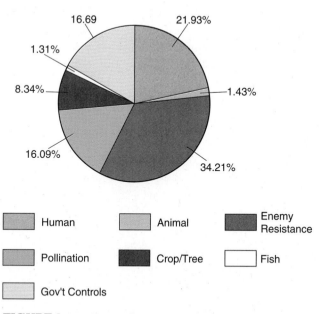

FIGURE 1 Losses Resulting From Pesticide Use
Source: U.S. Dept. of Agriculture; 1990.

Office of Research, the probability for developing cancer from pesticide ingestion is 200: 1 million. Since the 1990 census estimated the U.S. population at 240 million persons, the number of cancer cases resulting from pesticides should be roughly 40,000 per year. It is difficult to attach a monetary figure to this number as the value placed on each life is subjective.

The benefits of pesticide use revolve around greater crop yields. The cash sales for 1989 total $53,096,222 (U.S. Dept. of Agriculture, 1990, p. 12). For numerical breakdown of 1989 sales, see Appendix B.

In 1989 it is estimated that pesticide use resulted in a 21% increase in crop cash sales. Pesticides, therefore, were responsible for increasing crop sales by $11,150,207 (U.S.D.A., 1990, p. 18).

Another direct benefit of pesticide use is the capital that the pesticide industry generates. American farmers purchase about 40% of the pesticides sold by American pesticide companies. The remaining 60% is sold to foreign countries (U.S. Congress of Technology Assessment, p. 4). Last year, American pesticide companies reported a net gain of $20 billion, with 5% available as capital investment. As 40% of the market, American farmers account for $8 billion of the net gain, with $400 million available for reinvestment (p. 6).

Analysis. Because biocontrol methods are not used as extensively as pesticides, little information is available concerning the monetary values attached to indirect costs and benefits. The biggest cost associated with biocontrol is time. Biocontrol takes careful planning, and the results tend to be slower and less visible over time than pesticide applications. What one farmer can accomplish in one day with a chemical pesticide application may take the biocontrol farm two or three years of careful planning and monitoring to accomplish. Furthermore, pesticides do not require an extensive study into an area's pest population history in order to assess what types of natural methods will combat seasonal trends. The time factor is important because sudden outbreaks of crop pests cannot be handled in a timely manner with biocontrol. Entire crops could be lost if pesticides were completely banned, forcing many farms to withdraw from the market altogether.

Another cost is research and development. Chemical pesticide companies support the research within that field through profits. Biocontrol still remains largely within the hands of the government, which is unable to support the same amount of research as the chemical companies do. Biocontrol is subject to slower development than chemical pesticides.

The indirect benefits of biocontrol are largely long-term. Biocontrol, while unable to immediately lessen harmful pest populations, does bring about a long-term solution to pest problems. Most biocontrol methods involve only one application, and the results, while spread out, tend to be more effective than pesticides. This is due to the reduction in applications, and the elimination of pesticide tolerance. Biocontrols also eliminate the environmental hazards, as they are not harmful, in the form of pesticide residues, to either the beneficial insects, other animals, or humans.

FACTORS INHIBITING BIOCONTROL IMPLEMENTATION

Pesticides are still used as the primary method for pest control. This is due mainly to four actors that inhibit biocontrol implementation: government price-support programs, federal grading standards, farmer dependence on pesticides, and imperfect substitutability.

Government Price-Support Programs. Federal farm programs are designed mainly to support crop prices and farm income. The program supports the growth of specific crops. Then, commodity price supports are established by the government; direct payments and crop loans are made to the farms, and agricultural land is diverted from production through paid land diversions. In effect, the specialized cultivation of one or two crops is financially favored over crop rotation. With specialized cultivation comes a growth in the pest population that feeds on it. However, it is too costly for farmers to rotate crops because they could lose subsidy payments from the government. So, government support which encourages crop specialization also increases the likelihood of needing and using pesticides.

Federal Grading Standards. The government has established grading standards for all fruits and vegetables. These grading standards dictate that the most cosmetically attractive crops be marketed as raw produce. The rest, usually blemished and bruised, are used for canned and processed goods. But the price return is higher for raw produce. Certain pesticides also enhance the cosmetic value of fruits and vegetables, and are used heavily to meet federal standards. Alar, for example, helps apples to remain on trees until they ripen to a deep red color. However, this chemical, when processed, forms UDMH, a potent carcinogen (Goldstein, 1990, p. 50). But in order to receive a top price for produce, farmers must use more chemical pesticides to insure that their produce meets the high federal standards. Since there is no biocontrol method that is a direct alternative to growth/chemical pesticides, consumers must ingest this harmful carcinogen when eating raw fruits and vegetables.

Farmer Dependence on Pesticides. Each year the amount spent on pesticides increases. By 1991, this figure is expected to reach $25 mil-

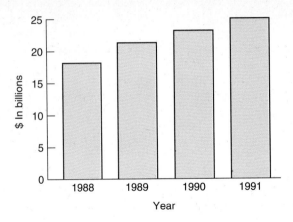

FIGURE 2   Yearly Pesticide Expenses

lion, due to increasing farmer dependence on pesticide use. (See Figure 2.)

DDT, the first pesticide, was developed during World War II as a method of chemical warfare. Since then, hundreds of pesticides have been developed and introduced into the market without sufficient research on their effects. Now, fifty years later, the farm owners are not the same group that practiced biocontrol before the introduction of chemical pesticides. In order for biocontrol methods to be used exclusively, the entire farming community would need to be reeducated as to how to maintain crop yields using alternative methods of pest control. As a government project, this would be too costly.

Imperfect Substitutability. Biocontrols do not perfectly substitute for each type of chemical used. As in the case of Alar, many chemical pesticides serve as growth enhancers as well. There is really little that the farmer can do beyond fertilizing to change the cosmetic appearance of produce using purely biocontrol methods. While biocontrol does offer ways to naturally reduce pest populations, it cannot accomplish all that pesticides do. Farmers will always be reluctant to adopt purely biocontrol methods until all facets of pesticide use are matched with an equivalent biocontrol method.

CONCLUSION

- Use the Conclusion to summarize the collected data, to interpret that data, and (if appropriate) to recommend a course of action in light of the findings and your interpretation of them.
- Do not introduce new information in the Conclusion.

CONCLUSION

SUMMARY OF FINDINGS

Biocontrol methods offer an economically feasible alternative to chemical pesticide use. Due to the high environmental hazards associated with pesticide use, biocontrol is much less costly and much less harmful to animals and humans. However, existing barriers to implementation of biocontrol methods, such as government farm policy, grading standards, farmer dependence on chemical use, and imperfect substitution, make it difficult for biocontrol to be developed as a viable alternative to pesticides.

INTERPRETATION

After reviewing the cost-benefit analysis of the two methods, it appears that biocontrol is the most economically efficient way of controlling pest populations. The benefits of biocontrol to both consumers and the environment far outweigh the costs associated with careful planning. Furthermore, pesticides are actually costing society much in the way of environmental contamination.

However, biocontrol methods will never be implemented as long as the government is the sole source for research and education funds. The hazards of pesticide use are of little concern to farmers and the government because they are not required to directly pay for the costs imposed on the environment. Biocontrol methods also require planning, which is something that today's farmers are not accustomed to. It would be too time-consuming and costly for the government to reeducate each farm owner.

RECOMMENDATIONS

The following recommendations show ways in which biocontrol methods can be implemented:

Public vs. Private Sector. If the market for biocontrol is moved from the public to the private sector, it will be subject to the same competition as found in the pesticide market. A competitive environment would foster extensive research and development in order to discover new methods for reducing pest populations. This type of market would also place the incentive to educate farmers in the hands of corporations where it could more efficiently be handled. Because private companies would have an interest in insuring that farmers use their methods, they would willingly educate farmers at their own expense. Biocontrol markets would then be generating the same amount of capital previously seen in the pesticide market.

External Costs. Farmers, as well as most corporate polluters, should be required to pay for the costs resulting from environmental contamination. By issuing a substantial tax on the purchase of each unit of chemical pesticide, the government will have an easy method of collecting payments for environmental damage. The tax will be passed on to the farmer in the form of higher pesticide prices, thus making biocontrol methods cheaper. Not only would this create a strong disincentive for farmers to use chemical pesticides, but it would also insure that society is not paying for the damage. It is more cost-efficient to place the responsibility on the party most able to control the damage. By forcing farmers to pay for damages caused by pesticides, the damage to the environment should decrease.

Lowering Grading Standards. The government should consider lowering the standard for raw produce so that pesticides are not encouraged for crop appeal. Farmers would not need to find an alternative to growth enhancements, making biocontrol a perfect substitute for pesticides. Prices could be determined by demand and not by appearance alone. This may be met with opposition by the public, as they are used to produce looking a certain way, but produce is a good that consumers will always purchase. Furthermore, fruit that looks bruised but is grown naturally will become the standard rather than the exception in produce markets as consumers become more environmentally conscious.

Include Rotation in Price-Support Systems. A variety of crops should be encouraged for price-support programs. This will insure that harmful pests do not return to a given area as their source of food diminishes. Furthermore, rotation provides a natural method of enriching soil as depleted nutrients have a chance to rebuild.

APPENDIX(ES)

- Use an Appendix to show research instruments, such as questionnaires used to obtain some of the data; to display documents such as maps that provide evidence of assertions or observations made.
- Label each Appendix A, B, C, and so on.

APPENDIX A

ANNUAL ESTIMATED ENVIRONMENTAL AND SOCIAL COST FROM PESTICIDE USE

1. Human Pesticide Poisonings $184,000,000

2. Animal Pesticide Poisonings and Contaminated Livestock Products 12,000,000

3. Reduced Natural Enemies and Pesticide Resistance   287,000,000

4. Honey Bee Poisonings and Reduced Pollination   135,000,000

5. Losses of Crops and Trees   70,000,000

6. Fishery and Wildlife Losses   11,000,000

7. Government Pesticide Pollution Controls   140,000,000

   Source: Pimenthal, 1987

APPENDIX B

CROP SALES FOR 1989

1. Fruits and Nuts $7,478,823

2. Vegetables and Melons   5,988,858

3. Wheat   34,615,195

4. Cotton   5,013,346

   Source: U.S. Congress, Office of Technology Assessment, 1990.

REFERENCES

The References or Works Cited page is the last section of the report; it includes the complete citations of every information source used in your report. Do not include uncited sources. See Appendix B for proper formatting of citations.

REFERENCES

Goldstein, Joan. (1990). *Demanding clean food and water.* New York: Plenum Press.

Pimental, David, and Perkins, John H. (Eds.). (1987). *Pest control: cultural and environmental aspects.* AAAS Select Symposium. Boulder: Westview Press.

U.S. Congress of Technology Assessment. (1990, March). *Technology, Public Policy, and the Changing Structure of American Agriculture.* OTA-F-285. Washington, DC: U.S. Government Printing Office.

U.S. Department of Agriculture. (1990). *U.S. agriculture in a global economy.* O-484-628. Washington, DC: U.S. Government Printing Office.

Acknowledgments

ART ACKNOWLEDGMENTS

Unless otherwise acknowledged, all photographs are the property of Scott, Foresman and Company.

Page 5 Copyright the British Museum / **Page 6** *Revolution in Time,* by David S. Landes, © 1983, Harvard College / **Page 16** Frank & Ernest reprinted by permission Newspaper Enterprise Association, Inc. / **Page 99** Frank & Ernest reprinted by permission Newspaper Enterprise Association, Inc. / **Page 178** Drawing by M. Twohy; © 1993 The New Yorker Magazine, Inc. / **Page 201** Von Braunschweig, Hieronymus, *The Vertuose Boke of Distyllacyon of the Waters of All Manner of Herbes,* 1527 / **Page 202** Elyton, Sir Thomas, *The Castel of Health,* 1536–1539 / **Page 215** Sierra Instruments, Inc. Calif. / **Page 221** Bodleian Library, Oxford, U.K. Copyright / **Page 221** *Leonardo & the Age of the Eye,* by Ritchie Calder, © 1970, Simon & Schuster, page 220 / **Page 228** Multiple-line graph from George Gaylord Simpson, *The Meaning of Evolution* Copyright © 1967 Yale University Press. Reprinted by permission / **Page 229** Area graph from *The Way Things Work.* Copyright © 1968 by International Thompson Publishing Service, Ltd. Reprinted by permission / **Page 230** Scatter graph by Lorelle M. Raboni, from "Paleoneurology and the Evolution of Mind," by Harry J. Jerison; *Scientific American,* Jan. 1976. Reprinted by permission of Scientific American / **Page 233** Multiple bar graph courtesy of Pacific Gas and Electric Co. / **Page 235** Pie chart from *The Scientist's Handbook for Writing Term Papers and Dissertations,* by Antoinette Wilkinson. Copyright © 1991 The American Psychological Association. Reprinted by permission / **Page 235** Drawing by M. Stevens; © 1992 The New Yorker Magazine, Inc. / **Page 237** Pie chart courtesy Pacific Gas and Electric Co. / **Page 239** Tree chart from George Gaylord Simpson, *The Meaning of Evolution.* Copyright © 1967 Yale University Press. Reprinted by permission / **Page 240** "The Water Meter" from *The Way Things Work* by David Macaulay. Compilation Copyright © 1988 by Dorling Kindersley Ltd., London. Illustration copyright © 1988 by David Macaulay. Text copyright © 1988 by David Macaulay and Neil Ardley. Reprinted by permission of Houghton Mifflin Company. All rights reserved. / **Page 242** Courtesy Lucas

TEXT ACKNOWLEDGMENTS

by permission / **Pages 455–458** Copyright © 1991 The San Jose Mercury News. Published in the *San Jose Mercury News*, Oct. 29, 1991: 1A. Reprinted by permission / **Pages 459–462** Copyright © 1990 *The Readers Digest, Inc.* Reprinted by permission / **Pages 464–465** Copyright © 1992 *Santa Clara Magazine*. Reprinted by permission / **Pages 472–474** Reproduced by permission of Paul Freitas / **Pages 475–479** Copyright © 1992 *New Scientist Syndication*. Published in *New Scientist,* June 6, 1992 Reprinted by permission / **Pages 483–485** Copyright © American Lung Association. Reprinted by permission / **Page 487** Reprinted by permission of Peter Drekmeier and Bay Area Action, 715 Colorado, #1, Palo Alto, CA 94303 / **Pages 489–490** Microsoft Project (Business Project Planning System software) Copyright © 1991 Microsoft Corp. Reprinted by permission / **Pages 492–493** Copyright © American Lung Association. Reprinted by permission / **Page 502** Copyright © 1991 The Health Source Corporation. *TopHealth* is published in English and Spanish editions by The Health Source Corp., P.O. Box 203, Newton, MA 02159 / **Page 508** Reproduced by permission of the Environmental Studies Program, Santa Clara University / **Page 509** Reproduced by permission of Dr. Kristina Irvin, D.C. / **Pages 556; 557–567** Letter of Transmittal and Report ("Biocontrol as a Substitute for Pesticides: An Economic Analysis") reproduced by permission of Lisa Bonnell-Sprenger / **Inside back cover** "Code for Communicators" used with permission from the Society for Technical Communication, Arlington, Virginia

Index